GEOMETRY

GEOMETRY

Harold R. Jacobs

W. H. FREEMAN AND COMPANY
San Francisco

The cover illustration is a lithograph titled
Ascending and Descending by Maurits Escher.
Reproduced by permission of the Escher Foundation,
Haags Gemeentemuseum, The Hague.

Library of Congress Cataloging in Publication Data

Jacobs, Harold R
    Geometry.

    Includes bibliographical references.
    1. Geometry. I. Title.
QA453.J26   516'.2   73-20024
ISBN 0-7167-0456-0

Printed in the United States of America

Contents

1
THE NATURE OF DEDUCTIVE REASONING

2
FUNDAMENTAL IDEAS: LINES AND ANGLES

3

SOME BASIC POSTULATES
AND THEOREMS

4

CONGRUENT TRIANGLES

5

TRANSFORMATIONS

6
INEQUALITIES

7
PARALLEL LINES

8
QUADRILATERALS

9
AREA

10
SIMILARITY

11
THE RIGHT TRIANGLE

12
CIRCLES

13

THE CONCURRENCE THEOREMS

14

REGULAR POLYGONS AND THE CIRCLE

15

GEOMETRIC SOLIDS

16
NON-EUCLIDEAN GEOMETRIES

A Letter to the Student

There is a legend that Pythagoras, a Greek mathematician of the sixth century B.C., wanted to see if he could teach someone geometry. After finding a somewhat reluctant student, Pythagoras agreed to pay him a penny for each theorem he learned. Because the student was very poor, he worked diligently. After a time, however, the student realized that he had become more interested in geometry than in the money he was accumulating. In fact, he became so intrigued with his studies that he begged Pythagoras to go faster, offering now to pay him back a penny for each new theorem. Eventually Pythagoras got all of his money back!*

What is there about geometry that the student could have found so fascinating? Perhaps it was the logical way in which Pythagoras was able to present it. Geometry was the first system of ideas developed by man in which a few simple statements were assumed and then used to derive more complex ones. Such a system is called *deductive*. The beauty of geometry as a deductive system has inspired men in other fields to organize their ideas in the same way. Sir Isaac Newton's *Principia*, in which he tried to present physics as a deductive system, and the philosopher Spinoza's *Ethics* are especially noteworthy examples. One eighteenth-century man of letters went so far as to say "a work of morality, politics, criticism . . . will be more elegant, other things being equal, if it is shaped by the hand of geometry."†

---

*Howard W. Eves, *In Mathematical Circles* (Prindle, Weber & Schmidt, 1969).

†Fontenelle in *Préface sur l'Utilité des Mathématiques et al Physique* (1729).

The study of geometry is also valuable because of its wide variety of applications to other subjects. We will consider, for example, how astronomers have used geometry to measure the distance from the earth to the moon, how artists have used it to develop the theory of perspective, and how chemists have used it to understand the structure of molecules.

Geometry was so named by the ancient Greeks, and it was one of them, Euclid, who systematized the ideas that we will study. We will also consider some interesting contributions to the subject that were made in India during the Dark Ages, when European scholarship was almost nonexistent, and in Europe during the Renaissance, when the pursuit of knowledge was revived. And we will briefly survey the "non-Euclidean" geometries developed in the nineteenth century and see how Einstein used them in his theory of the nature of space.

"When you come right down to it, Son, homework is the basis of civilization."

You will have many opportunities in your study of geometry to use your imagination. Since geometry is a logical system, however, you need to take the time to become thoroughly acquainted with the ideas contained within it. Although homework is hardly the basis of civilization, it is certainly the basis for success in geometry. We hope that you will find your study of the subject an enjoyable and rewarding endeavor.

Harold R. Jacobs

GEOMETRY

Old Euclid drew a circle
On a sand-beach long ago.
He bounded and enclosed it
With angles thus and so.
His set of solemn graybeards
Nodded and argued much
Of arc and of circumference,
Diameters and such.
A silent child stood by them
From morning until noon
Because they drew such charming
Round pictures of the moon.

VACHEL LINDSAY

INTRODUCTION

Euclid, the Surfer, and the Spotter

Euclid is one of the most famous mathematicians of all time, yet very little is known about him. He taught at the university at Alexandria, the main seaport of Egypt, in about 300 B.C.

The reason for Euclid's fame is a book he wrote, one of the most successful books ever written. It was called the *Elements*, which sounds like a book about chemistry, but is actually about geometry, algebra, and number theory. The *Elements* has probably been translated into more languages than any other book except the Bible. It was first printed in 1482 and since then more than a thousand editions of Euclid's book have been published.

Although very little of the mathematics in the *Elements* was original, what made the book unique was its logical organization of the subject, beginning with a few very simple principles and deriving from them everything else.

The first chapter of the *Elements* begins with an explanation of how to draw a triangle with three sides of equal length. Euclid's method requires the use of two tools: a straightedge for drawing straight lines and a compass for drawing circles. These tools have been used ever since in making geometric drawings called *constructions*.

A triangle having three sides of equal length is called *equilateral*. To construct an equilateral triangle, we begin by using the straightedge to draw a segment for one side. The segment is named AB in the figure at the left below. Next the radius of the

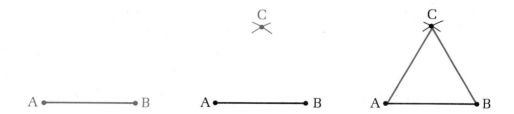

compass (the distance between pencil point and metal point) is adjusted so that it is equal to the length of the segment. Two arcs having this radius and A and B as their centers are drawn so that they intersect as shown in the second figure. Finally, two line segments are drawn from the point of intersection (labeled C in the figure) to points A and B to form the triangle.

Euclid not only told *how* to do this construction but also explained *why* it works. We will become acquainted with how he did this later in our study of geometry.

1. Use a straightedge and a compass to construct an equilateral triangle whose sides are each 12 centimeters long.

Next we will consider a couple of geometric problems involving equilateral triangles. To make them easier to understand, they will be presented in the form of an imaginary story.

The Puzzles of the Surfer and the Spotter

One night a ship is wrecked in a storm at sea and only two members of the crew survive. They manage to swim to a deserted tropical island where they fall asleep exhausted. After exploring the island the next morning, one of the men decides that he would like to stay there and spend the rest of his life surfing on the beaches. The other man, however, wants to escape and decides to use his time looking for a ship that might rescue him.

The island is overgrown with vegetation and happens to be in the shape of an equilateral triangle, each side being 12 kilometers (about 7.5 miles) long.

Wanting to be in the best possible position to spot any ship that might sail by, the man who hopes to escape (we will call him the "spotter") goes to one of the corners of the island. Since he doesn't know which corner is best, he decides to rotate from one to another, spending a day on each. He wants to build a shelter somewhere on the island and a path from it to each corner so that the sum of the lengths of the three paths is a minimum. (Digging up the vegetation to clear the paths is not an easy job.) Where should the spotter build his house?

Euclid, the Surfer, and the Spotter   3

The figure below is a scale drawing of the island in which 1 centimeter represents 1 kilometer. Suppose the spotter builds

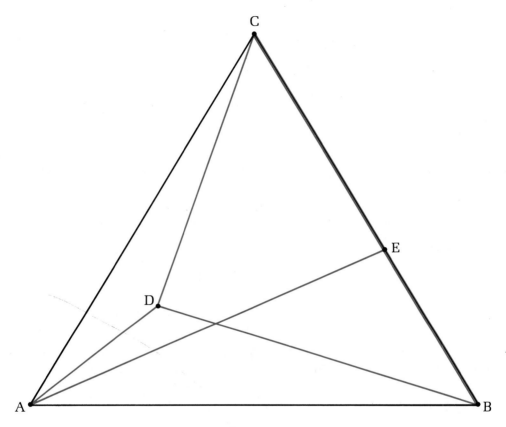

his house at point D. The three paths that he has to clear have the following lengths: DA = 4.4 km, DB = 9.1 km, and DC = 8.0 km. Check these measurements with your ruler, remembering that 1 cm represents 1 km. The sum of these lengths is 21.5 km.

If the spotter builds his house at point E, the path lengths are: EA = 10.4 km, EB = 5.0 km, and EC = 7.0 km, and their sum is 22.4 km. So point D is a better place for him to build than point E. But where is the best place?

2. Use the equilateral triangle you drew in the first exercise to represent the island. Choose several different points on it; for each point, measure the distance between it and each of the corners to the nearest 0.1 cm, and find their sums as illustrated for points D and E above.

3. On the basis of your work in Exercise 2, where do you think is the best place for the spotter to build his house? Also, how

many kilometers of path does he have to clear? (Remember that 1 cm on your map represents 1 km.)

4. Where do you think is the *worst* place on the island for the spotter to locate? How many kilometers of path would he have to clear from it?

Now we will consider the problem of where the surfer should build *his* house. He likes the beaches along all three sides of the island and decides to spend an equal amount of time on each. To make the paths from his house to each beach as short as possible, he constructs them so that they are perpendicular to the lines of the beaches. For example, in the figure below, if the surfer built his house at point D, the three paths to the beaches would be as shown. Path DX is perpendicular* to beach AC, path DY is

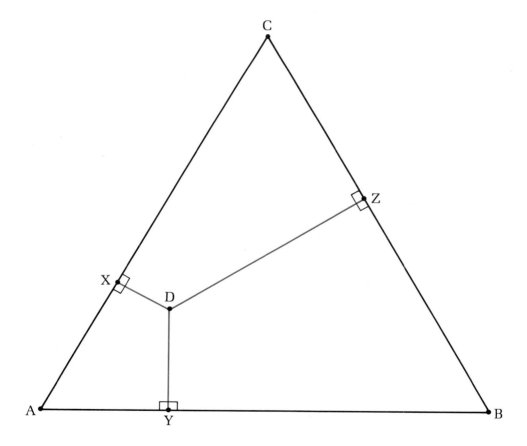

---

*Two lines that are perpendicular form right angles and these are shown by the small squares at the points of intersection, X, Y, and Z.

perpendicular to beach AB, and path DZ is perpendicular to beach BC. The lengths of the three paths are: DX = 1.5 km, DY = 2.8 km, and DZ = 6.1 km, so their sum is 10.4 km.

The surfer, like the spotter, wants to locate his house so that the sum of the lengths of the paths is a minimum. Where is the best place on the island for him?

5. Use a straightedge and compass to construct another equilateral triangle whose sides are 12 centimeters long. Choose several different points on it; for each point, measure the perpendicular distance from it to each of the sides to the nearest 0.1 cm, and find their sums as illustrated for point D above.

6. On the basis of your work in Exercise 5, where do you think is the best place for the surfer to build his house? Also, how many kilometers of path does he have to clear?

7. Where do you think is the worst place for the surfer to locate? How many kilometers of path would he have to clear from it?

Chapter 1

THE NATURE OF
DEDUCTIVE REASONING

"Help yourselves to the spaghetti, folks.
Peg will serve the meatballs."

DRAWING BY TOM HENDERSON; REPRINTED BY
PERMISSION OF THE AMERICAN LEGION MAGAZINE

Lesson **1**

Drawing Conclusions

One of the goals of studying geometry is to develop the ability to think critically. An understanding of the methods of deductive reasoning is fundamental in the development of critical thinking and so it is to this subject that we first turn our attention.

Misunderstandings are very common in everyday life. Like the couple in this cartoon, who have been told what is going to happen next but who are nevertheless probably in for a surprise, we often draw conclusions that are incorrect or at least questionable.

Exercises

Passages from several books are quoted on the following pages. Each is followed by a series of statements that are conclusions that might be drawn on the basis of accepting all of the information in the passage as literally true. Some of these conclusions are true, some are false, and some are questionable—that is, from the information in the passage, it cannot be definitely determined whether they are true or false.

Write the numbers of the statements on your paper and mark each "true," "false," or "not certain." Where you feel it is appropriate, briefly explain the basis for your answers.

"Mother used to send a box of candy every Christmas to the people the Airedale bit. The list finally contained forty or more names. Nobody could understand why we didn't get rid of the dog. I didn't understand it very well myself, but we didn't get rid of him. I think that one or two people tried to poison Muggs—he acted poisoned once in a while—and old Major Moberly fired at him once with his service revolver near the Seneca Hotel in East Broad Street—but Muggs lived to be almost eleven years old and even when he could hardly get around he bit a Congressman who had called to see my father on business."

JAMES THURBER, *My Life and Hard Times*

1. Muggs was an Airedale dog.

2. People who received boxes of candy from Mother at Christmas had been bitten by the dog.

3. At least one person had tried to poison Muggs.

4. The dog had bitten at least forty people.

5. Mother couldn't understand why we didn't get rid of Muggs.

6. Muggs tried to bite Major Moberly.

7. Major Moberly tried to kill Muggs.

8. Major Moberly missed Muggs when he fired at him.

9. The Seneca Hotel was in East Broad Street.

10. Muggs was eleven years old when he died.

11. Muggs died of old age.

" 'In answer to your question what we got out of English so far I am answering that so far I got without a doubt nothing out of English. Teachers were sourcastic sourpuses or nervous wrecks. Half the time they were from other subjects or only subs. . . .

Also no place to learn. Last term we had no desks to write only wet slabs from the fawcets because our English was in the Science Lab and before that we had no chairs because of being held in Gym where we had to squatt.

Even the regulars Mrs. Lewis made it so boreing I wore myself out yawning, and Mr. Loomis (a Math) hated teaching and us.' "

BEL KAUFMAN, *Up The Down Staircase*

12. The student writing this essay dislikes English.

13. The student writing this essay answered the question concerning what he (or she) had learned from his (or her) English class.

14. There are several spelling mistakes in this essay.

15. The student writing this essay is a poor speller.

16. The faucets in the science room leaked.

17. The gym had no bleachers since the students had to squat on the floor.

18. The school is overcrowded.

19. Mrs. Lewis was not one of the substitute teachers.

20. Mr. Loomis would rather teach math than English.

SET II

" 'Excellent!' said Gandalf, as he stepped from behind a tree, and helped Bilbo to climb down out of a thornbush. Then Bilbo understood. It was the wizard's voice that had kept the trolls bickering and quarrelling, until the light came and made an end of them.

The next thing was to untie the sacks and let out the dwarves. They were nearly suffocated, and very annoyed: they had not at all enjoyed lying there listening to the trolls making plans for roasting them and squashing them and mincing them. They had to hear Bilbo's account of what had happened to him twice over, before they were satisfied."

J. R. R. TOLKIEN, *The Hobbit*

1. Dwarves are not very large since they will fit inside sacks.

2. The trolls had been arguing with each other.

3. Gandalf was a wizard.

4. We know that the trolls had planned to eat the dwarves.

5. Trolls are fond of mince pie.

6. Bilbo had climbed into a thornbush.

7. Dwarves don't carry knives or else they would have cut their way out of the sacks.

8. Gandalf and Bilbo let the dwarves out of the sacks.

9. The dwarves hadn't been able to hear anything while they were tied up in the sacks.

10. Gandalf had been hiding behind a tree.

11. Maybe the dwarves deserved to have a good fright.

12. There were seven dwarves.

13. The sunlight changed the trolls into stone.

14. Gandalf fooled the trolls.

15. Bilbo knew all along what had been happening.

16. Bilbo told the dwarves what had happened to him more than once.

SET III

Study the following photograph carefully. Try to form some conclusions that you think seem reasonable.

Lesson 2
Conditional Statements

These three ads have something in common: each has a headline that begins with the word "if." Statements consisting of two clauses, one of which begins with the word "if" or "when" or some equivalent word, are called *conditional statements*. Such statements are often used when the purpose is to establish certain conclusions and so they are very common in the field of advertising. They are also important in mathematics in writing deductive proofs.

A conditional statement can be represented symbolically by "If a, then b," or, even more briefly, by

$$a \rightarrow b.$$

The letter a represents the "if" clause, or *hypothesis*, and the letter b represents the "then" clause, or *conclusion*. (The word "then," being understood, is usually omitted.) The symbols

$$a \rightarrow b$$

are read as "if a, then b," or as "a implies b." For example, in the first ad, a represents the words "Avis is out of cars" and b represents the words "we'll get you one from our competition." In the second ad, a stands for "life is discovered on Mars" and b for "it will come as news to you," and so on.

It is helpful in learning how to relate two or more conditional statements to each other to be able to represent them with circle diagrams. These are often called *Euler diagrams* after an eighteenth-century Swiss mathematician, Leonhard Euler, who first used them.

If you've seen the traffic in Paris, you ain't seen nothing yet.

CAUTION
KOALAS CROSS HERE

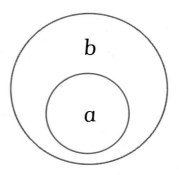

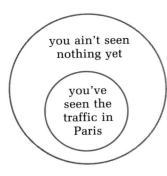

To represent a conditional statement with an Euler diagram, we draw two circles, one inside the other. The interior of the smaller circle represents *a*, the hypothesis, and the interior of the larger circle represents *b*, the conclusion. Notice that if a point is inside circle *a*, it is also inside circle *b*. Or, more briefly, "if *a*, then *b*," which is what the diagram is intended to represent. The headline of the last ad, represented by an Euler diagram, looks like the figure shown at the left.

Exercises

Beginning with this lesson, answers to many of the exercises are provided at the back of this book so that you can check some of your work as soon as you have finished it.

SET I

Conditional statements are not always written in the form "if *a*, then *b*." Rewrite each of the following sentences in "if-then" form. Be careful not to change the meanings of any of the sentences; for instance, if you write a true conditional statement in "if-then" form so that it turns out to be false, something is wrong.

Example. A baby sneezes when it gets pepper in its nose.

Answer. If a baby gets pepper in its nose, then it sneezes.

1. When you cross your eyes, I crack up.

2. Smokey the Bear wouldn't have to do commercials for a living if money grew on trees.

3. All surfers like big waves.

4. Licorice-flavored ice cream has a peculiar color.

5. A heavy object stored in the attic of a jungle mansion may crash down upon the occupants.

6. People who live in grass houses shouldn't stow thrones.

7. No ghost has a shadow.

SET II

1. Draw an Euler diagram to represent the following statement:

    All pelicans eat fish.

2. Several "if-then" statements are listed below. Which of them seem to be true if the diagram you have drawn represents a true statement?
    a) If a bird is a pelican, then it eats fish.
    b) If a creature eats fish, then it is a pelican.
    c) If a bird is not a pelican, then it doesn't eat fish.
    d) If a creature doesn't eat fish, then it is not a pelican.

3. Draw an Euler diagram to represent the statement:

    Pro basketball players are not midgets.

4. Which of the statements below are true if your diagram represents a true statement?
    a) If a fellow is a pro basketball player, then he is not a midget.
    b) If a fellow is not a pro basketball player, then he is a midget.
    c) If a fellow is not a midget, then he is a pro basketball player.
    d) If a fellow is a midget, then he is not a pro basketball player.

5. You know that the general conditional statement $a \rightarrow b$ is represented by the diagram shown here. Which of the following "if-then" statements does it represent?
    a) If b, then a.
    b) If not a, then not b.
    c) If a, then b.
    d) If not b, then not a.

    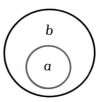

6. How many "if-then" statements does an Euler diagram represent?

BY PERMISSION OF JOHN HART AND FIELD ENTERPRISES, INC.

B.C. has made an analogy in this cartoon between flying and traveling by wheel. Peter said "If God had wanted you to fly, he would have given you wings."

1. What "if-then" statement is implied by B.C.'s analogy?

2. Can you write a different "if-then" statement that has the same meaning?

Lesson 3

Equivalent Statements

Lewis Carroll, the author of *Alice's Adventures in Wonderland* and *Through the Looking Glass*, was a mathematics teacher who wrote stories as a hobby. His books contain many amusing examples of both good and deliberately poor logic and, as a result, have long been favorites among mathematicians. Consider the following conversation held at the Mad Hatter's Tea Party.

"Then you should say what you mean," the March Hare went on.

"I do," Alice hastily replied; "at least—at least I mean what I say—that's the same thing, you know."

"Not the same thing a bit!" said the Hatter. "Why, you might just as well say that 'I see what I eat' is the same thing as 'I eat what I see'!"

"You might just as well say," added the March Hare, "that 'I like what I get' is the same thing as 'I get what I like'!"

"You might just as well say," added the Dormouse, who seemed to be talking in his sleep, "that 'I breathe when I sleep' is the same thing as 'I sleep when I breathe'!"

"It *is* the same thing with you," said the Hatter, and here the conversation dropped, and the party sat silent for a minute.

Carroll is playing here with pairs of related statements and the Hatter, the Hare, and the Dormouse are right: the sentences in each pair do not say the same thing at all. Consider the Dormouse's example. If we change his two statements into "if-then" form, we get

If I am sleeping, then I am breathing,

and

If I am breathing, then I am sleeping.

Although both statements may be true of the Dormouse, the first statement is true and the second statement is false for ordinary beings.

Notice that the hypothesis, "I am sleeping,' and the conclusion, "I am breathing," of the first statement are interchanged in the second. The second statement is called the *converse* of the first.

▶ The *converse* of a conditional statement is formed by interchanging its hypothesis and conclusion.

Symbolically, the converse of $a \rightarrow b$ is $b \rightarrow a$. We see that the converse of a true statement may be false. It is also possible that it may be true, but in either case a statement and its converse do not have the same meaning. In other words, accepting a statement as true does not require us to accept its converse as true. Euler diagrams for the Dormouse's sentences look like this.

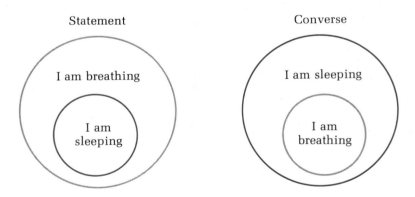

Statement   Converse

Notice that if a point is not inside the larger circle in one of these diagrams, then it is not inside the smaller circle. Because of this, each diagram illustrates *two* statements. The first one represents not only

If am sleeping, then I am breathing,

but also

> If I am not breathing, then I am not sleeping.

The second of these statements is called the *contrapositive* of the first.

▶ The *contrapositive* of a conditional statement is formed by interchanging its hypothesis and conclusion and denying both.

In symbols, the contrapositive of $a \rightarrow b$ is not $b \rightarrow$ not a. Since both a statement and its contrapositive are represented by the same diagram, they are said to be *logically equivalent*. One cannot be true and the other false; they are either both true or both false.

The second diagram above represents not only

> If I am breathing, then I am sleeping,

but also

> If I am not sleeping, then I am not breathing.

Since both statements are represented by the same diagram, they are also logically equivalent. The first is the converse of the Dormouse's original statement and the second is its *inverse*.

▶ The *inverse* of a conditional statement is formed by denying both its hypothesis and conclusion.

Symbolically, the inverse of $a \rightarrow b$ is not $a \rightarrow$ not b.

The relationships between statements considered in this lesson are summarized below.

$$
\text{logically equivalent} \begin{cases} \text{A conditional statement: } a \rightarrow b \\ \text{The contrapositive of the statement: not } b \rightarrow \text{not } a \end{cases}
$$

$$
\text{logically equivalent} \begin{cases} \text{The converse of the statement: } b \rightarrow a \\ \text{The inverse of the statement: not } a \rightarrow \text{not } b \end{cases}
$$

SET I

Following each of the numbered statements below are three lettered statements. Identify the relationship of each of the lettered statements to the numbered statement if possible. Write "converse," "inverse," "contrapositive," "original statement," or "none," as appropriate.

Exercises

1. If you live in Atlantis, then you need a snorkel.
    a) If you do not live in Atlantis, then you do not need a snorkel.
    b) If you need a snorkel, then you live in Atlantis.
    c) If you do not need a snorkel, then you do not live in Atlantis.

2. If you are over ninety, the Chop Chop Studio will give you free karate lessons.
    a) If the Chop Chop Studio won't give you free karate lessons, then you aren't over ninety.
    b) If you are ninety or less, the Chop Chop Studio will not give you free karate lessons.
    c) The Chop Chop Studio will give you free karate lessons if you are over ninety. (Careful.)

3. All Eskimos like pie.
    a) If someone likes pie, he is an Eskimo.
    b) If someone is not an Eskimo, he likes pie.
    c) A person who does not like pie is not an Eskimo.

4. Lady kangaroos do not need handbags.
    a) If a kangaroo is not a lady, it needs a handbag.
    b) If it needs a handbag, then it is not a lady kangaroo.
    c) A kangaroo does not need a handbag if it is a lady.

Write the indicated statement for each of the following sentences.

5. If the moon is full, the vampires are out. *Converse.*

6. If a giraffe has a sore throat, then gargling doesn't help much. *Contrapositive.*

7. If we have been receiving signals from Jupiter, it may not be wise to go there. *Inverse.*

8. You cannot comprehend geometry if you do not know how to reason deductively. *Converse.* (Careful.)

SET II

The electric circuit shown on the next page includes two light bulbs and a battery wired to a switchboard that has three positions. When the switch is turned to position 2, bulb *b* is on because electricity can flow around a complete circuit including it and the battery. Bulb *a*, however, is off because it is not part of a completed circuit.

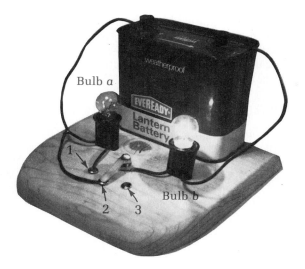

Bulb a

Bulb b

1 2 3

1. Copy and complete the following table.

Switch Position	Bulb a	Bulb b
1	▓	▓
2	Off	On
3	▓	▓

Now suppose you do not know which position the switch is in. Nevertheless, it is possible to conclude from this table that if bulb *a* is on, then bulb *b* is on. If we let *a* mean "bulb *a* is on" and *b* mean "bulb *b* is on," we can say this more briefly as $a \rightarrow b$. In other words, the two light bulbs can be thought of as representing the hypothesis and conclusion of the conditional statement $a \rightarrow b$.

2. If you do not know which position the switch is in, can you conclude that if bulb *b* is on, then bulb *a* is on? In other words, that $b \rightarrow a$?

3. What does this illustrate about the relationship of a statement and its converse?

Suppose we let not *a* mean "bulb *a* is off" and not *b* mean "bulb *b* is off."

4. If you do not know which position the switch is in, can you conclude that if bulb *a* is off, then bulb *b* is off? In other words, that not $a \rightarrow$ not *b*?

5. What does this illustrate?

6. If you do not know which position the switch is in, can you conclude that if bulb b is off, then bulb a is off? In other words, that not $b \longrightarrow$ not a?

7. What does this illustrate?

8. What simple change could be made in the board so that it would illustrate the statement $b \longrightarrow a$?

9. What other statement do you think it would then illustrate?

SET III

The advertising slogan "When you're out of Schlitz, you're out of beer" is a rather unusual one.

1. Is it necessarily true?

2. Write its converse, inverse, and contrapositive.

3. Which of these statements are true and which are false?

"*His logic certainly isn't my logic.*"

DRAWING BY ROSS; © 1970 THE NEW YORKER MAGAZINE, INC.

Lesson **4**

Valid and Invalid Deductions

Suppose that during a trial a lawyer claims that, from the evidence presented, the guilty person is obviously color-blind and that everyone on the jury accepts this as true. Then he produces proof that Mr. Black is color-blind. Must the jury conclude that Mr. Black is guilty? Suppose also that it is established that Miss White is not color-blind. Must Miss White be innocent?

If you think that Mr. Black is not necessarily guilty but that Miss White is definitely innocent, then your logic is probably better than that of the lady in the cartoon (assuming, that is, that her logic is her own peculiar kind).

One way to check these conclusions is with an Euler diagram.

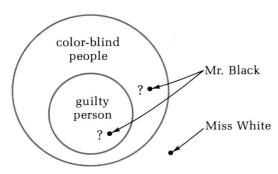

The conditional statement "If the person is guilty, then he is color-blind" is illustrated by the circles. Then we check to see in which regions the suspects belong: in both circles, in just the larger circle, or in neither. We can't tell where Mr. Black belongs, so no conclusion is justified. Since Miss White, however, is outside the larger circle, she cannot be inside the smaller one, and so she is not the guilty person.

Another way to check the two conclusions is as follows. The argument about Mr. Black is:

> If a person is guilty, then he is color-blind.
> Mr. Black is color-blind.
> Therefore, Mr. Black is guilty.

The first sentence in this argument is a conditional statement. The second and third sentences can be combined to form a second conditional statement:

> If a person is color-blind, then he is guilty.

What is the relationship of this statement to the first one? The hypothesis and conclusion of the second statement and those of the first are interchanged, so the second statement is the *converse* of the first. Since a statement and its converse are not logically equivalent, the second statement may be false even though the first statement is true. The conclusion "Mr. Black is guilty" does not necessarily follow from the other two statements, so it is *not a valid deduction*.

How about Miss White? The argument for her is:

> If a person is guilty, then she is color-blind.
> Miss White is not color-blind.
> Therefore, Miss White is not guilty.

Combining the second and third sentences to form the conditional statement

> If a person is not color-blind, then she is not guilty,

we see that it is the *contrapositive* of the first statement of the argument. So if we accept the first statement as true, we cannot help but also accept the second statement as true, since a statement and its contrapositive are logically equivalent. The deduction "Miss White is not guilty" is valid.

Exercises

In each of the following deductions, tell whether or not the third statement necessarily follows if the first two statements are accepted as true. If it does, write *valid* and, if it does not, name the error (*converse* error or *inverse* error) that would be made if the argument were accepted as correct.

1. If you see spots in front of your eyes, you're looking at a leopard.
   You're looking at a leopard.
   Therefore, you see spots in front of your eyes.

2. If you see strands in front of your eyes, your hair is too long.
   You don't see strands in front of your eyes.
   Therefore, your hair is not too long.

3. If you forgot your pencil, you may borrow one of mine.
   You forgot your pencil.
   Therefore, you may borrow a pencil from me.

4. If you brush your teeth with Brylcreem, you misunderstood the commercial.
   You didn't misunderstand the commercial.
   Therefore, you don't brush your teeth with Brylcreem.

5. All moths are attracted to candle flames.
   This insect is not a moth.
   Therefore, this insect is not attracted to candle flames.

6. All carbonated soft drinks contain bubbles.
   You are drinking something that is bubbly.
   Therefore, it is a carbonated soft drink.

7. No graduate of the White Elephant Memory School ever forgets.
   Eloise is very forgetful.
   Therefore, Eloise did not graduate from the White Elephant Memory School.

8. The students will stop paying attention if the class is boring.
   The class isn't boring.
   Therefore, the students will pay attention.

SET II

In each of the following exercises, two statements are given that are to be accepted as true. If possible, write a third statement that can be deduced from these statements. Otherwise, write "no deduction possible."

1. If I have reached the party to whom I am speaking, then I have dialed correctly.
   I have indeed reached the party to whom I am speaking.

2. If the Jolly Green Giant started turning blue, he should put on a sweater.
   The Jolly Green Giant has not started to turn blue.

3. If there is a fly in your soup, then you shouldn't be too quick to swallow each spoonful.
   You may swallow each spoonful quickly.

4. If I had a chimp for a nephew, then I'd be a monkey's uncle.
   I'm a monkey's uncle.

5. All Polaroid cameras take self-developing pictures.
   That camera is not a Polaroid.

6. All night owls hoot it up.
   Fred never gives a hoot.

7. No flying saucer can travel faster than the speed of light.
   The object hovering overhead is not a flying saucer.

8. His name ends in "o" if he is one of the Marx brothers.
   Groucho's name ends in "o."

SET III

The first sentence of the ad on the next page is a conditional statement that most people would probably accept as true. Compare it with the last sentence of the ad. Assuming that "you know just what to do" means that you will decide to buy a Volkswagen station wagon, is the ad's logic valid? Does this conclusion follow logically from what has been said before? Explain.

If the world looked like this,
and you wanted to buy a car that sticks out a little,
you probably wouldn't buy a Volkswagen Station Wagon.
    But in case you haven't noticed, the world doesn't look like this.
    So if you've wanted to buy a car that sticks out a little,
you know just what to do.

27

Lesson **5**

Arguments with Two Premises

"How is it you can all talk so nicely?" Alice said. . . . "I've been in many gardens before, but none of the flowers could talk."

"Put your hand down, and feel the ground," said the Tiger-Lily. "Then you'll know why."

Alice did so. "It's very hard," she said, "but I don't see what that has to do with it."

"In most gardens," the Tiger-Lily said, "they make the beds too soft—so that the flowers are always asleep."

LEWIS CARROLL, *Through The Looking Glass*

Perhaps the mums wouldn't talk even if they were awake! The argument that the Tiger-Lily is presenting to Alice is essentially this:

If the flower bed is too soft, the flowers are always asleep.
If the flowers are asleep, they don't talk.
Therefore, if the flower bed is too soft, the flowers don't talk.

If someone accepted the first two statements as being true, would he also have to accept the third statement as true? We can illustrate the first two statements, called the *premises* of the argument, with Euler diagrams.

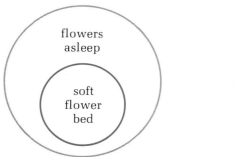
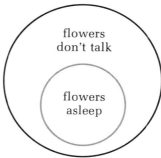

If we combine these two diagrams into one, we get the diagram on the left below. Omitting the second circle leaves the diagram shown at the right, which illustrates the third statement, called the *conclusion* of the argument.

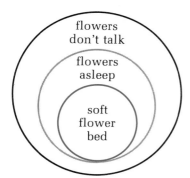
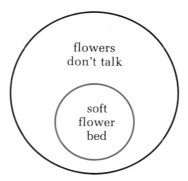

In more general terms, if we have two premises in which the conclusion of one is the same as the hypothesis of the other, such as

$$a \rightarrow b \quad \text{and} \quad b \rightarrow c,$$

then from them we can derive a third statement,

$$a \rightarrow c.$$

This statement, the conclusion, follows logically from the first two, *even though one or both of them may be false.*

Consider this argument:

If you live in Whangamata, then you live in Auckland.
If you live in Auckland, then you live in New Zealand.
Therefore, if you live in Whangamata, you live in New Zealand.

It is not necessary to look in an atlas to tell whether or not this is a valid argument. The conclusion is a logical consequence of the premises.

If either or both of the premises are false, then the conclusion may also be false, but this does not mean that the reasoning is incorrect. In other words, the truth or falsehood of the statements used has nothing to do with the validity of an argument. With this in mind, you should understand why the following argument is logically correct, even though none of the statements make any sense.

If a boy is sixteen, then he is tall for his age.
If a boy is tall for his age, he should stand on his head.
Therefore, if a boy is sixteen, he should stand on his head.

Here is an argument consisting of three true statements. Do you think it is a reasonable one?

If you are using this book, then you must be able to read.
If you are a geometry student, you must be able to read.
Therefore, if you are using this book, you are a geometry student.

An Euler diagram that correctly illustrates the premises but not the conclusion shows that the conclusion does not necessarily follow. This argument is not valid. The trouble is that the premises do not link together properly: the conclusion of one is not the hypothesis of the other.

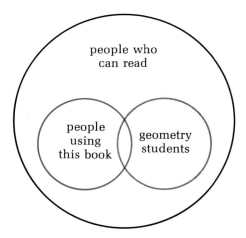

Exercises

Study each of the following arguments carefully to decide whether or not it is valid. Not all of the statements are in "if-then" form, so you may need to rewrite some of them before deciding. Write "valid" or "not valid" as your answers.

1. If the stones are rolling, they are not gathering moss.
   If the stones are not gathering moss, they are a smooth group of rocks.
   Therefore, if the stones are rolling, they are a smooth rock group.

2. If a penny has an Indian head on it, it is very old.
   If a penny has an Indian head on it, it is worth more than one cent.
   Therefore, if a penny is very old, it is worth more than one cent.

3. If you walk under a coconut tree, you will probably be hit on the head.
   If you visit Hawaii, then you will walk under coconut trees.
   Therefore, if you visit Hawaii, you will probably be hit on the head.

4. Blondes have more fun.
   A girl has more fun if she is saucy.
   Therefore, if a girl is a blonde, she is saucy.

"How do you like me as a blonde?"

COURTESY OF VIRGIL PARTCH

5. Tom would be a gardener if he had a green thumb.
   If Tom Thumb were a gardener, he would raise bonsai trees.
   Therefore, if Tom had a green thumb, he would raise bonsai trees.

6. All donkeys have long ears.
   All long-eared creatures are habitual eavesdroppers.
   Therefore, all donkeys are habitual eavesdroppers.

7. All movies directed by Alfred Hitchcock have suspenseful plots.
   The movie *North by Northwest* has a very suspenseful plot.
   Therefore, *North by Northwest* is a Hitchcock movie.

SET II

Does any conclusion follow from the following pair of premises?

If you can play the banjo while standing on your head, then you have practiced the banjo very hard.

If you can't play "Oh Susannah," then you haven't practiced the banjo very hard.

At first glance, the answer may appear to be no, since the conclusion of one premise is not the same as the hypothesis of the other. If we represent the information in the two premises symbolically, we have:

$$a \longrightarrow b$$
$$\text{not } c \longrightarrow \text{not } b$$

Since the contrapositive of a statement is always logically equivalent to the statement, we can replace the second statement, not $c \longrightarrow$ not b, with its contrapositive, $b \longrightarrow c$. Now we have $a \longrightarrow b$ and $b \longrightarrow c$, so we know that $a \longrightarrow c$.

Rewriting the second premise makes the conclusion more obvious:

If you can play the banjo while standing on your head, then you have practiced the banjo very hard.

If you have practiced the banjo very hard, then you can play "Oh Susannah."

Therefore, if you can play the banjo while standing on your head, you can play "Oh Susannah."

The following are pairs of premises from which it may or may not be possible to derive conclusions. In some cases, you will need to consider the contrapositive of one of the statements before being able to tell whether a conclusion is justified. Write either the conclusion statement or "no conclusion" as your answer.

1. If your children can copy their term papers, they will have more time to watch television.
   If you buy the *Encyclopedia Cribanana*, your children can copy their term papers from it.

2. If a lawn is made of Astroturf, then it is always green.
   A lawn is always green if it gets plenty of water.

3. If your name is in *Who's Who*, then you know what's what.
   If you're not sure of where's where, then you don't know what's what.

4. If you are afraid of earthquakes, then you shouldn't live in California.
   If you don't mind getting shaken up, then you are not afraid of earthquakes.

5. If you can leap tall buildings in a single bound, then you are a powerful jumper.
   If you are not a powerful jumper, then you're not from Calaveras County.

6. If you read *Mewsweek* magazine, then you like cats.
   If you are fond of mice, then you do not like cats.

7. All white flowers are fragrant.
   All carnations are white flowers.

8. All shoplifters are dishonest.
   No dishonest person is trustworthy.

SET III

If an argument is logically valid and the premises upon which it is based are true, then its conclusion is also true. Yet, on the surface, the following argument seems to contradict this.

Breadcrumbs are better than nothing.
Nothing is better than a big juicy steak.
Therefore, breadcrumbs are better than a big juicy steak.

Of all the people who agree with the first two statements in this argument, probably not one would accept its conclusion. Since the argument seems to be a logical one, can you explain what is wrong?

Lesson 6

Undefined Terms
and Definitions

BROOM-HILDA © 1973 THE CHICAGO TRIBUNE

The word "duck" is obviously being interpreted in more than one way by the characters in this cartoon. Some words have more than one meaning, perhaps entirely different, but ordinarily we can easily tell from the context in which such a word is being used which meaning is intended. Other words, even though they have just one definition in a dictionary, may mean slightly different things to different people. These different meanings depend upon each person's previous experiences with the word and, if the interpretations differ enough, misunderstandings or even disagreements may result. Variations in a word's meaning seldom cause much of a problem in everyday communication, but they are very serious if we are trying to reason in a precise way.

Consider the simple geometric figure shown here. Is it a square or is it a diamond? What is a square? Is it a figure having four equal sides? Or a figure having four right angles? A figure whose opposite sides are parallel? Perhaps it is a figure that has all of these properties and others; if so, would you have to think of all of them to convince yourself whether or not the figure pictured is a square?

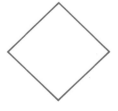

35

Before we are able to get anywhere in geometry, it should be evident that we need to agree upon precise definitions for the geometric terms we will use. Where shall we begin? What words should we define first? It would seem logical to start by defining the simplest terms, terms such as "point" and "line." However, there is a basic problem in trying to do this. For example, how shall we define "point"? Suppose we say "A point is a dot." If someone wants to know what a "dot" is, we might say "A dot is a tiny spot." But what if they want us to define "tiny spot"? Somehow this chain seems about as futile as an often quoted statement of Gertrude Stein, "A rose is a rose is a rose."

Putting it in the form of a pun, what's the point of continuing in this fashion? Let's forget about trying to define the word "point" and call it an *undefined term*. This may seem like giving up, but to define a word we need to use other words. But what about the other words—do we define them too? If we do, then we need still other words. The chain has to end somewhere (unless we come back to where we began, which is what a dictionary has to do), so we choose some especially simple terms as a basis for building definitions of the rest. Among these simple terms, which we will leave undefined, are "point," "line," "plane," and "set."

Exercises

SET I

1. If you look up the word "kiwi" in *Webster's New Collegiate Dictionary*, you will find the definition: "an apteryx." Two definitions of "apteryx" are given, one of which is: "a kiwi." If this were the only definition for "apteryx," could you learn from this dictionary what a kiwi is if you do not know what an apteryx is?

2. If someone wanted to learn how to read Spanish, he might buy a Spanish dictionary. Can you explain why the dictionary

may be of no help at all? (Assume that the dictionary is a good one and that the person has no trouble reading it.)

3. The following dialogue is from a scene in a classroom in the Laurel and Hardy film *Pardon Us*.

> Teacher: What is a comet?
> One of the students: A star with a tail on it.
> Teacher: Can anyone give us an example?
> Laurel: Rin-Tin-Tin?

What is the reason for Laurel's unexpected answer?

4. You are away from home when your cat knocks a flowerpot onto the floor. Nobody is there to hear the crash when the pot breaks. Suppose the cat is deaf so that even it hears nothing. Does the breaking flowerpot make any sound? Most people would probably say that it does. How could someone argue that it does *not*?

SET II

1. Is it possible to define every geometric term by using simpler geometric terms?

2. Euclid, the author of the most famous of all geometry books, tried to define the term "point" as "that which has no part." What do you think he meant by this?

3. If we were to accept Euclid's definition of "point," what word would then either require definition or have to be accepted as an undefined term?

4. An eighteenth-century French mathematician, d'Alembert, once said: "It is the disgrace of geometry that it does not supply me with a definition of line." It is possible to define the term "line" but, if we do, we must then replace it with one or more other undefined terms. Try to write a clear, concise definition to explain your concept of a line.

5. If we defined "plane" as a "flat surface," we would then have to either accept "flat" and "surface" as undefined terms or else provide definitions for these words. Try to write a short definition for each.

SET III

"I don't know what you mean by 'glory,'" Alice said.

Humpty Dumpty smiled contemptuously. "Of course you don't —till I tell you. I meant 'there's a nice knock-down argument for you!'"

"But 'glory' doesn't mean 'a nice knock-down argument,'" Alice objected.

"When *I* use a word," Humpty Dumpty said in rather a scornful tone, "it means just what I choose it to mean—neither more nor less."

"The question is," said Alice, "whether you *can* make words mean so many different things."

"The question is," said Humpty Dumpty, "which is to be master—that's all."

LEWIS CARROLL, *Through The Looking Glass*

There are two sides to every argument and, in this one between Humpty Dumpty and Alice, there is something to be said for each side. Can you explain why?

Lesson **7**

More on Definitions

BY PERMISSION OF JOHN HART
AND FIELD ENTERPRISES, INC.

B.C.'s predicament in trying to learn the meaning of "ecology" from *Wiley's Dictionary* reminds us that it is impossible to define everything without going around in circles.

After some simple terms such as "point," "line," "plane," and "set" have been accepted as undefined, we can begin to define other terms by using them. For example, "space" can be defined as "the set of all points." In saying this, we mean that the word "space" and the phrase "the set of all points" have exactly the same meaning. Not only can we say,

> Space is the set of all points,

we can also say,

> The set of all points is space.

Notice that each of these statements is the converse of the other. You know that, in general, the converse of a true statement is not necessarily true. It is a consequence of the nature of a definition, however, that its *converse is always true.*

Another example illustrated by means of Euler diagrams will help make this clearer. We have already defined "an equilateral triangle" as "a triangle having three sides of equal length." In "if-then" form, this says that

If a triangle is equilateral, then it is a triangle having three sides of equal length.

The converse of this statement is:

If a triangle has three sides of equal length, then it is an equilateral triangle.

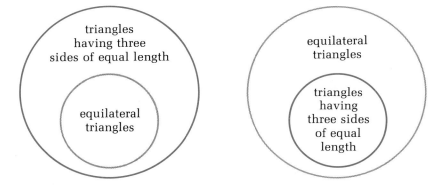

Euler diagrams of these two statements seem to be different, but since "equilateral triangle" and "triangle having three sides of equal length" mean the same thing, the two circles in each figure are not really different at all. They are the same and we can redraw both figures as one.

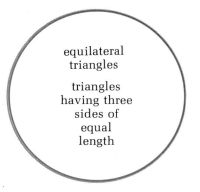

When both a statement and its converse are true, there is a convenient way to combine the two into one. It is by means of the phrase "if and only if." When we say

$$a \text{ if and only if } b,$$

we mean both

$$if\ a, then\ b \quad \text{and} \quad if\ b, then\ a.$$

We can represent the phrase "if and only if" by the symbol $\leftrightarrow$. To write $a \leftrightarrow b$ means that both $a \rightarrow b$ and $b \rightarrow a$ are true.

Look back at the two "if-then" statements about equilateral triangles on the facing page. We can combine them into a single statement by saying

A triangle is equilateral if and only if it is a triangle having three sides of equal length.

Mathematicians usually abbreviate the phrase "if and only if" by writing "iff."

Exercises

SET I

1. You have learned in this lesson that the following statement is true:

   If a statement is a definition, then its converse is true.

   Does it necessarily follow that if its converse is not true, a statement cannot be a definition? Explain.

Decide which of the following true statements are good definitions of the italicized words by determining whether or not their converses are true.

2. If something is *cold*, then it has a low temperature.

3. If a creature is an *ant*, then it is a social insect.

4. A *mandolin* is a stringed musical instrument.

5. A *kitten* is a young cat.

1. Compare the following two sentences:

   If it is your birthday, then you get some presents.
   Only if it is your birthday, do you get some presents.

   a) Do both sentences say the same thing?
   b) Which sentence does this Euler diagram illustrate?
   c) Write the sentence illustrated by the diagram in "if-then" form.

2. The following questions refer to the two Euler diagrams below.

   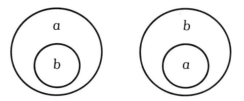

   a) In which diagram is a point inside circle *a only if* it is inside circle *b*?
   b) If you have answered this correctly, the diagram you have chosen illustrates the general statement *a only if b*. Use the diagram to rewrite this statement in "if-then" form.
   c) The statement *a if and only if b* means the same thing as

      a if b   and   *a only if b*.

   Show that the phrase "if and only if" implies that both a statement and its converse are true by rewriting each of the two statements

      a if b   and   *a only if b*

   in "if-then" form.

3. Rewrite the following two statements as a *single* statement using the phrase "if and only if."

   If an automobile is a jalopy, then it is a junky old car.
   If an automobile is a junky old car, then it is a jalopy.

4. Rewrite the following statement as two "if-then" statements, one of which is the converse of the other.

   A person is a goof-off if and only if he habitually shirks responsibility.

5. The *rectangle* is a basic geometric figure. One way to define it is:

A rectangle is an equiangular quadrilateral.

From this definition we can conclude not only that

All rectangles are equiangular quadrilaterals,

but also something else. What?

6. The following statement is a definition of *parallel planes*:

Two planes are parallel if they do not intersect.

Which of the following statements must also be true?
a) If two planes are parallel, then they do not intersect.
b) If two planes are not parallel, then they intersect.
c) If two planes intersect, then they are not parallel.

7. You know that the word *line* is among those terms that we will leave undefined. Suppose instead that we tried to define it in terms of the words *point* and *set*, both having been accepted as undefined. Why is the following definition unsatisfactory?

A line is a set of points.

SET III

It is a well-known rumor that the Upmost Crust Country Club is so snobbish that only those who are descendants of first-class passengers on the *Mayflower* may become members. Franklin Benjamin The Twenty-Second can trace his family tree all the way back to one of its first-class passengers. He is crustfallen when he finds out that the folks of Upmost Crust won't have him.

Can the rumor about those who are eligible for membership be true?

Lesson **8**

Postulates about the Undefined Terms

In order to be a good football player, you have to know the game's rules. *The Official N.C.A.A. Football Guide* lists 197 of them, many of which contain more than one part. The rules of football are, of course, arbitrary, and some of them have been changed since the game was first invented. In order for the game to be playable, however, it is important that the rules be both *sufficient* and *consistent*. By sufficient, we mean that they tell what to do in every situation that might occur and, by consistent, that they neither contradict each other nor lead to contradictions.

Geometry, or any deductive system, is very much like a game. Before playing the game, it is necessary to accept some basic rules, which we will call *postulates*. The postulates in geometry are man-made, just as the rules of football are, and what the subject will be like depends upon the nature of the postulates used. In fact, many different sets of postulates have been invented for geometry, so that there are actually many different geometries rather than just one. In this course, we will spend most of our time studying (perhaps "playing with" is more appropriate) the geometry called Euclidean, named after Euclid. When we use the word geometry, we will be referring to Euclidean geometry. For many centuries, it was the only geometry known, because it took man a long time to realize that more than one set of rules were possible.

You will be happy to know that geometry has very few rules compared with football. We will need to supplement them, however, with some of the rules of algebra with which you are already familiar. For example, you know that any real number is equal to itself (the reflexive rule of equality) and that the order in which two numbers are added is not important (the commutative rule of addition). The rules, or postulates, of algebra concern numbers and operations performed on them.

With what do the postulates of geometry deal? With sets of points and their relationships. A logical way to begin is with some postulates about the undefined terms, "point," "line," and "plane." Since we have no definitions to tell what these words mean, we will give them some meaning by assuming some relationships between them.

To see how this works, let's consider some relationships between *lines* and *points*. The term *line* will always refer in our geometry to a line that is *straight* and that *extends without end* in both directions. We will represent a line like this, the arrowheads indicating that the line does not end where we stop drawing it.

Consider a single point in space. How many lines can pass through, or contain, it? The drawing at the left below shows ten, all necessarily on the surface of the page, but surely there are many, many more in other directions—an unlimited number, in fact.

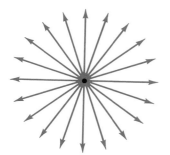

Now consider *two* points in space.* How many lines can contain them? If we didn't think of a line as being straight, the answer would again be an unlimited number, as the right-hand figure on page 45 suggests. Or, if a line is straight but of limited length, the answer would still be an unlimited number, as the figure below, which shows two lines through two points, suggests.

However, if a line is straight and infinite in extent, there seems to be *only one line through two given points*. We shall assume that this is true, and call it a postulate.

▶ **Postulate 1**
For any two points, there is exactly one line that contains them.

This is often said more briefly in the form: Two points determine a line.

Exercises   SET I

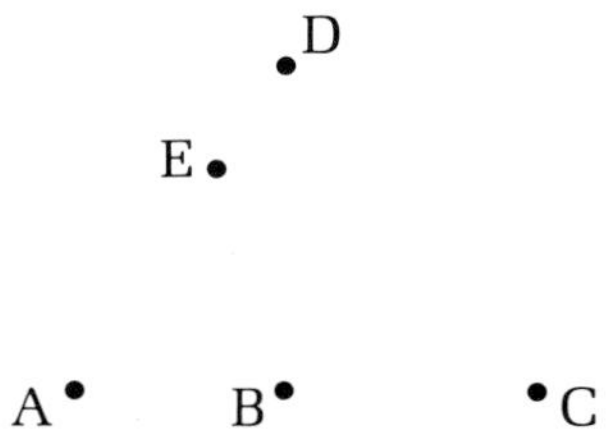

.D

E •

A •   B •   • C

1. The following questions refer to the figure shown here. It represents five points, each named with a capital letter. How many lines do you think can be drawn
   a) through point A? (Ignore the other points in answering this.)
   b) that contain both points A and B?
   c) that contain points A, B, *and* C?
   d) that contain points B, C, *and* D?

▶ **Definition**
Points that lie on the same line are called *collinear points*.

2. Name two sets of at least three points each in the figure for Exercise 1 that seem to be collinear.

3. Restate the definition of "collinear points," using the phrase "if and only if."

---

*Whenever we refer to *two* things in our geometry, we will always mean that they are not the same thing.

4. The statement "Three points determine a line" is false. Make a true statement out of it
   a) by changing just one word.
   b) by adding one word without changing anything else.

5. The following questions refer to the figure shown here. It represents four lines, each named with a small letter. On the assumption that none of the lines intersect except as shown in the figure, how many points do you think can be chosen
   a) on line *a*? (Ignore the other lines in answering this.)
   b) that are on both lines *a* and *b*?
   c) that are on both lines *c* and *d*?
   d) that lie on lines *a*, *b*, and *c*?
   e) that lie on lines *a*, *b*, and *d*?

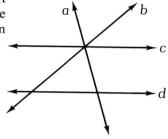

▶ Definition
Lines that contain the same point are called **concurrent lines.**

6. Name a set of at least three lines in the figure above that seem to be concurrent.

7. The statement "Two lines determine a point" is false. Make a true statement out of it by adding just one word.

Although a line contains an unlimited number of points, mathematicians like to be as conservative as possible in what they assume. Instead of saying that a line has that many points, let's simply claim that it contains at least two.

▶ Postulate 2
A line contains at least two points.

8. The first figure shown here represents two lines that do not intersect. On the basis of Postulate 2, what is the minimum number of points that we can claim they have together?

9. The second figure represents two lines that do intersect in a point named P. Each line contains this point. According to Postulate 2, how many *more* points can we say must be on line *c* and how many more points on line *d*? What, then, is the minimum number of points that they have together?

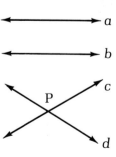

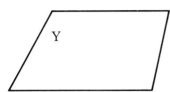

Next we will consider some relationships between *planes* and *points*. When we use the term *plane* in geometry, we mean a *flat surface* that has *no boundaries*. To represent a plane, we draw a rectangle that lies in the plane. The figure labeled X might represent the plane of this page. If we are not looking at the plane "straight on," then a rectangle that lies in it will appear distorted. The figure labeled Y is meant to represent a horizontal, or level, plane. Even though our drawings have boundaries, remember that planes do not.

This photograph shows a model in which points are represented by points of thumbtacks and a plane by a stiff sheet of clear plastic. Three thumbtacks are needed to support the sheet; if one or two of them were removed, the sheet would tip and slide off.

1. The sheet would also tip and slide off if one of the thumbtacks were moved to a different position with respect to the other two. What is it?

2. What relationship must three points have, then, to determine a plane?

We will state this as another postulate.

▶ Postulate 3
For any three noncollinear points, there is exactly one plane that contains them.

3. Remember that points that lie on the same *line* are called *collinear*. What do you think is a logical term to describe points that lie in the same *plane*?

4. Make a drawing to illustrate a plane and two points that lie in it. Draw the line determined by the two points.

5. If two points lie in a plane, what relationship does the line that contains those points have to the plane?

6. Make a drawing to illustrate a plane that is intersected by a line in just one point.

SET III

The figure at the left below shows three noncollinear points and all of the lines (three) determined by these points. The figure at the right shows four points, no three of which are collinear, and all of the lines (six) they determine.

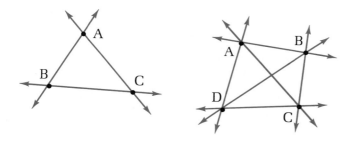

1. Make a drawing to illustrate five points, no three of which are collinear, and all of the lines they determine. How many lines are there in all?

2. Do the same for six points, no three of which are collinear.

3. Without making a drawing, can you figure out how many lines are determined by *ten* points, no three of which are collinear?

Lesson 9

Direct Proof: Arguments with Several Premises

COURTESY OF
THE ADVERTISING COUNCIL

A fish died
because
        it couldn't breathe
because
        its gills got clogged with silt
because
        mud ran into the river
because
        there was nothing to trap the rain
because
        there was a forest fire
because
        someone got careless with fire.

        So please, be careful with fire . . .
because.

The chain of events described in this ad might be expressed as follows:

If you are careless with fire,
    then there will be a forest fire.

If there is a forest fire,
    then there will be nothing to trap the rain.

If there is nothing to trap the rain,
    then the mud will run into the river.

If mud runs into the river,
    then the gills of the fish will get clogged with silt.

If the gills of a fish get clogged with silt,
    then it can't breathe.

If a fish can't breathe,
    then a fish will die.

Therefore, if you are careless with fire,
    then a fish will die.

Notice how the conclusion of each statement matches the hypothesis of the following one. Using letters to represent each hypothesis and conclusion makes the underlying pattern more obvious:

$$a \rightarrow b, \quad b \rightarrow c, \quad c \rightarrow d, \quad d \rightarrow e, \quad e \rightarrow f, \quad f \rightarrow g.$$

Therefore, $a \rightarrow g$.

This argument is a simple example of a *direct proof*. Its pattern suggests the idea of a chain reaction, as the photograph of the dominoes in the ad illustrates. The knocking over of each domino by the one preceding it corresponds to the linking together of the hypothesis of each statement with the conclusion of the previous one.

The first six statements in the argument, called its *premises*, lead to the last statement, called its *conclusion*. The conclusion statement might be considered a *theorem*.

▶ *Theorems* are statements that are proved by means of other statements.

Here is an example of a simple theorem about numbers, and how it might be proved on the basis of some other statements about numbers.

Theorem.

If a number is odd, then it can be written as $2n + 1$ (where n is a whole number).

Proof.

If a number is odd,
   it is 1 more than an even number.

If a number is 1 more than an even number,
   it is 1 more than a multiple of 2.

If a number is 1 more than a multiple of 2,
   it can be written as $2n + 1$ (where n is a whole number).

If we accept the three statements used in the proof as true, then we must accept the theorem that is derived from them as true. Notice that we make no claim about the *absolute* truth of the theorem; it is true *relative* to the statements upon which it is based. To make the meaning of this clearer, suppose someone knows how to reason deductively but does not know whether the three statements in our proof are true or not. He can still recognize, nevertheless, that the theorem is a logical consequence of these statements.

Exercises   SET I

Each of the following exercises consists of a "theorem" and a proof in which one or more of the statements has been omitted. By studying the relationships of the statements given, write the missing statements.

1. *Theorem.*

   If you could count to one million without stopping, it would take you more than a week to do it.
   Proof.

   If you could count to one million without stopping, it would take you at least one million seconds.

   If something takes at least one million seconds, it takes more than sixteen thousand minutes.

   If something takes more than sixteen thousand minutes, it takes more than two hundred hours.

   (What statement belongs here?)

   If something takes more than eight days, it would take you more than a week to do it.

2. *Theorem.*

If the electricity was off during the night, you will be late to school.

Proof.

If the electricity was off during the night, your clock will be slow.

If your clock is slow, you won't realize what time it is.

(What statement belongs here?)

3. *Theorem.*

If there is a total eclipse of the sun, the temperature can be determined without a thermometer.

Proof.

(What statement belongs here? Look at the theorem and the next statement.)

If the sky becomes dark, crickets think that it is night.

(What statement belongs here?)

If crickets start chirping, it is possible to estimate the temperature by counting the number of chirps per minute.

If the temperature is estimated by counting cricket chirps, it can be determined without a thermometer.

4. What theorem is proved by the following statements?

Proof.

If the moon were made of green cheese, mice would make eager astronauts.

If mice were eager astronauts, sooner or later NASA would send some on a lunar mission.

If mice were sent on a lunar mission, the eyes of the entire world would be watching them on television.

If the eyes of the entire world watched some mice on television, it would be one giant peep for mousekind.

5. What is wrong with the following proof?

Theorem.

If all three sides of a triangle are equal, then each of its angles has a measure of 60°.

Proof.

If all three sides of a triangle are equal, then the triangle is equilateral.

If a triangle is equiangular, then it is also equilateral.

If a triangle is equiangular, then each of its angles has a measure of 60°.

1. The following logic exercise was written by Lewis Carroll.

   Theorem.
   Babies cannot manage crocodiles.
   Proof.
   Babies are illogical.
   Nobody is despised who can manage a crocodile.
   Illogical persons are despised.

   Carroll deliberately made the proof difficult to follow by not stating his sentences in "if-then" form and by not stating them in logical order. Make the proof more understandable by rewriting it, doing both of these things.

2. The following proof looks incorrect but is actually valid. Show why there is nothing wrong with it by replacing one of the statements with one that is logically equivalent.

   Theorem.
   If the opposite sides of a quadrilateral are equal, then its diagonals bisect each other.
   Proof.
   If the opposite sides of a quadrilateral are equal, then they are parallel.
   If a quadrilateral is not a parallelogram, then its opposite sides are not parallel.
   If a quadrilateral is a parallelogram, then its diagonals bisect each other.

3. Reorganize the proof below so that it is easier to follow. (It should be rewritten on your paper.)

   Theorem.
   If there is no Great Pumpkin, Snoopy won't have pie for dinner.
   Proof.
   If Lucy plays a trick on Charlie Brown, he will be upset.
   If Linus is mistaken, Lucy is pleased.
   If Lucy becomes rambunctious, she plays a trick on Charlie Brown.
   If Linus is not mistaken, there is a Great Pumpkin.
   If Charlie Brown doesn't feed Snoopy, Snoopy won't have pie for dinner.
   If Lucy is pleased, she becomes rambunctious.
   Charlie Brown forgets to feed Snoopy if he is upset.

Can you rearrange the following statements in logical order? If so, what theorem do they prove?

If I have trouble with a proof, it is not easy.
If I study a proof without getting dizzy, it is one I understand.
If a proof is not arranged in a logical order, I can't understand it.
A proof is giving me trouble if I get dizzy while studying it.
This proof is not arranged in a logical order.

"Don't take this lesson too seriously because tomorrow's lesson will contradict what you've learned today."

Lesson **10**

Indirect Proof

A series of lessons in a subject that contradicted each other would make that subject very confusing. Yet, in reasoning deductively in geometry, it is sometimes helpful to try to arrive at contradictions deliberately!

In the last lesson you became acquainted with the method that we will use to prove most of the theorems of interest in this course. Called the *direct method* of proof, it consists of a sequence of conditional statements such as

$$a \rightarrow b, \quad b \rightarrow c, \quad c \rightarrow d, \quad d \rightarrow e$$

that "link together" to yield a theorem, which in this case is

$$a \rightarrow e.$$

It is not always convenient, or even possible, to prove a theorem directly in this manner. When this is the case, another method is useful, called the *indirect method*. It is this method that involves reasoning to a contradiction.

To get an idea of how indirect reasoning works, consider the following puzzle.

Emerson, Lake, and Palmer are different heights. Who is the tallest and who is the shortest if *only one* of the following statements is true?

1. Emerson is the tallest.
2. Lake is not the tallest.
3. Palmer is not the shortest.

One way of starting to figure this puzzle out would be to assume that the first statement is the one that is true, which would mean that the other two are false. This results in the following three statements:

1. Emerson is the tallest.
2. Lake is the tallest.
3. Palmer is the shortest.

(Note that if it is false that "Lake is not the tallest," then Lake is the tallest.) The first two statements contradict the information that Emerson, Lake, and Palmer are *different* heights. Something is wrong. It is reasonable to conclude that, if there is nothing wrong with the puzzle itself, our assumption that the first statement is true must be wrong. Therefore, it is either the second or the third statement that is true.

On the assumption that the second statement is the true one, we get:

1. Emerson is not the tallest.
2. Lake is not the tallest.
3. Palmer is the shortest.

Again we come to a contradiction. One of the three *has* to be the tallest, yet these statements eliminate all three. So, our assumption that the second statement is true is apparently wrong.

That leaves just one more possibility if the puzzle has a solution at all; namely, that the third statement is the true one. We have:

1. Emerson is not the tallest.
2. Lake is the tallest.
3. Palmer is not the shortest.

This time there is no contradiction. Evidently Lake is the tallest of the three and Emerson is the shortest.

The indirect method consists of considering all of the possibilities and then eliminating all but one by showing that they lead to logical contradictions. As to which statement is true in the puzzle about Emerson, Lake, and Palmer, there are three possibilities to consider. We showed that two of these yield contradictory results.

You will find in applying the indirect method to geometric proofs that it is usually necessary to consider just two possibilities. After eliminating one of them, we can conclude that the other is true.

Exercises

SET I

The basic pattern in proving a theorem, say $a \rightarrow b$, indirectly is to begin by assuming *not b*. It is by reasoning from *this* statement that we hope to arrive at a contradiction. The statements b and *not b* are called *opposites* of each other.

1. Write the opposites of the following statements.
   a) Seven is a prime number.
   b) Mr. Spock does not like contradictions.
   c) A line contains at least two points.

To prove a theorem indirectly, we begin by assuming the opposite of its conclusion.

2. Write the opposite of the conclusion of each of the following theorems.
   a) If a number is odd, its square is odd.
   b) If two lines intersect, they intersect in no more than one point.
   c) In a plane, two lines perpendicular to a third line are parallel to each other.

Here is an example of an indirect proof.

Theorem. If today is Groundhog Day, Uncle Porky will do some shadow tricks.
Proof. Suppose Uncle Porky won't do any shadow tricks. If Uncle Porky will not do any shadow tricks, then today is not a holiday. And if today is not a holiday, it certainly can't be Groundhog Day. However, today *is* Groundhog Day. So, Uncle Porky will do some shadow tricks.

3. The following questions refer to this proof.
   a) What is the relationship of the opening statement in the proof to the theorem being proved?
   b) What does the reasoning based upon this statement lead to?
   c) What does this result indicate?

Indirect proofs can be very simple. For example, it is sufficient to know that poison ivy has leaves in groups of three to prove that the plant in this photograph is not poison ivy.

4. The following questions refer to such a proof.
    a) With what assumption would we begin the proof?
    b) What conclusion follows from this assumption?
    c) What does this conclusion contradict?

    Since our initial assumption led to a contradiction, it must be false. In other words, the statement "The plant in this photograph is poison ivy" is false.
    d) What statement, then, must be true?

SET II

You have been using the indirect method in your reasoning since the time you first began to think. Simple examples, such as the poison-ivy problem, seem to be a part of our "common sense." The following problems are a little more challenging.

1. A backward geometry student named Dilcue is taking a true-false quiz of five questions. From past experience, he is certain of the following facts:

    If the first answer is true, the next one is false.
    The last answer is always the same as the first answer.

    When he sees the quiz, he is positive that the second answer is true. On the assumption that all of this information is correct, write an indirect proof that the last answer is false.

2. Lorelei's boy friend has given her a "diamond" ring but she isn't certain that it is genuine. On the assumption that each of the following statements is true, can you prove anything about Lorelei's "diamond"?

> If a stone is a diamond, its index of refraction is more than 2.
> The stone in Lorelei's ring has an index of refraction of 2.4
> If a stone is not a diamond, its hardness is less than 10.
> If a stone's hardness is less than 10, it can be scratched by corundum.
> The stone in Lorelei's ring cannot be scratched by corundum.

SET III

Another puzzle about Emerson, Lake, and Palmer. They are arguing about how many friends Tarkus has. Emerson claims that Tarkus has at least a hundred friends. Lake says that Tarkus certainly doesn't have that many, whereas Palmer remarks that Tarkus must have at least one friend.

If what *only one* of the three is saying is true, how many friends does Tarkus actually have?

Lesson 11

Some Theorem Proofs

The picture shown here is a painting by a French artist, Francois Morellet. Titled *Four Superimposed Webs*, it consists of a set of intersecting lines drawn over a dark background. Someone has described the painting in these words: "Circular shapes and concentric rings, some with bright, some with dark centers, make their appearance and then disappear again. Their appearance is so unstable that the eye cannot hold any one of them for long but is quickly drawn away to others clamoring for attention. The whole surface becomes a field studded with stars."*

This painting includes illustrations of the three basic undefined terms of geometry: point, line, and plane. In Lesson 8, we made up some postulates about these undefined terms, which are listed here. A fourth postulate, suggested by one of the exercises in that lesson, is included in the list.

▶ **Postulate 1**
For any two points, there is exactly one line that contains them.

*Cyril Barrett in *Op Art* (The Viking Press, 1970), p. 50.

▶ **Postulate 2**

A line contains at least two points.

▶ **Postulate 3**

For any three noncollinear points, there is exactly one plane that contains them.

▶ **Postulate 4**

If two points lie in a plane, then the line that contains them lies in the plane.

The information in these postulates can be used to prove other relationships between points, lines, and planes. In other words, *if we accept these four statements as being true, we can show by deductive reasoning that other statements must also be true.*

For example, consider Postulate 3. It is illustrated again by the "tacks and plastic" photograph. In effect, the postulate says that

three tacks, not all in line with each other, are sufficient to determine the position of the plastic sheet. There are also other ways. For instance, the edge of a ruler and a golf tee will work. In more

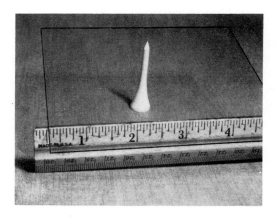

abstract terms, it seems that:

> For a line and a point not on the line,
> there is exactly one plane that contains them.

Can this statement be proved on the basis of our postulates? In "if–then" form, it says:

> If there is a line and a point not on the line,
> then there is exactly one plane that contains them.

To prove this, we must show through a sequence of statements that the hypothesis implies the conclusion. Our proof will look something like this:

> If there is a line and a point not on the line, then *b*.
> If *b*, then *c*.
> If *c*, then there is exactly one plane that contains them.

Of course, the proof may have more than three steps as shown here, or it might even have only two.

The figure at the right illustrates the hypothesis of the theorem: "If there is a line and a point not on the line." Do any of our postulates permit us to draw any conclusions from this? Postulates 1, 3, and 4 assume that we have two or more points and there is only one in the figure, point A. Postulate 2, however, says that a line contains at least two points, so we may choose two points, B and C, on our line. Postulate 3 says that for any three noncollinear points, there is exactly one plane that contains them. So we know that exactly one plane contains points A, B, and C. Does it also contain the line through B and C? Yes, because Postulate 4 says, if two points lie in a plane, then the line that contains them lies in the plane.

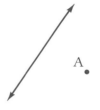

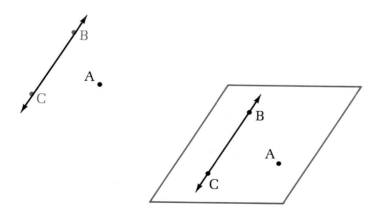

Exercises

1. The statement "For a line and a point not on the line, there is exactly one plane that contains them" is a theorem, because we have proved it by means of our postulates and the definition of "collinear points." The proof is rewritten here. Give the number of the postulate or identify the definition that justifies each statement.

   Theorem.
   For a line and a point not on the line, there is exactly one plane that contains them.

   Proof.
   a) If there is a line, there are two points on the line.
   b) If there are two points on a line and a third point *not* on the line, the three points are noncollinear. (Hint: The reason is a definition.)
   c) If three points are noncollinear, then there is exactly one plane that contains the points.
   d) If a plane contains two points, it also contains the line through those points.

2. The following theorem is easy to prove by the *indirect method*.

   Theorem.
   If two lines intersect, they intersect in no more than one point.

   a) With what assumption should we begin?
   b) If the two lines intersect in more than one point, then they intersect in at least two points. But this contradicts one of our postulates. The figure below will help you decide which one it is. (The lines in the figure have been drawn curved, but this is *not* the contradiction. We have no postulate that says that lines are straight.)

   c) What does the fact that we have come to a contradiction indicate?
   d) What follows from the fact that our initial assumption is false?

1. The figures below illustrate the ideas used in proving another theorem:

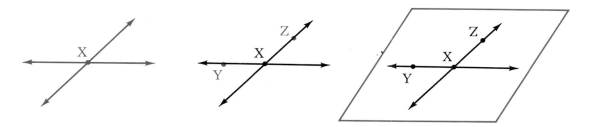

Theorem.
If two lines intersect, there is exactly one plane that contains them.

The reasons for the statements in the following proof of this theorem are either postulates, definitions, or previously proved theorems. Identify the reason for each.

Proof.
a) If two lines intersect, they intersect in no more than one point.
b) If two lines intersect in no more than one point, there is an extra point on each line.
c) If there is an extra point on each line, then the three points are noncollinear.
d) If three points are noncollinear, then there is exactly one plane that contains them.
e) If a plane contains two points, it also contains the line through those points.

2. One more theorem that we can prove by the indirect method is:

Theorem.
If a line intersects a plane that does not contain it, they intersect in no more than one point.

a) With what assumption should we begin?
b) The figure shown here illustrates this assumption. Which one of our postulates does it contradict?
c) What have we proved to be false?
d) What follows from this conclusion?

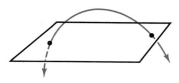

The following theorem can be proved by means of one of the postulates below.

Theorem.   If two dogs have some fleas, they share no more than one flea.

Postulate 1.   For any two fleas, there is exactly one dog that has them.
Postulate 2.   A dog has at least two fleas.
Postulate 3.   If two fleas live in a doghouse, then the dog that has them lives in the doghouse.

1. Do you think it would be best to try to prove the theorem *directly* or *indirectly*?

2. Can you figure out a proof?

3. A perceptive person might say that we have proved this before. Can you explain why?

Chapter 1/Summary and Review

The following check list includes the most important concepts in the Introduction and Chapter 1.

Basic Ideas

Postulates

1. Two points determine a line.   46
2. A line contains at least two points.   47
3. Three noncollinear points determine a plane.   48
4. If two points lie in a plane, then the line that contains them lies in the plane.   62

Exercises

1. Write in "if-then" form:
   a) All limericks have five lines.
   b) I will make a fortune when I perfect my perpetual motion machine.

2. "In *that* direction," the Cat said, waving its right paw round, "lives a Hatter: and in *that* direction," waving the other paw, "lives a March Hare. Visit either you like: they're both mad."

   "But I don't want to go among mad people," Alice remarked.

   "Oh, you can't help that," said the Cat: "we're all mad here. I'm mad. You're mad."

   "How do you know I'm mad?" said Alice.

   "You must be," said the Cat, "or you wouldn't have come here."

   Alice didn't think that proved it at all. . . .

   LEWIS CARROLL, Alice in Wonderland

   The Cat is claiming that
   If you are not mad, then you wouldn't have come here.

   Which of the following statements do you think the Cat would agree with?
   a) If you are not here, then you are not mad.
   b) If you are mad, then you are here.
   c) If you are here, then you are mad.

3. The following pairs of statements appear in the book *Knots* by the psychiatrist R. D. Laing.* What is the relationship of the second statement to the first one in each case?
   a) I never got what I wanted.
      I always got what I did not want.
   b) What I want, I can't get.
      What I get, I don't want.
   c) I get what I deserve.
      I deserve what I get.

4. Write the indicated statement for each of the following sentences.
   a) Where there's smoke, there's fire. *Converse.*
   b) If it isn't an Eastman, it isn't a Kodak. *Contrapositive.*

---

*Pantheon Books, 1971.

c) If Evel Knievel tried to jump over the Grand Canyon, he
   would land in the Colorado River. *Inverse.*

BY PERMISSION OF JOHN HART AND FIELD ENTERPRISES, INC.

5. Peter's argument in this cartoon is:

> All apes have tails.
> You do not have a tail.
> Therefore, you are not an ape.

a) Is this argument valid?
b) Does it prove what B.C. asked?

SET II

1. "Euclid gathered together the
   geometric knowledge of his time,
   and arranged it
   not just in a hodge-podge manner,
   but,
   he started with what he thought were
   self-evident truths
   and then proceeded to
   PROVE all the rest by
   LOGIC.
   A splendid idea, as you will admit.
   And his system has served
   as a model
   ever since."

   LILLIAN LIEBER, *The Education of T. C. Mits*

a) What are Euclid's "self-evident truths" called?
b) What are the rest of the truths proved by logic called?
c) What is another name for the logic used in geometric
   proofs?

2. Why are the terms *point*, *line*, and *plane* left undefined?

3. Comment on the good and bad points of the following statement:

    The surface of a window is a plane.

4. Sherlock Holmes once said to Watson:

    "How often have I said to you that when you have eliminated the impossible, whatever remains, *however improbable, must be the truth?*"*

    What method of proof is Holmes describing?

5. Distracted father: Snooks, stop making that same noise!
    Baby Snooks: This ain't the same noise, daddy. It's another one just like it!

    Explain the basis for disagreement in this conversation.

6. Notice that the following statement contains the phrase *only if*.

    A person enjoys gambling only if he likes to take a chance.

    a) Draw an Euler diagram to illustrate it.
    b) Write a statement in "if-then" form that says the same thing.

SET III

1. Can you rewrite the following proof so that it is easier to follow?

    If a rabbit's name is Harvey, he is invisible.
    A rabbit will not be taken seriously if he is thought to be imaginary.
    If a rabbit is over six feet tall, his name is Harvey.
    If only a rabbit's best friends can see him, everyone else will think he is imaginary.
    If a rabbit is invisible, then only his best friends can see him.

2. What theorem does it prove?

---

*Sir Arthur Conan Doyle, *The Sign of the Four*.

Chapter 2

FUNDAMENTAL IDEAS:
LINES AND ANGLES

Lesson 1

The Distance Between
Two Points

BY PERMISSION OF JOHN HART AND FIELD ENTERPRISES, INC.

Before building his tree tunnel, Peter tried to use some simple geometry to figure out how high it should be. Unfortunately, he didn't think his method through very carefully. If Peter had only *added* his height to the height of his wheel, his head would have cleared the top of the tunnel with some room to spare. Not realizing that his confused idea of distance relationships was at fault, he put the blame on his tape measure instead.

The concept of distance is a very basic one in geometry. In fact, it was used in the puzzles of the surfer and the spotter with which we began our study of the subject. In this lesson, we will consider in more detail what distance is and how it is measured.

To measure the distance between two points on a line, we need a ruler. A ruler provides us with a way of assigning *numbers* to the *points* on the line. From the numbers associated with two points, we can figure out *another number*, the *distance* between them.

Peter, in the adjoining figure, is measuring the distance from the top of his head, A, to the bottom of his feet, B. He has lined up his tape-measure ruler with these two points and sees that they correspond to the numbers 70 and 10, respectively. From this he can conclude that the distance between them is

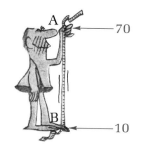

$$70 - 10 = 60.$$

If the numbers on the ruler are one inch apart, Peter is 60 inches tall.

In the figure at the right, the ruler has been moved so that the numbers that correspond to A and B have changed. The distance between the two points, nevertheless, is still the same because

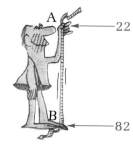

$$82 - 22 = 60.$$

We will represent the distance between points A and B by the symbol AB. In each case, then, AB = 60.

Of course, Peter can move his ruler to many other positions so that the numbers that correspond to points A and B are different in each case. However, for a given unit of measurement such as the inch, the distance between the two points never changes: it will always be 60. We will state this observation as a postulate.

▶ **Postulate 5** (The Distance Postulate)
To every pair of points there corresponds a unique positive number.

To find this number, the *distance* between the two points, we will assume that we have an infinite ruler.

▶ **Postulate 6** (The Ruler Postulate)
The points on a line can be placed in a one-to-one correspondence with the real numbers so that number differences measure distances.

In making a one-to-one correspondence between the points on a line and the real numbers, we will call the number that corresponds to each point its *coordinate*. According to our Distance Postulate, the distance between two points is always *positive*. Therefore, in calculating it from the coordinates of the two points, we must always subtract the smaller coordinate from the larger.
Here is a restatement of this conclusion in symbols:

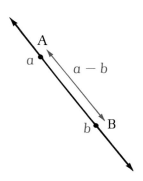

If points A and B have coordinates a and b, and a is greater than b, then

$$AB = a - b.$$

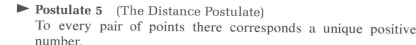

Exercises

Points A, N, and T are on the same line.

1. What word describes their relationship?

2. If the distance from A to N is twice the distance from N to T (written: AN = 2NT),
   a) find AN if NT = 3 units.
   b) find NT if AN = 10 units.

3. In the figure below, the coordinates of F and Y are 4 and 11, respectively, and FL = 2.

   a) What is the coordinate of L?
   b) Find LY.

4. In the figure below, the coordinates of B, U, and G are b, u, and g, respectively. Suppose the coordinates increase from left to right.

   a) What distance does $g - u$ represent?
   b) Express BU in terms of b and u.

5. Find the following distances in the figure below.
   a) AT.
   b) NA.
   c) GN.
   d) GT.

G   N A   T

−33   −7 3   77

1. What seems to be wrong with the figure below?

2. Copy the figure below and, using the following clues, name as many of the points as you can.

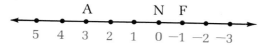

a) The coordinate of L is −2.
b) AN = NY. (Assume that A and Y are different points.)
c) The point R corresponds to 4.
d) The coordinate of O is the negative of the coordinate of F.
e) The coordinate of D is one larger than the coordinate of R.
f) GF = 3.

3. The points A, P, H, I, and D are spaced evenly along a line.

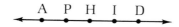

If the coordinates of
a) A and P are 0 and 3, respectively, find the coordinate of D.
b) A and D are 5 and 15, respectively, find the coordinate of H.
c) P and H are 1 and −3, respectively, find the coordinate of D.

4. Points T, I, C, and K are collinear.
a) If the coordinate of I is 6 and IC = 4, what numbers could be the coordinate of C? Draw a figure to illustrate.
b) If the coordinate of I is 0 and the coordinates of T and K are negatives of each other, what can you say about TI and IK?

A Distance Mystery. Use your ruler to draw a horizontal line segment that is 2 inches long. Mark its midpoint with a dot. Then draw directly downward from this point a vertical line segment that is 2 inches long to form a T-shaped figure.
There is something very peculiar about the result. What is it?

Lesson 2

Betweenness of Points

The map of the United States shown here appears to be "upside down." This is because we are used to thinking of the North Pole as the "top" of the world. If the South Pole were chosen as the top instead, the directions of north and south as customarily shown on maps would probably be reversed. Perhaps this seems like idle speculation, but if civilization had developed in the southern hemisphere rather than the northern, it is likely that our maps would now be drawn like the one shown above. At any rate, we are so conditioned to looking at maps with north on top that this one doesn't "look right" even though it is a perfectly good one.

Look at the points on the map representing the cities of Boston, New York, and Philadelphia. Which city is between the other two? The answer to this question seems fairly obvious: New York is. Now look at the points representing Salt Lake City, Las Vegas, and San Francisco. Which one of these cities is between the other two? In this case, the answer is not so obvious.

The idea of "betweenness" of points is a very basic one in geometry. In defining the term, we will limit it to sets of points that are *collinear* so as to avoid situations that might be confusing. In the figure at the left below, for instance, the points shown in brown are between points X and Y; the rest of the points are not.

In the figure at the right, point B is between points A and C. Notice that the distance from A to C is equal to the sum of the distance from A to B and the distance from B to C; that is, $AC = AB + BC$.

► Definition (Betweenness of Points)
Point B is **between** points A and C iff A, B, and C are collinear and

$$AB + BC = AC.$$

We will use the symbol A-B-C to mean that B is between A and C.

Now that we have defined the word *between*, we can use it to define other important terms. For example, look at the figure shown here. The set of the two points A and B and all of the points on the line between them is called the line segment AB.

► Definition
A *line segment* is the set of two points and all the points between them.

A line segment is a *set of points*, whereas its length is a *number*.

► Definition
The *length of a line segment* is the distance between its endpoints.

For simplicity, we will refer to line segments that have equal lengths as "equal segments."

► Definition
The *midpoint of a line segment* is the point on the segment that divides it into two equal segments.

If B is the midpoint of line segment AC in the figure shown here, then $AB = BC$.

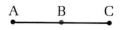

For any two points A and B, we will want to distinguish between the *line* they determine, the *line segment* they determine, and the *length* of the line segment they determine. To do this, we will use the symbols

$$\overleftrightarrow{AB}, \overline{AB}, \text{ and } AB,$$

respectively.

Symbol	Meaning	Illustration
$\overleftrightarrow{AB}$	The *line* through A and B.	 A B
$\overline{AB}$	The *line segment* with endpoints A and B.	 A B
AB	The *length* of the line segment $\overline{AB}$.	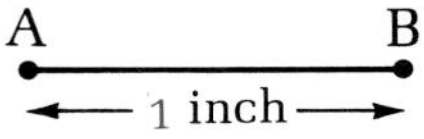 A B ← 1 inch → AB = 1

Exercises

1. In the adjoining figure, point A is between points J and R.
   a) According to the definition of betweenness of points, what relationship must the three points have?
   b) Write an equation that includes the distances JA, AR, and JR.
   c) The symbol J-A-R means that A is between J and R. What other way of writing it indicates the same thing?

   J   A   R

2. In the figure shown here, U is the midpoint of $\overline{MG}$.
   a) Write an equation containing the lengths MU and UG that follows from this fact.
   b) Since the midpoint of a line segment is also between its endpoints, we can write another equation containing the lengths MU, UG, and MG. What is it?

   M   U   G

3. Point C is between points A and N.
   a) Write a three-letter symbol to represent this fact.
   b) Write an equation that follows from this fact.
   c) Can we conclude that $AC = CN$?

4. What do each of the following statements describe?
   a) It is the set of points C and U and all points P such that C-P-U.
   b) It is the point O on $\overline{PT}$ such that $PO = OT$.

5. Points V, A, and T are collinear. Which point is between the other two if
   a) TV + VA = TA?
   b) TA = 1, VA = 2, and VT = 3?

6. The points in the figure below are collinear.
   a) Is point O between points B and L?

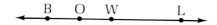

   b) Explain your answer.

7. Points T, U, and B are collinear.
   a) How many lines do they determine?
   b) How many line segments do they determine? Name them.

SET II

1. Read the definition of *line segment* again. Could this definition have been stated before we had defined betweenness of points?

2. If point E is between points K and G, we can write the following equation:

   $$KE + EG = KG.$$

   Two other equations, each containing all three lengths, can also be written. What are they?

3. If four points are collinear, and in the order shown here, we can write the equation, PA + AI + IL = PL. This follows from the definition of betweenness of points, but is not quite the same thing because we have more than one point between two others.

   Show that this equation is correct by doing the following:
   a) Write an equation based upon the fact that A is between P and L.
   b) Write an equation based upon the fact that I is between A and L.
   c) Tell why the two equations you have written imply the equation

   $$PA + AI + IL = PL.$$

4. Points U, R, and N are collinear and the coordinate of U is 5. If UR = 4, UN = 3, and RN = 7, what are the coordinates of R and N if
   a) the coordinate of R is larger than the coordinate of U?
   b) the coordinate of R is smaller than the coordinate of U?

5. Why is the following statement incorrect?

   J is the midpoint of UG.

6. It is often said that "the shortest distance between two points is a straight line." Although its meaning seems fairly clear, this statement is actually incorrect.
   a) Why is it incorrect?
   b) With what phrase could the last three words be replaced so that the statement makes more sense?

SET III

Five crows are sitting in a row on a fence. Can you figure out the order in which they are seated from the following clues?

1. Clyde is the same distance from Alvin that Alvin is from Benny.

2. Ed is seated between Donald and Alvin.

3. Benny is sitting next to Ed.

4. Ed is not seated between Benny and Donald.

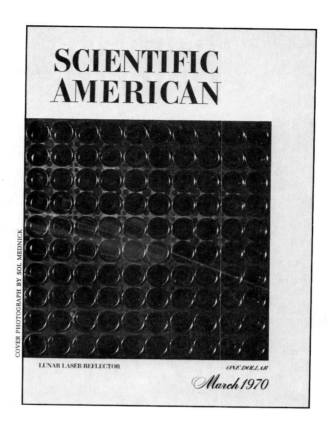

COVER PHOTOGRAPH BY SOL MEDNICK

Lesson 3

Rays and Angles

It has long been known that the distance from the earth to the moon is approximately 240,000 miles. In fact, the Greek astronomer Hipparchus used some basic geometry to figure this out back in the second century B.C.! Since then, the accuracy to which the distance is known has been improved to the extent that we can now tell the distance at any moment (it is actually continually changing) to the *nearest six inches*! This is possible because of a lunar laser reflector left on the moon by astronaut Edwin Aldrin during the first landing of men on the moon in 1969. It is pictured on this cover of *Scientific American*.

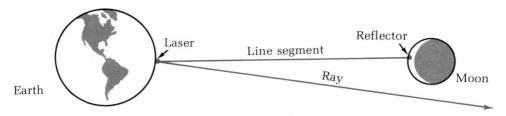

Earth — Laser — Line segment — Ray — Reflector — Moon

The method includes bouncing a laser beam from the earth off the moon and back. Since the path of the beam from the earth to the moon seems to be part of a line bounded by two endpoints, it is a good physical model of a *line segment*. If the laser were beamed away from the moon and if it went infinitely far off into space, it would illustrate a *ray*. A ray, in contrast to a line segment, has only one endpoint.

To get an idea of just how a ray may be precisely defined, imagine the laser beam slowly being turned away from the moon until it just misses it and shoots off into space. The ray shown starts at point A and passes the moon at point B. It includes all of the points on the line $\overleftrightarrow{AB}$ that are between A and B and all of those points on the line that are on the other side of B with respect to A.

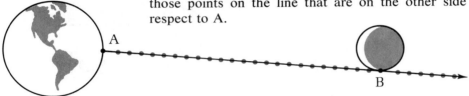

These ideas are included in the following definition.

► Definition
A *ray* $\overrightarrow{AB}$ is the set of points A, B, and all points P such that A-P-B or A-B-P.

Notice the symbol used to represent a ray in this definition. The endpoint of a ray is always named first, followed by the name of any other point on it.

If a point is chosen on a line as shown in this figure, then it is the endpoint of two rays that point in opposite directions. Choosing an extra point on each ray enables us to name them ray $\overrightarrow{AB}$ and ray $\overrightarrow{AC}$.

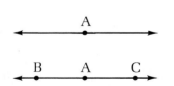

► Definition
Rays $\overrightarrow{AB}$ and $\overrightarrow{AC}$ are *opposite rays* iff A is between B and C.

Notice that, according to this definition, opposite rays have the same endpoint. Two rays that have a common endpoint and that are *not* opposite rays form an *angle*.

Some examples
of angles

▶ Definition
An *angle* is a pair of rays that have the same endpoint but do not lie on the same line.

The rays are called the *sides* of the angle and their common endpoint is called the *vertex* of the angle. Angles can be named in several different ways. For example, the angle shown here can be named ∡BAC or ∡CAB (where the vertex is named in the middle). Since no other angle is shown that has the same vertex, we can also simply call it ∡A. Finally, we can name it with the number written inside: ∡1.

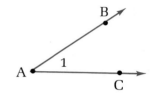

SET I

Exercises

1. In the figure below, points M, E, and W are collinear.

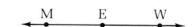

   a) What are rays $\overrightarrow{EM}$ and $\overrightarrow{EW}$ called?
   b) According to our definition of an angle, do these rays form an angle?

2. The four points in the figure shown here are collinear.

   a) What is the endpoint of $\overrightarrow{KB}$?
   b) Name $\overrightarrow{KB}$ in two other ways.
   c) According to our definition, are $\overrightarrow{AR}$ and $\overrightarrow{RA}$ opposite rays?
   d) Give a simple name for the set of points that $\overrightarrow{AR}$ and $\overrightarrow{RA}$ have in common.

3. Name the angle shown here in three different ways.

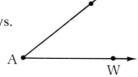

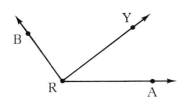

4. It is possible to answer the following questions about ∡YAP without looking at a figure.
   a) What point is its vertex?
   b) What rays are its sides?

5. The figure shown here contains three angles. Since they all have the same vertex, R, it is necessary to name each with three letters in order to tell which is which. Name the three angles.

SET II

1. What do each of the following statements describe?
   a) It is the set of points L and O and all points W such that L-W-O or L-O-W.
   b) It is the set of points L and O and all points W such that O-W-L or O-L-W.

2. In the figure shown here, G, U, and T are collinear, as are R, U, and N. Which of the following are correct names for angles whose sides are shown in the figure?
   a) ∡GNR.
   b) ∡U.
   c) ∡RUT.
   d) ∡N.
   e) ∡UGR.

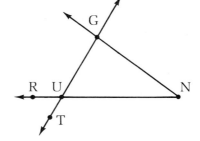

3. Draw figures to help in answering the following questions. How many angles are formed by
   a) two intersecting lines?
   b) three concurrent lines?

4. Suppose that H, O, W, and L are four points such that $\overleftrightarrow{HO}$ contains W and $\overline{WL}$ contains O. Which of the following statements must be true?
   a) The four points are collinear.
   b) W is between H and O.
   c) $\overrightarrow{LW}$ contains O.
   d) $\overrightarrow{OL}$ and $\overrightarrow{OH}$ are opposite rays.

". . . he discovered at last that with great concentration and self-control he was able to change direction and bend wherever he chose. So he did, and made an angle."

These words from the book *The Dot and the Line*, by Norton Juster, describe how a line learns to turn itself into an angle.

1. In doing this, is it possible for it to be both a line and an angle at the same time? Explain.

2. Could the line turn itself into a ray? Explain.

3. Is it possible, if the line bends itself at several points simultaneously, for it to become more than one angle at a time? Explain.

Lesson **4**

Angle Measurement

BY PERMISSION OF JOHN HART AND FIELD ENTERPRISES, INC.

The ancient Babylonians chose 360 as their number for measuring angles more than four thousand years ago. The unit based upon this number, the *degree*, has been used to measure angles in geometry ever since.

Imagine a ray that rotates in a plane about its endpoint as shown in the adjoining figure. The measure of one rotation is 360°. Now let's consider the set of all the rays corresponding to the positions of this rotating ray. All of these rays have the same endpoint and lie in the same plane. We will refer to this set of rays as a *rotation of rays*. Some of them are shown here.

These rays form many different angles. To measure these angles, it seems reasonable that we should number the rays in the same way that we numbered the points on a line so that we could measure distances. For a given coordinate system on a line, exactly one real number corresponds to each point.

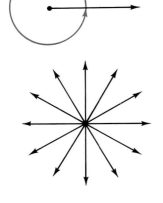

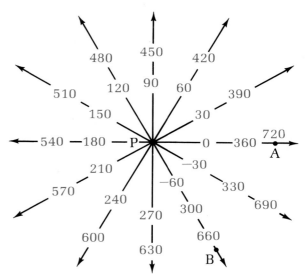

In a comparable system for a rotation of rays, an unlimited number of real numbers correspond to each ray. A few of these are illustrated in the rotation of rays shown above. If ray $\overrightarrow{PA}$ is numbered 0, it can also be numbered 360, 720, and so forth. Ray $\overrightarrow{PB}$ can be numbered −60, 300, 660, and so forth.

Our problem is this. For simplicity, we would like every angle to have a unique measure just as every line segment has a unique length. But which numbers corresponding to the sides of an angle do we choose? Is the measure of ∡APB 360 − 300 = 60°? Or is it 300 − 0 = 300°? Or perhaps 660 − 0 = 660°?

To avoid this difficulty, we will limit our set of rays to those included in half a rotation. We will refer to such a set of rays as a *half-rotation of rays*. The figure at the right below illustrates some of the rays in such a set. Rays $\overrightarrow{PA}$ and $\overrightarrow{PC}$ are opposite rays: if we assign the number 0 to $\overrightarrow{PA}$, then the number 180 can be assigned to $\overrightarrow{PC}$ and the rest of the rays in the half-rotation can be matched with the numbers between 0 and 180. This is, in fact, the basis for the design of a *protractor*, the instrument used to measure angles. It is numbered from 0 to 180 to correspond to the rays in a half-rotation.

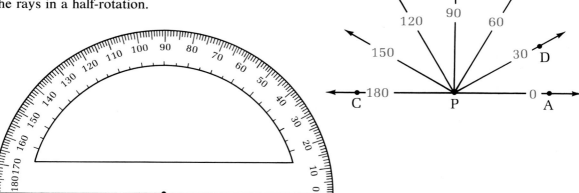

Angles can now be measured by number differences without confusion in the same way that lengths are. The measure of ∡APD in the half-rotation figure is $30 - 0 = 30°$. The measure of ∡CPD in the same figure is $180 - 30 = 150°$.

We are now ready to state some postulates concerning the measurement of angles that are comparable to Postulates 5 and 6 for measuring lengths.

► **Postulate 7** (The Angle Measure Postulate)
To every angle there corresponds a unique number between 0 and 180.

This number is called the *measure* of the angle.

► **Postulate 8** (The Protractor Postulate)
The rays in a half-rotation can be placed in a one-to-one correspondence with the real numbers from 0 to 180 inclusive so that number differences measure angles.

In making such a correspondence, we will call the number that corresponds to each ray its *coordinate*. It follows from our Angle Measure Postulate that the measure of an angle is always positive. Therefore, in calculating it from the coordinates of the angle's sides, we must always subtract the smaller coordinate from the larger.

To distinguish between an angle (a set of points) and its measure (a number), we will use the symbol ∡ when we mean the *angle*, and the symbol ∠ when we mean its *measure*. In symbols, we can write:

If the sides of ∡APB have coordinates a and b, and a is greater than b, then

$$\angle APB = a - b.$$

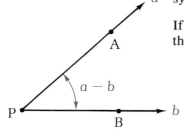

Angles are classified according to their measures as follows:

► Definitions
An *acute angle* is an angle whose measure is less than 90°.
A *right angle* is an angle whose measure is 90°.
An *obtuse angle* is an angle whose measure is greater than 90°.

1. According to *The Random House Dictionary of the English Language*, a synonym for "acute" is "sharp-witted" and a synonym for "obtuse" is "dull." Could this help someone remember the types of angles to which these words refer?

Exercises

2. Obtuse Ollie thinks that ⦠O is larger than ⦠X.

   a) Why does he think this?
   b) How could Acute Alice set him right?

3. Find the measure of each of the following angles in the figure shown here and classify each as acute, right, or obtuse.
   a) ⦠PST.
   b) ⦠TSH.
   c) ⦠ESN.
   d) ⦠RSA.

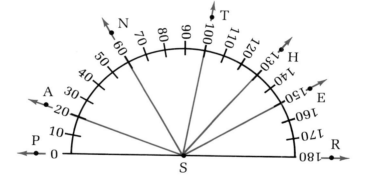

4. Without using a protractor, try to make drawings of angles whose measures are approximately

   $$30°, 45°, 90°, 110°, \text{ and } 150°.$$

   Then use your protractor to see how close you came to each number.

5. It is sometimes convenient to subdivide the degree unit for measuring angles into smaller units. One degree is equal to 60 minutes and one minute is equal to 60 seconds. The symbol for minute is ' and the symbol for second is ".
   a) What is the measure of a right angle in minutes?
   b) The Angle Measure Postulate is stated in terms of degrees. Restate it in terms of minutes.
   c) How many seconds are contained in 1 degree?

Use a protractor to find the measures of the angles in the figures below as accurately as you can. Express each measure to the nearest degree.

1.

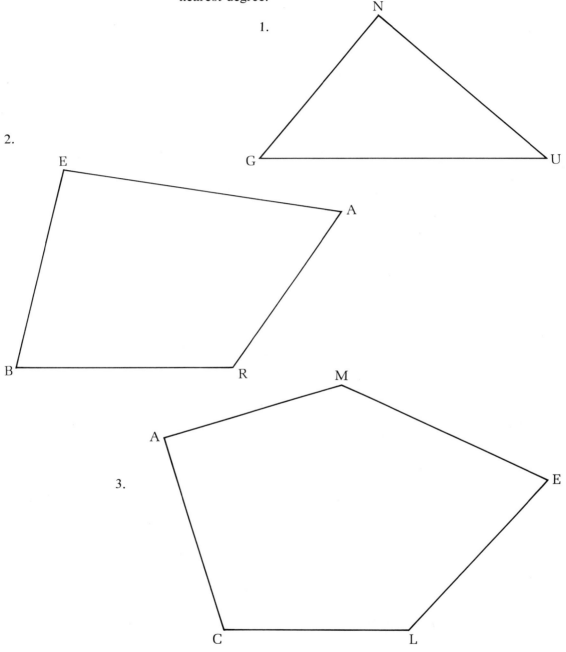

2.

3.

4. Find the sum of the measures of the angles in figure
   a) GNU.
   b) BEAR.
   c) CAMEL.

5. There is a relationship between the sum of the measures of the angles of a figure and the number of sides that it has. What do you think is the sum of the measures of the angles of a figure that has
   a) six sides?
   b) ten sides?

6. Use a straightedge and a protractor to draw a triangle each of whose angles has a measure of 60°. What seems to be true?

7. Use a straightedge and a protractor to draw a four-sided figure whose angles, reading in order around the figure, have measures of 35°, 145°, 35°, and 145°, respectively. What seems to be true?

___

SET III

Although the scale on the protractor shown here looks peculiar, it is possible to measure angles correctly with it. Trace the figure, cut it out, and see if you can figure out how this can be.

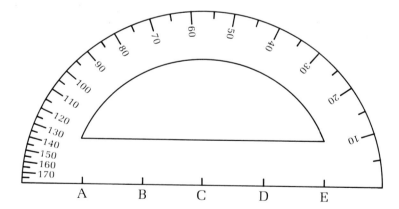

Lesson 5

Complementary and Supplementary Angles

The word "angle" has several different meanings in everyday language. One of its uses as a verb is to mean "to fish with hook and line." An angler, then, is not a mathematician who studies angles; he is a fisherman.

One of the three anglers shown in the picture below has just hooked a large fish. As a result, the angle that his line and pole makes is much larger than the angle of either of the other two. The measures of the three angles have been chosen to illustrate a couple of important angle relationships. Notice that $\angle A + \angle B = 90°$; such angles are said to be *complementary*.

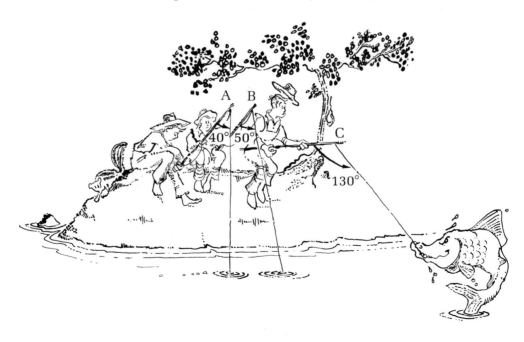

► Definition

Two angles are *complementary* iff the sum of their measures is 90°.

Angles A and B are called *complements* of each other. Since the sum of the measures of two complementary angles is always 90°, we can find the measure of the complement of an acute angle by subtracting the measure of the angle from 90°. It is easy to show that this fact follows directly from our definition. Here is a direct proof in the form: $a \rightarrow b$, $b \rightarrow c$, therefore, $a \rightarrow c$.

If ∡X is the complement of ∡Y, then $\angle X + \angle Y = 90°$.
If $\angle X + \angle Y = 90°$, then $\angle X = 90° - \angle Y$.
Therefore, if ∡X is the complement of ∡Y,
    then $\angle X = 90° - \angle Y$.

The conclusion in this proof, that the measure of a complement of an angle is found by subtracting the measure of the angle from 90°, is an example of a *corollary*.

► A *corollary* is a theorem that can be easily proved as a consequence of a definition, postulate, or another theorem.

Another basic angle relationship is illustrated by the picture of the anglers. Since $\angle B + \angle C = 180°$, ∡B and ∡C are called *supplementary*.

► Definition

Two angles are *supplementary* iff the sum of their measures is 180°.

Such angles are called *supplements* of each other and a corollary that follows from this definition is:

The measure of the supplement of an angle is found by subtracting the measure of the angle from 180°.

A third basic relationship that can exist between angles is that they have equal measures. For simplicity, we will refer to such angles as "equal angles."

► Definition

The *bisector of an angle* is the ray or line that divides it into two equal angles.

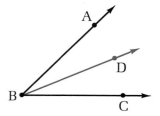

If $\overrightarrow{BD}$ bisects ∡ABC in the figure shown here, then

$$\angle ABD = \angle DBC.$$

Exercises

1. State the definition of complementary angles in the form of two "if-then" statements, one of which is the converse of the other.

2. Obtuse Ollie says: "∠E = 50°, ∠F = 60°, and ∠G = 70°. Therefore, they are supplementary."
   a) Why does he think this?
   b) Acute Alice, having learned the definition of supplementary angles, knows that this isn't right. Why?

3. The supplement of an *acute* angle is *obtuse*. What can you say about
   a) the supplement of a right angle?
   b) the complement of an acute angle?
   c) the complement of an obtuse angle?

4. Find the measure of each of the following angles. Remember that one degree is equal to 60 minutes and that one minute is equal to 60 seconds.
   a) The complement of an angle whose measure is 11°.
   b) The complement of an angle whose measure is 1′.
   c) The supplement of an angle whose measure is 70°.
   d) The supplement of an angle whose measure is 70″.

5. If an acute angle is bisected, the angles formed are also acute. What kind of angles are formed when an obtuse angle is bisected? (Careful.)

SET II

1. In the figure below, numbers have been assigned to some of the rays as assumed in the Protractor Postulate. Angles AOB and BOC are complementary, and $\overrightarrow{OC}$ bisects ∢BOD.
   a) Find ∠BOD. (In other words, what is the measure of ∢BOD?)
   b) Find ∠BOC.
   c) What number corresponds to $\overrightarrow{OC}$?
   d) Find ∠AOB.
   e) What number corresponds to $\overrightarrow{OA}$?

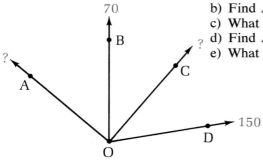

2. The measures of the supplement and complement of a 20°
angle are 160° and 70°, respectively. Find the measures of
the supplement and complement of
   a) a 75° angle.
   b) a 1° angle.
   c) In what way are the results for the 20° angle in the example
      and those for the 75° angle and 1° angle alike?
   d) Use algebra to show why the difference of the supplement
      and the complement of every acute angle is the same
      number. Let $m =$ the measure of the acute angle.

3. Copy and complete the following proof of this corollary: The
   measure of a supplement of an angle is found by subtracting
   the measure of the angle from 180°.
   If ∡X is the supplement of ∡Y, then ▨▨▨▨.
   If ▨▨▨▨, then $\angle X =$ ▨▨▨▨.
   Therefore, if ∡X is the supplement of ∡Y, then $\angle X =$ ▨▨▨▨.

SET III

A Clock Puzzle. You know that minutes and seconds are names
of units of time as well as angle measure. Consider the minute
hand of a clock. During 1 minute of time, it moves through 1/60
of a revolution. Since one revolution has a measure of 360°, the
minute hand moves

$$\frac{1}{60} \cdot 360° = 6°$$

during 1 minute of time. Since a degree is equal to 60 minutes, it
also moves

$$6 \cdot 60' = 360',$$

or *360 angular minutes* during *1 minute of time.*

Can you figure out through how many *angular seconds* the
second hand of a clock moves during *1 second of time?*

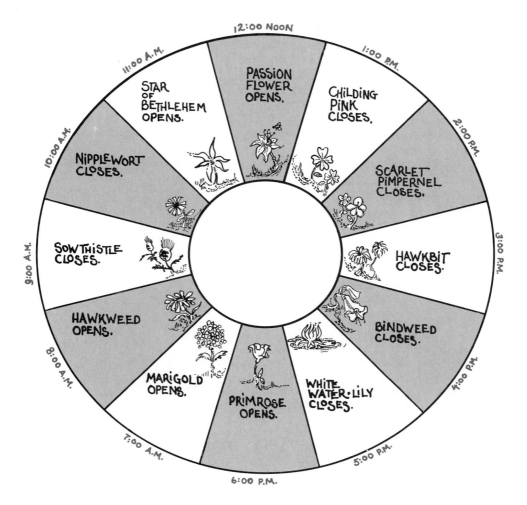

Lesson 6

Betweenness of Rays

There are many flowers that regularly open and close at the same times each day. The Swedish naturalist Carolus Linneaus used this fact to design a flower garden in the shape of a clock that could be used to tell time. It is shown here. If, for example, the marigolds are open but the hawkweed is still closed, it is between seven and eight o'clock in the morning.

Do you think it is correct to say that, on this flower clock, the water-lilies are planted between the bindweed and the primroses?

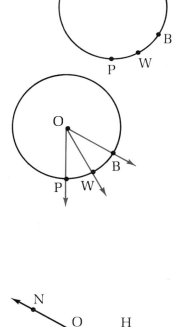

If we use the word "between" in the sense that we have defined "betweenness of points," the answer is no. In the simplified drawing of the clock shown here, points B, W, and P are not collinear. So W is not between P and B.

Now imagine a ray with its endpoint at the center of the clock turning as an hour hand might. Three of its positions are shown by $\overrightarrow{OB}$, $\overrightarrow{OW}$, and $\overrightarrow{OP}$ in the second figure. Is $\overrightarrow{OW}$ between $\overrightarrow{OB}$ and $\overrightarrow{OP}$? The answer to this question depends upon what we mean by "betweenness of rays."

▶ Definition (Betweenness of Rays)
Ray $\overrightarrow{OB}$ is *between* rays $\overrightarrow{OA}$ and $\overrightarrow{OC}$ iff $\overrightarrow{OA}$, $\overrightarrow{OB}$, and $\overrightarrow{OC}$ lie in the same plane and $\angle AOB + \angle BOC = \angle AOC$.

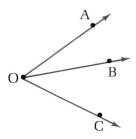

According to this definition, we can say that in the flower clock diagram, $\overrightarrow{OW}$ is between $\overrightarrow{OB}$ and $\overrightarrow{OP}$. Which ray is between the other two in the adjoining figure? The figure satisfies one criterion of our definition of betweenness of rays — that the three rays lie in the same plane — but does not satisfy the other criterion. The sum of the measures of no two of the angles is equal to the measure of the third. How about the next figure? If points S, O, and H are collinear, is $\overrightarrow{OC}$ between $\overrightarrow{OS}$ and $\overrightarrow{OH}$? Again the answer is no; in this case, the figure contains only two angles. Rays $\overrightarrow{OS}$ and $\overrightarrow{OH}$ lie on the same line, so "∡SOH" is not an angle and we cannot write $\angle SOC + \angle COH = \angle SOH$. We will give a special name to ∡SOC and ∡COH, however. We will call them a *linear pair*.

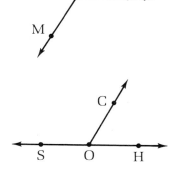

▶ Definition
Two angles are a *linear pair* iff they have a common side and their other sides are opposite rays.

The sum of the measures of the two angles in a linear pair is evidently 180°, which means that they are supplementary. We will assume this as a postulate.

▶ **Postulate 9** (The Linear Pair Postulate)
If two angles are a linear pair, then they are supplementary.

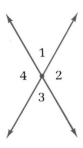

Two lines that intersect form two pairs of "opposite" angles that are called vertical angles. In the figure shown here, ∡1 and ∡3 are one pair of vertical angles; ∡2 and ∡4 are the other pair. Notice that the word "vertical" does not have the same meaning when applied to angles as it does when referring to lines. To define vertical angles precisely, we will use the idea of opposite rays.

▶ Definition
Two angles are *vertical angles* iff the sides of one are rays opposite the sides of the other.

Exercises   SET I

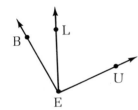

1. Coplanar rays, like coplanar points, lie in the same plane. In the adjoining figure, $\overrightarrow{EB}$, $\overrightarrow{EL}$, and $\overrightarrow{EU}$ are coplanar.
   a) Which ray is between the other two?
   b) Write an equation expressing the relationship of ∠BEL, ∠LEU and ∠BEU.

2. Rays $\overrightarrow{RO}$, $\overrightarrow{RS}$, and $\overrightarrow{RE}$ are coplanar and ∠SRO + ∠ORE = ∠SRE.
   a) Draw a figure to illustrate this.
   b) Which ray is between the other two?

3. In the figure at the left below, points O, L, and D are collinear.
   a) What are ∡OLG and ∡GLD called?
   b) What measure relationship do ∡OLG and ∡GLD have?
   c) Is $\overrightarrow{LG}$ between $\overrightarrow{LO}$ and $\overrightarrow{LD}$? Explain.

4. In the figure at the right, points W, H, and T are collinear.
   a) Name *two* linear pairs in the figure.
   b) Are there any vertical angles in the figure?
   c) Which of the four rays in the figure is between two others?

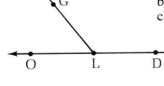

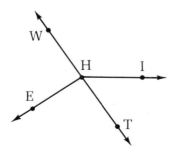

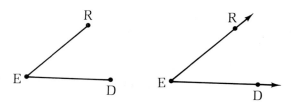

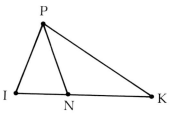

Even though the figure at the left above is not an angle because its sides are segments, it *determines* the angle shown in the figure at the right above.

5. In the first adjoining figure, points I, N, and K are collinear. The segments in the figure determine seven angles. Name them.

6. In the second adjoining figure, points B, O, and N are collinear, as are points R, O, and W.
   a) Name the two pairs of vertical angles determined by the segments in the figure.
   b) Name the four linear pairs determined by the segments in the figure.

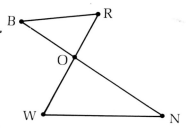

SET II

1. According to the definition of supplementary angles, if the sum of the measures of two angles is 180°, then they are supplementary. Is the converse of this statement true?

2. According to the Linear Pair Postulate, if two angles are a linear pair, then they are supplementary.
   a) State the converse of this.
   b) Is the converse true?

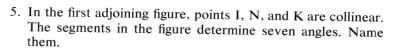

3. In the adjoining figure, $\overrightarrow{RY}$ is between $\overrightarrow{RG}$ and $\overrightarrow{RA}$. In addition to saying that

$$\angle GRY + \angle YRA = \angle GRA,$$

we can write two other equations. One of them is $\angle GRY = \angle GRA - \angle YRA$. What is the other?

4. Copy and complete the following equations concerning the measures of the angles determined by the adjoining figure. Assume that O, E, and A are collinear.
   a) $\angle RAO + \angle OAN = \angle$ ▓▓▓▓.
   b) $\angle ORG + \angle$ ▓▓▓▓ $= \angle ORA$.
   c) $\angle OGN - \angle OGR = \angle$ ▓▓▓▓.
   d) $\angle OEG + \angle GEA =$ ▓▓▓▓.

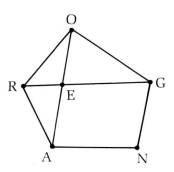

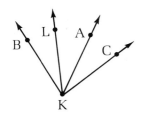

5. Rays $\overrightarrow{KB}$, $\overrightarrow{KL}$, $\overrightarrow{KA}$, and $\overrightarrow{KC}$ in the figure shown here are coplanar.
   a) Write an equation based upon the fact that $\overrightarrow{KL}$ is between $\overrightarrow{KB}$ and $\overrightarrow{KC}$.
   b) Write another equation that can be used with the equation you have just written to show that
   $$\angle BKL + \angle LKA + \angle AKC = \angle BKC.$$

6. Rays $\overrightarrow{PL}$, $\overrightarrow{PU}$, and $\overrightarrow{PM}$ are coplanar. Make a drawing for which it is true that
   a) $\angle LPU + \angle UPM + \angle MPL = 360°$.
   b) $\angle LPU + \angle UPM + \angle MPL = 180°$.

SET III

1. Can you make a drawing illustrating three rays having the same vertex in space such that they form three right angles?

2. Can you make a comparable drawing illustrating the same situation in a plane?

"Have you ever put something down and then not been able to find it?"

COURTESY OF VIRGIL PARTCH

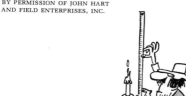

BY PERMISSION OF JOHN HART
AND FIELD ENTERPRISES, INC.

Lesson 7

Some Consequences of the Ruler
and Protractor Postulates

According to the Ruler Postulate, the points on a line can be put in a one-to-one correspondence with the real numbers so that number differences measure distances. In the adjoining figure, point P has been chosen on line ℓ. It separates the line into two parts: the points on each side, together with point P, form two opposite rays.

$$\xleftarrow{\hspace{2cm}} \bullet \xrightarrow{\hspace{2cm}} \ell$$
P

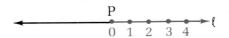

Suppose we set up a point and number correspondence (called a *coordinate system*) so that the number 0 is paired with P and the positive numbers with the points on the ray pointing to the right. Then it is apparent that on this ray there is exactly one point at any given distance from point P. Although this observation is a very simple one, we will find it useful in our future work. Since it is a direct consequence of the Ruler Postulate, we will call it a *corollary* to it.

▶ **The Unique Point Corollary**
There is exactly one point on a ray at any given distance from the endpoint of the ray.

A second consequence of the Ruler Postulate concerns a special point on a line segment.

▶ **The Midpoint Corollary**
A line segment has exactly one midpoint.

These conclusions about rays and line segments result from the way in which the Ruler Postulate numbers points. Some comparable conclusions about angles follow from the way in which the Protractor Postulate numbers rays. According to the Protractor Postulate, the rays in a half-rotation can be placed in a one-to-one correspondence with the real numbers from 0 to 180 inclusive so that number differences measure angles.

To recall what we mean by a half-rotation, consider the figure below. Just as a point separates a line into two parts, so does a line separate a plane into two parts. The points on each side, together with the line, form two *half-planes*. The line is called the *edge* of each half-plane. If we choose a point on the line, the set of all rays in one of the half-planes that have this point for their endpoint is called a half-rotation. The figure shows some of the rays in the half-rotation from point P in the upper half-plane.

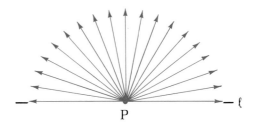

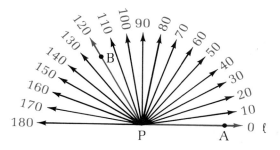

Now suppose we set up our ray and number coordinate system so that the number 0 is paired with $\overrightarrow{PA}$, the ray on the edge of the half-plane that points to the right. Then in the upper half-plane, there is exactly one ray that forms with $\overrightarrow{PA}$ an angle having any given measure. For example, $\overrightarrow{PB}$ is the ray that forms a 120° angle with $\overrightarrow{PA}$. This observation is a direct consequence of the Protractor Postulate. We will state it here for future use.

▶ **The Unique Ray Corollary**
If a ray is on the edge of a given half-plane, there is exactly one ray in the half-plane that forms with the given ray an angle having a given measure.

A second consequence of the Protractor Postulate concerns a special ray in an angle.

▶ **The Angle Bisector Corollary**
An angle has exactly one bisector.

SET I

Exercises

1. In the figure below, points A and B are 3 centimeters apart. How many points that are 4 centimeters from point A are there on
   a) $\overleftrightarrow{AB}$?
   b) $\overrightarrow{AB}$?
   c) $\overline{AB}$?

2. A line through the midpoint of a segment *bisects* the segment or contains the segment. In the adjoining figure, if I is the midpoint of $\overline{SL}$, then line k is a bisector of $\overline{SL}$.
   a) How many midpoints does a line segment have?
   b) How many bisectors can a line segment have?

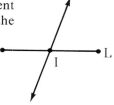

Lesson 7: Consequences of the Ruler and Protractor Postulates   103

3. In the figure below, point E lies on $\overleftrightarrow{HM}$ and $\overleftrightarrow{HM}$ lies in plane P. How many rays with E as their endpoint can be chosen so that they form a 10° angle with $\overrightarrow{EM}$
   a) if the rays are not limited to plane P?
   b) if the rays must lie in plane P?
   c) if the rays must lie in the upper half-plane of plane P with respect to $\overleftrightarrow{HM}$?

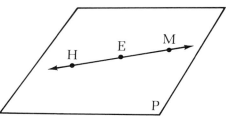

SET II

1. Later in the course, it will be necessary in developing proofs to add points or lines to the figures illustrating them. The corollaries to the Ruler and Protractor Postulates presented in this lesson will often be used to justify these additions.

   For practice, see if you can either name or state the reason that allows each of the indicated steps in completing the drawing of the sailboat.
   a) Let M be the midpoint of $\overline{AB}$.
   b) Draw $\overrightarrow{MX}$ in the half-plane above $\overleftrightarrow{AB}$ so that $\angle AMX = 95°$.
   c) Choose C and D on $\overrightarrow{MX}$ so that MC = 0.5 cm and MD = 2 cm.
   d) Draw $\overrightarrow{DY}$ and $\overrightarrow{DZ}$ in the half-planes to the left and right of $\overleftrightarrow{DC}$ as shown so that $\angle CDY = 45°$ and $\angle CDZ = 30°$.
   e) Choose E and F on $\overrightarrow{DY}$ and $\overrightarrow{DZ}$ so that DE = MD and DF = DC.
   f) Draw $\overleftrightarrow{ME}$ and $\overleftrightarrow{CF}$.

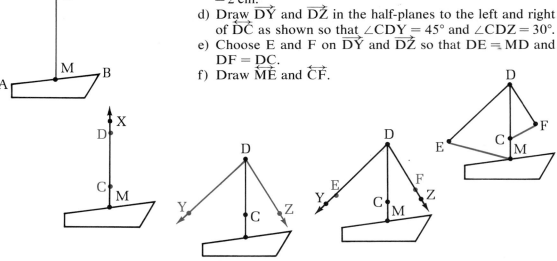

2. Use indirect reasoning to explain why both $\overrightarrow{NA}$ and $\overrightarrow{NY}$ cannot bisect ∡RNO.

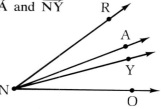

3. Use indirect reasoning to explain why, if L and A are different points on $\overrightarrow{FX}$, FL cannot be equal to FA.

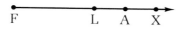

SET III

You know that the figures of geometry, such as points, lines, rays, and segments, are mathematical concepts and not physical objects. The figure shown here, for instance, is not a line segment but merely illustrates one. It is literally a strip of brown ink, consisting of trillions of atoms.

A •————————————• B

If we thought of geometry only in terms of physical objects, we might conceive of a point as an atom and a line as an infinite "string" of atoms that is one atom thick. Any given line segment, then, would consist of a certain number of atoms.

At first glance, it might seem helpful to think of geometric figures in this fashion. If we did, however, we would immediately get into serious logical difficulties. As an example of this, consider the Midpoint Corollary in this lesson. Would it be correct to say that every line segment has exactly one midpoint? Explain.

Chapter 2 / Summary and Review

You should be familiar with the following concepts introduced in Chapter 2.

Basic Ideas

Postulates

5. *The Distance Postulate.* To every pair of points there corresponds a unique positive number. 73
6. *The Ruler Postulate.* The points on a line can be placed in a one-to-one correspondence with the real numbers so that number differences measure distances. 73

   The Unique Point Corollary. There is exactly one point on a ray at any given distance from the endpoint of the ray. 102

   The Midpoint Corollary. A line segment has exactly one midpoint. 102
7. *The Angle Measure Postulate.* To every angle there corresponds a unique number between 0 and 180. 88
8. *The Protractor Postulate.* The rays in a half-rotation can be placed in a one-to-one correspondence with the real numbers from 0 to 180 inclusive so that number differences measure angles. 88

   The Unique Ray Corollary. If a ray is on the edge of a given half-plane, there is exactly one ray in the half-plane that forms with the given ray an angle having a given measure. 103

   The Angle Bisector Corollary. An angle has exactly one bisector. 103
9. *The Linear Pair Postulate.* If two angles are a linear pair, then they are supplementary. 97

SET I

Exercises

1. Complete the following definition statements by using *as few* words or symbols as you can.
   a) If $\measuredangle A$ and $\measuredangle B$ are complementary, then ||||||||.
   b) If $\measuredangle C$ is acute, then ||||||||.
   c) If D is the midpoint of $\overline{EF}$, then ||||||||.
   d) If points G and H have coordinates g and h and h is greater than g, then $GH =$ ||||||||.
   e) If points I, J, and K are collinear and $IJ + JK = IK$, then ||||||||.
   f) $\overline{LM}$ is the set of points L and M and all points N such that ||||||||.
   g) $\overrightarrow{OP}$ is the set of points O and P and all points Q such that either |||||||| or ||||||||.
   h) If $\overrightarrow{RS}$ and $\overrightarrow{RT}$ are opposite rays, then ||||||||.

2. The following questions concern the naming of angles.
   a) Name the vertex and sides of ∡OWL.
   b) Under what condition can an angle be named with a single letter?
   c) Which of the following might also be names for ∡OWL?

   $$∡W \quad ∡L \quad ∡WOL \quad ∡VWX$$

3. In the figure at the right below, $\overrightarrow{NW}$ bisects ∡SNA.
   a) Write the equation that follows from the fact that $\overrightarrow{NW}$ is between $\overrightarrow{NS}$ and $\overrightarrow{NA}$.
   b) Write the equation that follows from the fact that $\overrightarrow{NW}$ bisects ∡SNA.

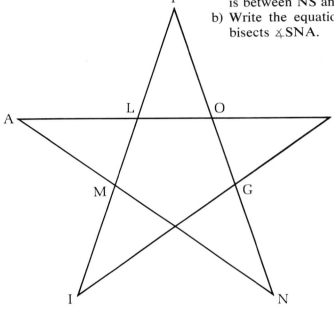

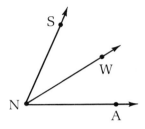

4. Use your protractor to measure each of the following angles in the star figure.
   a) ∡A.
   b) ∡FLO.
   c) ∡LMN.

5. The following questions refer to the same figure. You may assume that points that look collinear actually are.
   a) Name two relationships that ∡OGI and ∡NGI have.
   b) What is the relationship between ∡AMF and ∡IMN?

6. The following questions are about the angles in a linear pair.
   a) If one angle in a linear pair has a measure of 30°, what is the measure of the other angle?
   b) If the two angles of a linear pair are equal, what kind of angles are they?
   c) Can the two angles of a linear pair be complementary?
   d) Can both angles in a linear pair be obtuse?

7. Give a reason for adding each part to the figure at the left below to give the figure at the right.
   a) Let T be the midpoint of $\overline{SO}$.
   b) Choose R on $\overrightarrow{KO}$ so that KR = ST.
   c) Draw $\overleftrightarrow{TR}$.

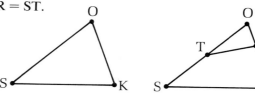

SET II

1. The five labeled points in the figure below are collinear. Copy the figure and add the following information.

   The coordinates of O and N are −17 and 31, respectively.
   OB = 11 and BI = 24.
   O is the midpoint of $\overline{RI}$.

   Now figure out each of the following numbers:
   a) ON.
   b) The coordinate of B.
   c) OI.
   d) The coordinate of I.
   e) RO.
   f) The coordinate of R.

2. Make drawings of three coplanar rays having the same end-point such that they form:
   a) three acute angles.
   b) two acute angles and one obtuse angle.
   c) one acute angle and two obtuse angles.
   d) three obtuse angles.
   e) only two angles, one of which is acute and the other obtuse.

3. Suppose that $\overrightarrow{DO}$, $\overrightarrow{DV}$, and $\overrightarrow{DE}$ are coplanar rays and that $\angle ODV = 75°$ and $\angle VDE = 30°$. What numbers could be the measure of $\angle ODE$? Illustrate your answer.

4. Points J, A, and Y are collinear, JA = 7, and AY = 5. If the coordinate of A is −3 and the coordinate of Y is more than that of J, what numbers could be the coordinates of J and Y? Illustrate your answer.

5. Find the measure in *minutes* of an angle whose complement has a measure of 30,000 *seconds*.

A tripod is a stand having three adjustable legs. Suppose that after the one in this cartoon is captured and brought back for the picture, it adjusts its legs so that one of them is between the other two. In terms of the diagram, suppose $\overline{CO}$ is between $\overline{CR}$ and $\overline{CW}$.

Since we have not defined betweenness of *segments*, we might explain this in terms of either betweenness of rays or betweenness of points.

1. Under what conditions is $\overrightarrow{CO}$ between $\overrightarrow{CR}$ and $\overrightarrow{CW}$?

2. Under what conditions is O between R and W?

If the tripod adjusted its legs in this fashion, it would probably fall over.

3. Why?

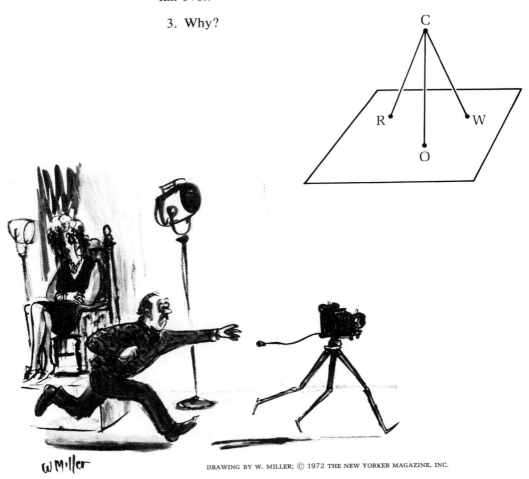

DRAWING BY W. MILLER; © 1972 THE NEW YORKER MAGAZINE, INC.

SOME BASIC POSTULATES
AND THEOREMS

Lesson 1

Postulates of Equality

At this point in your study of geometry you know that it, like arithmetic, is a very precise subject. In reasoning about numbers such as lengths of line segments and measures of angles, we want to avoid situations in which "the answers are mostly a matter of opinion." To accomplish this, we will review some ideas about equality, most of which you have already used in solving equations in algebra.

112

If the symbols a and b represent real numbers, what do we mean by the statement $a = b$? Simply that a and b *represent the same number.* The statements in the list below are direct consequences of this idea. Although you may have proved some of them as theorems in your algebra course, we will refer to these statements as postulates. Since they are not geometric in nature, we will not include them in our list of postulates about geometric figures.

In the interest of simplicity, we will state these postulates symbolically. Each has been given a name with which you may identify it when you use it. The letters in each postulate represent real numbers.

▶ **Postulate 1**   (The Reflexive Postulate)
Any number is equal to itself. In symbols: $a = a$.

▶ **Postulate 2**   (The Symmetric Postulate)
If $a = b$,   then $b = a$.

▶ **Postulate 3**   (The Transitive Postulate)
If $a = b$ and $b = c$,   then $a = c$.

▶ **Postulate 4**   (The Substitution Postulate)
If $a = b$,   then a can be substituted for b (and b for a).

▶ **Postulate 5**   (The Addition Postulate)
If $a = b$,   then $a + c = b + c$.

▶ **Postulate 6**   (The Subtraction Postulate)
If $a = b$,   then $a - c = b - c$.

▶ **Postulate 7**   (The Multiplication Postulate)
If $a = b$,   then $ac = bc$.

▶ **Postulate 8**   (The Division Postulate)
If $a = b$ and $c \neq 0$,   then $\dfrac{a}{c} = \dfrac{b}{c}$.

▶ **Postulate 9**   (The Square Roots Postulate)
If $a = b \geq 0$,   then $\sqrt{a} = \sqrt{b}$.

For the square roots postulate to be meaningful, it is important to understand that the symbol $\sqrt{n}$ represents the *positive* square root of n. For example, 9 has *two* square roots: 3 and -3. The symbol $\sqrt{9}$, however, represents only *one* of them: 3.

Here is an example of how some of the postulates can be put to use. Suppose that the two sticks shown in the figure at the left below have equal lengths and that they are placed together as shown in the top figure at the right. The distance between the left ends, A and C, must be equal to the distance between the right ends, B and D. Can you explain why?

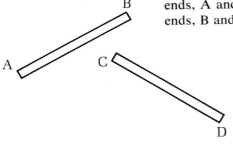

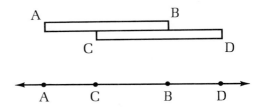

We might represent the sticks as two overlapping segments on a line. From the hypothesis that AB = CD, we want to show that AC = BD. Here is an argument to this effect:

By hypothesis, AB = CD.
AB = AC + CB because C is between A and B, and
CD = CB + BD because B is between C and D.
So AC + CB = CB + BD by the substitution postulate.
Therefore, AC = BD by the subtraction postulate.

Although this result is a simple one and may seem rather obvious, we will eventually use the equality postulates to draw some conclusions that will probably surprise you.

Exercises

SET I

1. Some of the postulates of equality are stated below in terms of lengths of line segments and measures of angles. Copy and complete each one.
   a) The symmetric postulate: If ∠E = ∠D, then ▓▓▓▓.
   b) The multiplication postulate: If JA = CK, then 2JA = ▓▓▓▓.
   c) The square roots postulate: If $DO^2 = UG^2$, then ▓▓▓▓.
   d) The reflexive postulate: AL = ▓▓▓▓.
   e) The subtraction postulate: If DA = VE, then DA − LE = ▓▓▓▓.
   f) The transitive postulate: If MI = KE and KE = NT, then ▓▓▓▓.

Chapter 3: SOME BASIC POSTULATES AND THEOREMS

2. The following exercises will give you some practice in using the substitution postulate. Write the conclusion of each statement.
   a) If $DA - NI = EL$ and $NI = CK$, then ▨▨▨▨.
   b) If $3\angle P + \angle E = 90°$ and $\angle T = \angle E$, then ▨▨▨▨.
   c) If $BR^2 - 1 = AN^2$ and $AN = DY$, then ▨▨▨▨.
   d) If $\dfrac{PH}{GA} = \dfrac{IL}{RY}$ and $NE = IL$, then ▨▨▨▨.

You have used the postulates of equality many times in solving algebraic equations. Name the postulate being used in each step in the solutions of the following equations.

3. The equation to be solved: $2x + 1 = 7$.
   a) If $2x + 1 = 7$, then $2x = 6$. Why?
   b) If $2x = 6$, then $x = 3$. Why?

4. The equation to be solved: $3(x - 4) = 5x$.
   a) If $3(x - 4) = 5x$, then $3x - 12 = 5x$. Why?
   b) If $3x - 12 = 5x$, then $-12 = 2x$. Why?
   c) If $-12 = 2x$, then $2x = -12$. Why?
   d) If $2x = -12$, then $x = -6$. Why?

5. The equation to be solved: $\frac{1}{2}x(x + 2) = x + 8$. Solve for $x > 0$.
   a) If $\frac{1}{2}x(x + 2) = x + 8$, then $\frac{1}{2}x^2 + x = x + 8$. Why?
   b) If $\frac{1}{2}x^2 + x = x + 8$, then $\frac{1}{2}x^2 = 8$. Why?
   c) If $\frac{1}{2}x^2 = 8$, then $x^2 = 16$. Why?
   d) If $x^2 = 16$, then $x = 4$ (since we are to solve for $x > 0$ only). Why?

SET II

Name the definition or postulate that justifies each of the conclusions about the geometric figures shown.

1. a) If J-H-N, then $JH + HN = JN$.
   b) If $OH = HN$, then $HN = OH$.
   c) If $OH^2 = 4$, then $OH = 2$.
   d) If $JO = ON$, then $JO - OH = ON - OH$.

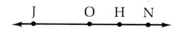

2. a) $PU = PU$.
   b) If U is the midpoint of $\overline{AL}$, then $AU = UL$.
   c) If $\angle A = \angle L$ and $\angle L = 75°$, then $\angle A = 75°$.
   d) If $AL = 2AU$ and $AU = UL$, then $AL = 2UL$.

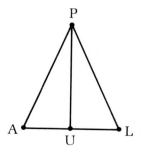

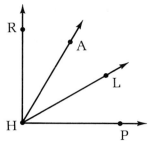

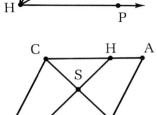

3. a) If $\angle RHL = \angle AHP$, then $\frac{1}{2}\angle RHL = \frac{1}{2}\angle AHP$.
   b) If $\overrightarrow{HA}$ is between $\overrightarrow{HR}$ and $\overrightarrow{HP}$, then $\angle RHA + \angle AHP = \angle RHP$.
   c) If $\angle RHA = \angle LHP$, then $\angle RHA + \angle AHL = \angle AHL + \angle LHP$.
   d) If $\overrightarrow{HL}$ bisects $\angle AHP$, then $\angle AHL = \angle LHP$.

4. a) If $CA = CR$ and $CR = ER$, then $CA = ER$.
   b) $\angle ACR = \angle HCS$.
   c) If $LS = SH$, then $\dfrac{LS}{LH} = \dfrac{SH}{LH}$.
   d) If $\dfrac{CH}{HS} = \dfrac{RL}{LS}$ and $CS = HS$, then $\dfrac{CH}{CS} = \dfrac{RL}{LS}$.

SET III

Consider the following three statements:

> Snail A is as slow as itself.
> If snail A is as slow as snail B,
>   then snail B is as slow as snail A.
> If snail A is as slow as snail B and snail B is as slow
>   as snail C, then snail A is as slow as snail C.

These three statements sound like the reflexive, symmetric, and transitive postulates of equality. A relation that is reflexive, symmetric, and transitive is called an *equivalence relation*.
   Which of the following do you think are equivalence relations? Explain.

1. is older than.

2. is the same color as.

3. is a friend of.

PHOTOGRAPH BY SOL MEDNICK

Lesson 2

Two Bisection Theorems

The man in this picture is holding a photograph of himself that has been cut into two parts. Has the photograph been *bisected* by the cut? No, because the lower part is larger than the upper one. To bisect means to divide into two *equal* parts.

A *line segment* is bisected by its midpoint, or by any line, ray, or segment that intersects it in its midpoint. An *angle* is bisected by the ray or line that divides it into two equal angles.

What is the relationship between the lengths of each part of a bisected line segment and the whole segment? The parts are half the length of the whole segment. Although this relationship is prefectly obvious and seems to need no proof, it is possible to show that it is a logical consequence of the definition of midpoint and the postulates of equality. Remember that statements proved by means of other statements are called *theorems*. We will prove this relationship as our first theorem.

▶ **Theorem 1**
The midpoint of a line segment divides it into segments that are half as long.

117

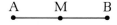

A   M   B

The adjoining figure illustrates the theorem. In terms of this figure, the hypothesis and conclusion of the theorem are:

Hypothesis: M is the midpoint of $\overline{AB}$.
Conclusion: $AM = \frac{1}{2}AB$ and $MB = \frac{1}{2}AB$.

The object of our proof is to show that the conclusion follows logically from the hypothesis. To make it easier to follow, we will write it in two columns with the statements in the left column and the reasons that justify them in the right column.

Proof.

Statements	Reasons
1. M is the midpoint of $\overline{AB}$.	Hypothesis.
2. $AM = MB$.	The midpoint of a line segment divides it into two equal segments.
3. $AM + MB = AB$.	Definition of betweenness of points.
4. $AM + AM = AB$.	Substitution postulate.
5. $2AM = AB$.	Substitution postulate.
6. $AM = \frac{1}{2}AB$.	Multiplication postulate.
7. $MB = \frac{1}{2}AB$.	Substitution postulate (steps 2 and 6).

Notice how, with the exception of the third statement in the proof, each successive statement follows from the one before. The third statement is based upon the observation that, in the figure, point M is between points A and B. To keep our proofs from getting very complicated, we will agree that *certain basic relationships may be assumed on the basis of the figures used to illustrate them.* These relationships will be limited to the following:

> *collinearity of points,*
> *betweenness of points,* and
> *betweenness of rays.*

In proving Theorem 1, this means that we could assume not only that M is the midpoint of $\overline{AB}$, but also that A, M, and B are collinear and that A-M-B.

We have now proved that the midpoint of a line segment divides it into segments that are half as long. A comparable theorem for angles is:

▶ **Theorem 2**
The bisector of an angle divides it into angles that are half as large.

The proof of this theorem is very much like that for Theorem 1 and is left for you to complete as an exercise.

One of the goals of studying geometry will be to learn how to write proofs on your own. At first, to become more familiar with how proofs are developed, you will be asked to complete a few that have been written for you. In each of the following proofs, state the missing reasons.

Exercises

1.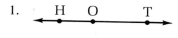

Hypothesis: H-O-T.
Conclusion: $HO = HT - OT$.

Proof.

Statements	Reasons
1. H-O-T.	Hypothesis.
2. $HO + OT = HT$.	a) Why?
3. $HO = HT - OT$.	b) Why?

2.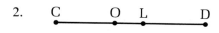

Hypothesis: $CO = LD$.
Conclusion: $CL = OD$.

Proof.

Statements	Reasons
1. $CO = LD$.	Hypothesis.
2. $CO + OL = OL + LD$.	a) Why?
3. $CL = CO + OL$.	b) Why?
4. $CL = OL + LD$.	c) Why? (See steps 2 and 3.)
5. $OL + LD = OD$.	d) Why?
6. $CL = OD$.	e) Why? (See steps 4 and 5.)

3.

Hypothesis: D is the midpoint of both $\overline{TP}$ and $\overline{IE}$; $TP = IE$.

Conclusion: $DE = DP$.

Proof.

Statements	Reasons
1. D is the midpoint of $\overline{TP}$.	Hypothesis.
2. $DP = \frac{1}{2}TP$.	a) Why?
3. $TP = IE$.	b) Why?
4. $DP = \frac{1}{2}IE$.	c) Why?
5. D is the midpoint of $\overline{IE}$.	Hypothesis.
6. $DE = \frac{1}{2}IE$.	d) Why?
7. $DE = DP$.	e) Why?

Lesson 2: Two Bisection Theorems   119

4. 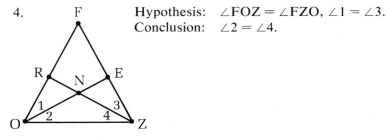 Hypothesis: ∠FOZ = ∠FZO, ∠1 = ∠3.
Conclusion: ∠2 = ∠4.

Proof.

Statements	Reasons
1. ∠FOZ = ∠FZO.	Hypothesis.
2. ∠FOZ = ∠1 + ∠2; ∠FZO = ∠3 + ∠4.	a) Why?
3. ∠1 + ∠2 = ∠3 + ∠4.	b) Why?
4. ∠1 = ∠3.	c) Why?
5. ∠1 + ∠2 = ∠1 + ∠4.	d) Why?
6. ∠2 = ∠4.	e) Why?

SET II

The proof of Theorem 2 is identical in format to that of Theorem 1. A figure, hypothesis, conclusion, and the first three steps are shown below. Copy them and complete the proof, using the proof of Theorem 1 in this lesson as a guide.

Theorem 2.
The bisector of an angle divides it into angles that are half as large.

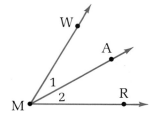 Hypothesis: $\overrightarrow{MA}$ bisects ∡WMR.
Conclusion: ∠1 = ½∠WMR and
∠2 = ½∠WMR.

Proof.

Statements	Reasons
1. $\overrightarrow{MA}$ bisects ∡WMR.	Hypothesis.
2. ∠1 = ∠2	Why?
3. ∠1 + ∠2 = ∠WMR.	Why?
(The rest of the proof is left to you.)	

An Experiment with an Unexpected Result. Cut out a strip of paper about 10 inches long and 1 inch wide. Make it into a loop, turn one end over, and tape the two ends together as shown in the first photograph. Now cut the loop along its center as shown in the second photograph. You know that to bisect is to cut or divide into two equal parts. Does cutting all the way around the loop bisect it?

Lesson **3**

Some Angle Relationship
Theorems

The artist in this cartoon has an outlandish source of inspiration. As you can see, his specialty is drawing "lines" that suddenly change direction. These "lines" are actually sets of line segments that intersect at their endpoints. Each pair of intersecting segments determines an angle and in the figure at the top of the next page the three angles thus determined are lettered A, B, and C.

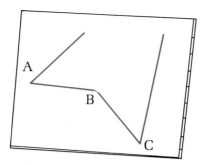

Suppose, in the figure, that ∡A and ∡B are supplementary and ∡B and ∡C are also supplementary. In what way are ∡A and ∡C related?

If "supplementaryness" of angles were transitive, like equality of numbers, we could say that ∡A and ∡C are supplementary. This, however, is not true of the angles in the picture. Suppose ∡A has a measure of 50°. Remembering that ∡A and ∡B are supplementary, what must be the measure of ∡B? Since ∡B and ∡C are also supplementary, what is the measure of ∡C?

The fact that it has the same measure as ∡A suggests the following theorem:

▶ **Theorem 3**
Supplements of the same angle (or equal angles) are equal.

Since the proofs of both versions of this theorem are very much alike, we will prove just the first one. The adjoining figure shows three angles, two of which are supplements of the third. The proof will be based upon the definition of supplementary angles.

Hypothesis: ∡A and ∡C are supplementary;
           ∡B and ∡C are supplementary.
Conclusion: $\angle A = \angle B$.

Proof.

Statements	Reasons
1. ∡A and ∡C are supplementary.	Hypothesis.
2. $\angle A + \angle C = 180°$.	If two angles are supplementary, the sum of their measures is 180°.
3. ∡B and ∡C are supplementary.	Hypothesis.
4. $\angle B + \angle C = 180°$.	Same reason as step 2.
5. $\angle A + \angle C = \angle B + \angle C$.	Substitution postulate.
6. $\angle A = \angle B$.	Subtraction postulate.

A similar theorem holds for complementary angles:

▶ **Theorem 4**
Complements of the same angle (or equal angles) are equal.

Since the proof of this theorem is very much like that for Theorem 3, it is left for you to complete as an exercise.

In an earlier lesson, you guessed that vertical angles are equal in measure. We will prove this as:

▶ **Theorem 5**
Vertical angles are equal.

The figure below shows two lines that intersect to form two pairs of vertical angles: $\angle AOC$ and $\angle DOB$ are one pair; $\angle AOD$ and $\angle COB$ are the other. How can we prove that $\angle AOC = \angle DOB$ and that $\angle AOD = \angle COB$?

Let's consider $\angle AOC$, one of the angles in the first pair. Notice that $\angle AOC$ and $\angle AOD$ are a linear pair, so they are supplementary. Furthermore, $\angle DOB$ and $\angle AOD$ are a linear pair, so they are also supplementary. What does Theorem 3 say about two angles that are supplements of the same angle?

Here is a proof of Theorem 5 in two-column form.

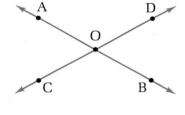

Hypothesis: $\angle AOC$ and $\angle DOB$ are vertical angles.
Conclusion: $\angle AOC = \angle DOB$.

Proof.

Statements	Reasons
1. $\angle AOC$ and $\angle DOB$ are vertical angles.	Hypothesis.
2. $\overrightarrow{OC}$ and $\overrightarrow{OD}$ are opposite rays; also $\overrightarrow{OA}$ and $\overrightarrow{OB}$ are opposite rays.	If two angles are vertical angles, the sides of one are rays opposite the sides of the other.
3. $\angle AOC$ and $\angle AOD$ are a linear pair; $\angle AOD$ and $\angle DOB$ are a linear pair.	If two angles have a common side and their other sides are opposite rays, they are a linear pair.
4. $\angle AOC$ and $\angle AOD$ are supplementary; $\angle AOD$ and $\angle DOB$ are supplementary.	If two angles are a linear pair, they are supplementary.
5. $\angle AOC = \angle DOB$.	Supplements of the same angle are equal.

1. Suppose that ∢O is supplementary to ∢U and ∢U is supplementary to ∢T.
   a) What can you conclude about ∢O and ∢T?
   b) Why?

2. According to Theorem 5, if two angles are vertical, then they are equal.
   a) What is the converse of this theorem?
   b) Is it true?
   c) What is the contrapositive of this theorem?
   d) Is it true?

3. Suppose that ∢N is complementary to ∢E, ∠E = ∠A, and ∢A is complementary to ∢R.
   a) What can you conclude about ∢N and ∢R?
   b) Why?

4. Suppose that ∢I and ∢N are two vertical angles that are also complementary.
   a) What can you conclude about ∢I and ∢N since they are vertical angles?
   b) Why?
   c) What can you conclude about ∢I and ∢N since they are complementary?
   d) Why?
   e) What can you conclude about ∢I and ∢N, using both of these facts?

SET II

State the missing reasons in each of the following proofs.

1.   Hypothesis: ∢1 and ∢2 are vertical angles;
                     $\overrightarrow{DI}$ bisects ∢NDH.
       Conclusion: ∠1 = ∠3.

Proof.

Statements	Reasons
1. ∢1 and ∢2 are vertical angles.	Hypothesis.
2. ∠1 = ∠2.	a) Why?
3. $\overrightarrow{DI}$ bisects ∢NDH.	Hypothesis.
4. ∠2 = ∠3.	b) Why?
5. ∠1 = ∠3.	c) Why?

Lesson 3: Angle Relationship Theorems   125

2.

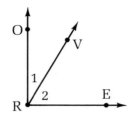

Hypothesis: ∡ORE is a right angle.
Conclusion: ∡1 and ∡2 are complementary.

Proof.

Statements	Reasons
1. ∡ORE is a right angle.	Hypothesis.
2. ∠ORE = 90°.	a) Why?
3. ∠ORE = ∠1 + ∠2.	b) Why?
4. ∠1 + ∠2 = 90°.	c) Why?
5. ∡1 and ∡2 are complementary.	d) Why?

3. *Theorem 4.*

Complements of equal angles are equal.

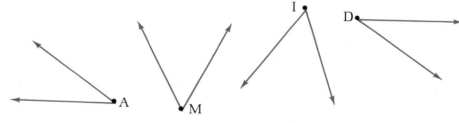

Hypothesis: ∡A and ∡M are complementary;
∡I and ∡D are complementary;
∠M = ∠I.
Conclusion: ∠A = ∠D.

Proof.

Statements	Reasons
1. ∡A and ∡M are complementary; ∡I and ∡D are complementary.	Hypothesis.
2. ∠A + ∠M = 90°; ∠I + ∠D = 90°.	a) Why?
3. ∠A + ∠M = ∠I + ∠D.	b) Why?
4. ∠M = ∠I.	Hypothesis.
5. ∠A + ∠I = ∠I + ∠D.	c) Why?
6. ∠A = ∠D.	d) Why?

If you looked at a geometry book used by students in a foreign country, you would probably be able to guess what many of the proofs were about even though you didn't understand the language. The reason is that most of the symbols used in mathematics are the same throughout the world. For example, look at the passage from a Chinese geometry book shown here. Can you figure out what it says?

定理 3. 對頂角相等.

［設］ 二直線 *AB,CD* 相交於 *O* 點.

∠*AOD* 同 ∠*BOC* 是對頂角,

∠*AOC* 同 ∠*BOD* 是對頂角.

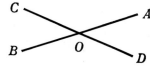

［求證］ ∠*AOD* = ∠*BOC*, ∠*AOC* = ∠*BOD*.

［證］ ∠*AOD* + ∠*AOC* = 2∠*R* ⎫
⎬
∠*BOC* + ∠*AOC* = 2∠*R* ⎭

　　(二鄰角的外邊成一直線,則二角互為補角).

∴　∠*AOD* + ∠*AOC* = ∠*BOC* + ∠*AOC*

　　　　　　　　　　　　(凡平角必等).

∴　∠*AOD*　　　 = ∠*BOC*　(等量減去同量).

仿此　　　　∠*AOC* = ∠*BOD*.

Lesson **4**

Theorems about Right Angles

This picture is a copy of a sketch made by the Dutch artist Maurits Escher for his lithograph *Belvedere*. The young man holds a strange cubelike structure in his hands that is not like any ordinary cube. It is based upon the drawing shown at the top of the next page. The arrows indicate the two places in which the edges appear to cross each other. Which edges are in front and which are in back?

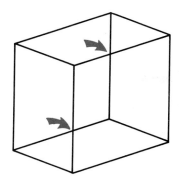

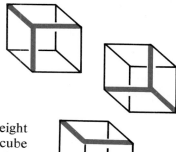

A cube contains 24 right angles in all (three at each of its eight corners). The sides of these angles contain the edges of the cube and the first two adjoining drawings show the only real relationships in space that they can have. Escher has chosen the third, an impossibility.

It seems obvious that, no matter what relative positions two right angles may have, they are always equal in measure. It is easy to prove that this is a direct consequence of our definition of a right angle.

► **Theorem 6**

Any two right angles are equal.

What relationship do the *sides* of a right angle have to each other? They are *perpendicular*. We will first define the word perpendicular in terms of lines and then extend it to include rays and segments.

► **Definition**

Two lines are **perpendicular** iff they form a right angle.

Rays and segments are perpendicular iff the lines that contain them are perpendicular. The symbol for "perpendicular" is ⊥. To indicate that two lines, $\overleftrightarrow{AB}$ and $\overleftrightarrow{CD}$, are perpendicular, we write $\overleftrightarrow{AB} \perp \overleftrightarrow{CD}$.

Two more useful theorems about right angles are stated below.

► **Theorem 7**

If two lines are perpendicular, they form four right angles.

► **Theorem 8**

If the two angles of a linear pair are equal, then each is a right angle.

Exercises

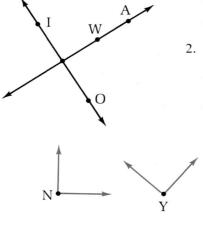

1. The two lines in the adjoining figure are perpendicular.
   a) Write this fact symbolically.
   b) It does not make sense to write IO ⊥ WA. Why not?
   c) Is it true that $\overline{\text{IO}}$ ⊥ $\overline{\text{WA}}$?

2. It is certainly not true that any two acute angles are equal, or that any two obtuse angles are equal. The reason that any two right angles are equal is based upon our definition of a right angle. Copy and complete the following proof of Theorem 6: Any two right angles are equal.

   Hypothesis: ∡N and ∡Y are right angles.
   Conclusion: ∠N = ∠Y.

   Proof.

Statements	Reasons
1. ∡N and ∡Y are right angles. (There are two more steps.)	Why?

3. In proving Theorem 6, it might be misleading to draw the figure shown here, even though it does contain two right angles. Why?

4. a) Draw a figure containing two lines, *c* and *a*, such that *c* ⊥ *a*.
   b) If *c* ⊥ *a* can you conclude that *a* ⊥ *c*?
   c) To what postulate of equality is this property of perpendicular lines comparable?
   d) Add a third line, ℓ, to the figure so that *a* ⊥ ℓ.
   e) Is perpendicularity of lines transitive? (In other words, if *c* ⊥ *a* and *a* ⊥ ℓ, is *c* ⊥ ℓ?)

1. The fact that "two perpendicular lines form four right angles" is called a theorem because it can be proved by means of other facts we already know. How is this statement different from our definition of perpendicular lines?

2. Rather than writing a complete proof of Theorem 7, answer the following questions about the accompanying figure to explain how it follows from other postulates and theorems.

Theorem 7.

If two lines are perpendicular, they form four right angles.

By the definition of perpendicular lines, we know that they form one right angle, say ∢1.
a) Why must ∢3 also be a right angle?
b) What relationships do ∢1 and ∢2 have?
c) Why must ∢2 be a right angle?
d) Why must ∢4 also be a right angle?

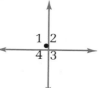

3. Copy and complete the following proof of Theorem 8.

Theorem 8.

If the two angles in a linear pair are equal, then each is a right angle.

Hypothesis: ∢1 and ∢2 are a linear pair;
$\angle 1 = \angle 2$.
Conclusion: ∢1 and ∢2 are right angles.

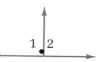

Proof.

Statements	Reasons
1. ∢1 and ∢2 are a linear pair.	Hypothesis.
2. ∢1 and ∢2 are supplementary.	Why?
3. $\angle 1 + \angle 2 = 180°$.	Why?
4. $\angle 1 = \angle 2$.	Why?
5. $\angle 1 + \angle 1 = 180°$, so $2\angle 1 = 180°$.	Why?
6. $\angle 1 = 90°$.	Why?
7. ∢1 is a right angle.	Why?
8. $\angle 2 = 90°$.	Why? (See steps 4 and 6.)
9. ∢2 is a right angle.	Same reason as step 7.

SET III

PHOTOGRAPH BY C. F. COCHRAN

This is a photograph of a wooden crate that is even stranger than the "cube" in Escher's picture. On the assumption that the photograph has not been "doctored," can you explain how such a crate could be built?

Lesson 5

Some Original Proofs

Abraham Lincoln considered the study of geometry an excellent way to sharpen the mind. In his *Short Autobiography* (in which he referred to himself in the third person), Lincoln wrote the following:

"He studied and nearly mastered the six books of Euclid since he was a member of Congress. He began a course of rigid mental discipline with the intent to improve his faculties, especially his powers of logic and language. Hence his fondness for Euclid, which he carried with him on the circuit till he could demonstrate with ease all the [theorems] in the six books; often studying far into the night, . . . while his fellow-lawyers, half a dozen in a room, filled the air with interminable snoring."

You are now to the point in your study of geometry at which you have some idea of how Euclid organized his logical system, beginning with definitions and postulates and deriving theorems from them by means of deductive reasoning. As we proceed, we will need to introduce more definitions and postulates, but the emphasis will now be upon developing logical proofs. Some of the proofs will be for theorems, which should be committed to memory since they will provide the basis for later proofs. Others will concern relationships in specific geometric figures.

In writing your own proofs, you should begin by making a large neat drawing of the figure. Then you should copy the hypothesis and conclusion, which we will call "given" and "prove" in problems that are not theorems. You will find it helpful at first to organize the proof itself in two columns, with statements in the left column and reasons in the right. This format is not a logical necessity but is useful in keeping track of our ideas. Later we will write many of our proofs in paragraph form.

Two more finished proofs are presented here as examples. Go over each one carefully to see how they are organized.

Example 1.   Given: $\overrightarrow{SA}$ bisects $\angle MSR$; $\angle 1 + \angle R = 90°$.
                  Prove: $\angle 2$ and $\angle R$ are complementary.

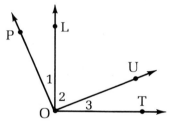

Proof.

Statements	Reasons
1. $\overrightarrow{SA}$ bisects $\angle MSR$.	Given.
2. $\angle 1 = \angle 2$.	If an angle is bisected, it is divided into two equal angles.
3. $\angle 1 + \angle R = 90°$.	Given.
4. $\angle 2 + \angle R = 90°$.	Substitution postulate.
5. $\angle 2$ and $\angle R$ are complementary.	If the sum of the measures of two angles is 90°, the angles are complementary.

Example 2.   Given: $\overrightarrow{OP} \perp \overrightarrow{OU}$; $\angle LOT$ is a right angle.
                  Prove: $\angle 1 = \angle 3$.

Proof.

Statements	Reasons
1. $\overrightarrow{OP} \perp \overrightarrow{OU}$.	Given.
2. $\angle POU$ is a right angle.	If two lines are perpendicular, they form right angles.
3. $\angle LOT$ is a right angle.	Given.
4. $\angle POU = \angle LOT$.	Any two right angles are equal.
5. $\angle POU = \angle 1 + \angle 2$.	Definition of betweenness of rays.*
6. $\angle 1 + \angle 2 = \angle LOT$.	Substitution postulate.
7 $\angle LOT = \angle 2 + \angle 3$.	Same as step 5.
8. $\angle 1 + \angle 2 = \angle 2 + \angle 3$.	Same as step 6.
9. $\angle 1 = \angle 3$.	Subtraction postulate.

---

*Except for the definitions of betweenness of points and betweenness of rays, which are somewhat clumsy to quote, each definition that you use as a reason should be stated as a complete sentence.

Exercises

Supply a reason for each statement in the following proof.

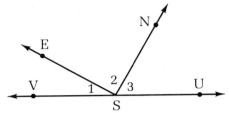

Given: $\overrightarrow{SV}$ and $\overrightarrow{SU}$ are opposite rays; $\overrightarrow{SE} \perp \overrightarrow{SN}$.
Prove: ∢1 and ∢3 are complementary.

Proof.

Statements

1. $\overrightarrow{SV}$ and $\overrightarrow{SU}$ are opposite rays.
2. ∢1 and ∢ESU are a linear pair.
3. ∢1 and ∢ESU are supplementary.
4. $\angle 1 + \angle ESU = 180°$.
5. $\angle ESU = \angle 2 + \angle 3$.
6. $\angle 1 + \angle 2 + \angle 3 = 180°$.
7. $\overrightarrow{SE} \perp \overrightarrow{SN}$.
8. ∢2 is a right angle.
9. $\angle 2 = 90°$.
10. $\angle 1 + 90° + \angle 3 = 180°$.
11. $\angle 1 + \angle 3 = 90°$.
12. ∢1 and ∢3 are complementary.

In each of the following exercises, copy the figure, given, and prove on your paper. Then write a proof in two-column form. The par is the number of steps the proof can be completed in.

1. Given: ∢2 and ∢3 are a linear pair;
   ∢1 and ∢3 are supplementary.
   Prove: $\angle 1 = \angle 2$.
   Par 4.

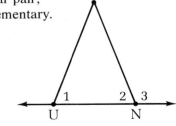

2. Given: $\angle 1$ and $\angle 3$ are vertical angles;
   $\angle 2 = \angle 3$.
   Prove: $\overrightarrow{NA}$ bisects $\angle SNT$.
   Par 5.

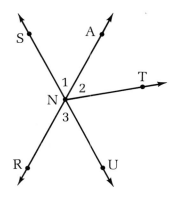

3. Given: $\overrightarrow{HE}$ and $\overrightarrow{HR}$ are opposite rays;
   $\angle 1 = \angle 2$.
   Prove: $\overleftrightarrow{AT} \perp \overleftrightarrow{ER}$.
   Par 5.

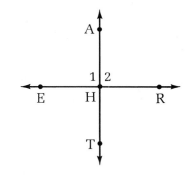

4. Given: $\overline{OT} \perp \overline{CE}$; $\angle OTE = \angle M$.
   Prove: $\angle M$ is a right angle.
   Par 6.

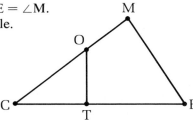

SET III

This problem appears in a Portuguese geometry book. Can you
translate the problem and explain briefly how it can be proved?

$\angle AOC = 1 \angle$ recto; demonstrar que
o $\angle 1$ e o $\angle 3$ são complementares.

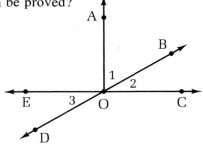

Chapter 3/Summary and Review

You should be sure that you know the definitions of the following terms introduced in Chapters 2 and 3. They are frequently used as reasons in proofs.

Betweenness of points   77
Betweenness of rays   97
Bisector of an angle   93
Complementary angles   93
Linear pair   97
Midpoint of a line segment   77
Perpendicular lines, rays, segments   129
Right angle   88
Supplementary angles   93
Vertical angles   98

Also, the Linear Pair Postulate   97

You should also know how to use the following postulates about numbers in equations.

Postulates of Equality   113

1. Reflexive
2. Symmetric
3. Transitive
4. Substitution
5. Addition
6. Subtraction
7. Multiplication
8. Division
9. Square Roots

Theorems

1. The midpoint of a line segment divides it into segments that are half as long.   117
2. The bisector of an angle divides it into angles that are half as large.   118
3. Supplements of the same angle (or equal angles) are equal.   123
4. Complements of the same angle (or equal angles) are equal.   124
5. Vertical angles are equal.   124
6. Any two right angles are equal.   129
7. If two lines are perpendicular, they form four right angles.   129
8. If the two angles in a linear pair are equal, then each is a right angle.   129

SET I

Exercises

1. Complete each of the following statements about the figure below and name the postulate of equality that justifies each.
   a) If ∠H = ∠R and ∠R = ∠E, then ▨▨▨.
   b) If 2HT = 2TR, then HT = ▨▨▨.
   c) If HA = HT + TA and TA = TE, then ▨▨▨.
   d) If ∠ETA = ∠HTR, then ∠HTR = ▨▨▨.

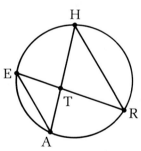

2. In the adjoining figure, $\overrightarrow{OJ}$ and $\overrightarrow{OK}$ are opposite rays.
   a) What relationships do ∡JOE and ∡EOK have?
   b) Write the equation that follows from the fact that $\overrightarrow{OR}$ is between $\overrightarrow{OJ}$ and $\overrightarrow{OE}$.
   c) If ∠ROE = ∠EOK, what can you conclude about $\overrightarrow{OE}$ and ∡ROK?
   d) If ∡JOR is a right angle, what can you conclude about $\overrightarrow{OR}$ and $\overrightarrow{OJ}$?

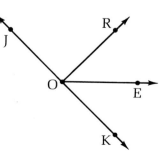

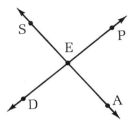

3. State the missing reasons in the following proof.

Given: ∡SED and ∡PEA are vertical angles;
        ∡SED and ∡PEA are supplementary.
Prove: $\overleftrightarrow{SA} \perp \overleftrightarrow{DP}$.

Proof.

Statements	Reasons
1. ∡SED and ∡PEA are vertical angles.	Given.
2. ∠SED = ∠PEA.	a) Why?
3. ∡SED and ∡PEA are supplementary.	Given.
4. ∠SED + ∠PEA = 180°.	b) Why?
5. ∠PEA + ∠PEA = 180°, so 2∠PEA = 180°.	c) Why?
6. ∠PEA = 90°.	d) Why?
7. ∡PEA is a right angle.	e) Why?
8. $\overleftrightarrow{SA} \perp \overleftrightarrow{DP}$.	f) Why?

4. The following definition is incorrectly stated and, as a result, it is false.

    Two angles are complementary if their measures are 90°.

    Explain why. (Hint: Look at the cartoon.)

DRAWING BY CHAS. ADDAMS; © 1956 THE NEW YORKER MAGAZINE, INC.

5. We have proved two theorems whose conclusions mention right angles. State each as a complete sentence.

SET II

Write a complete proof for each of the following.

1. Given: $\overline{CU} \perp \overline{LB}$.
   Prove: $\angle CUL = \angle CUB$.
   Par 3.

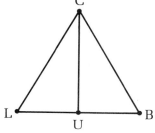

2. Given: $\angle 1$ and $\angle 2$ are a linear pair;
   $\angle J$ and $\angle A$ are supplementary;
   $\angle 1 = \angle J$.
   Prove: $\angle 2 = \angle A$.
   Par 5.

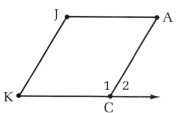

3. Given: $\angle KIG$ and $\angle GIN$ are complementary.
   Prove: $\overrightarrow{IK} \perp \overrightarrow{IN}$.
   Par 6.

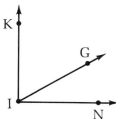

4. Write, without using any references, a complete proof of the following theorem:

   Complements of the same angle are equal.

Dilcue, the backward geometry student, was asked to prove the following theorem:

If the two angles in a linear pair are equal,
then each is a right angle.

Although he managed to write all the correct statements in the correct order, *every one of his reasons is wrong.* Instead of merely stating the correct reasons, can you explain why each of Dilcue's is wrong? If you can, you have a good understanding of what constitutes a logical proof.

Hypothesis: ∡1 and ∡2 are a linear pair;
$\angle 1 = \angle 2$.

Conclusion: ∡1 and ∡2 are right angles.

Proof.

Statements	Reasons
1. ∡1 and ∡2 are a linear pair.	They're not vertical angles.
2. ∡1 and ∡2 are supplementary.	If the sum of the measures of two angles is 180°, they are supplementary.
3. $\angle 1 + \angle 2 = 180°$.	This is obvious from looking at the figure.
4. $\angle 1 = \angle 2$.	Any two right angles are equal.
5. $\angle 1 + \angle 1 = 180°$, so $2\angle 1 = 180°$.	Addition postulate.
6. $\angle 1 = 90°$.	A right angle has a measure of 90°.
7. ∡1 is a right angle.	Perpendicular lines form right angles.
8. $\angle 2 = 90°$.	Same as step 6.
9. ∡2 is a right angle.	If the two angles in a linear pair are equal, then each is a right angle.

Chapter 4

CONGRUENT TRIANGLES

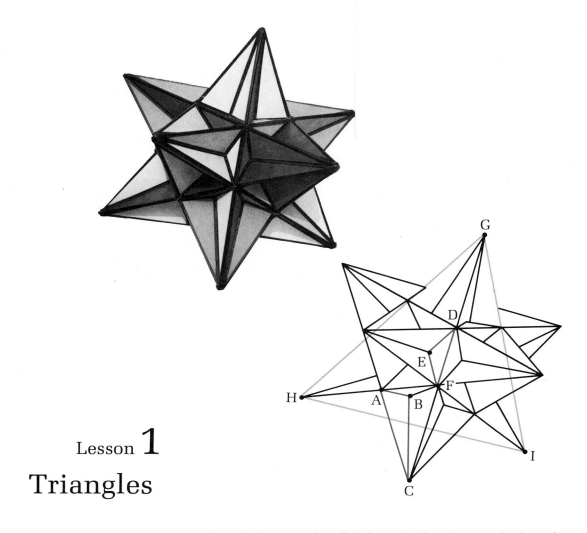

Lesson 1

Triangles

The beautiful geometric solid shown in the photograph above is one that Euclid never got to see. It was discovered in 1809 by a French mathematician, Louis Poinsot, and is called a "great icosahedron."

The edges of the great icosahedron form a large number of triangles having several different shapes. Although the arrangement of these triangles in the solid is very complex, the triangles themselves are very simple figures.

A triangle is determined by any three noncollinear points and consists of the three line segments that join the points.

▶ Definition
A *triangle* is the union of the three line segments determined by three noncollinear points.

The line segments are called the *sides* of the triangle and the points are called its *vertices* (the plural of *vertex*). We will use the symbol △ to mean "triangle": the vertices of △ABC are points A, B, and C, and its sides are $\overline{AB}$, $\overline{BC}$, and $\overline{AC}$. Every triangle determines three angles, called the *angles* of the triangle. The angles of △ABC are ∡A, ∡B, and ∡C.

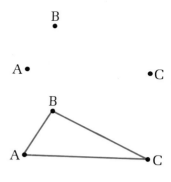

Triangles are given special names with respect to the relative lengths of their sides. Most of the triangles in the great icosahedron have the shape of △ABC in the figure on the facing page. It has no equal sides and is called *scalene*. The solid also contains some triangles that have two equal sides, called *isosceles*, such as △DEF. And, not so obvious at first glance, there are some triangles that have three equal sides, called *equilateral*, such as △GHI.

► Definitions
A triangle is:
   scalene iff it has no equal sides;
   isosceles iff it has at least two equal sides;
   equilateral iff it has three equal sides.

Triangles are also classified according to the measures of their angles.

► Definitions
A triangle is:
   acute iff it has three acute angles;
   right iff it has a right angle;
   obtuse iff it has an obtuse angle;
   equiangular iff it has three equal angles.

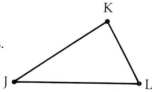

The words "included" and "opposite" are used to describe the position relationships of the sides and angles of a given triangle. For example, in △JKL, side $\overline{JK}$ is included by ∡J and ∡K and is *opposite* ∡L. ∡K is *included* by sides $\overline{KJ}$ and $\overline{KL}$ and is *opposite* side $\overline{JL}$.

► Definitions
A *side* of a triangle is *included* by the two angles whose vertices are its endpoints and is *opposite* the third angle.

An *angle* of a triangle is *included* by the two sides of the triangle that lie on the sides of the angle and is *opposite* the third side.

Every triangle is said to contain six parts: its sides and its angles.

The sides and angles of two types of triangles, isosceles triangles that have exactly two equal sides and right triangles, are given special names. They are illustrated in the figures shown here and defined below.

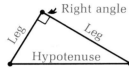

An isosceles triangle

► Definitions

In an *isosceles triangle* that has two equal sides, the equal sides are called its *legs* and the third side is called its *base*. The angle included by its legs is called the *vertex angle* and the angles that include the base are called *base angles*.

A right triangle

In a *right triangle*, the sides that include the right angle are called the *legs* and the side opposite the right angle is called the *hypotenuse*.

Exercises

SET I

1. The following questions refer to △ASP, in which SA = SP.
   a) What kind of triangle is it?
   b) Name its legs.
   c) Name its base.
   d) Name its vertex angle.
   e) Name its base angles.

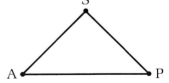

2. The following questions refer to △BOA, which has no equal sides and one obtuse angle.
   a) What kind of triangle is it? (Use two words.)
   b) Name the side opposite ∡A.
   c) Name the sides that include ∡B.
   d) Name the angle opposite side $\overline{BA}$.
   e) Name the angles that include side $\overline{OA}$.

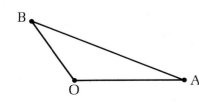

3. The following questions refer to △VIE, in which ∢I is a right angle.
   a) What kind of triangle is it?
   b) Name its legs.
   c) Name its hypotenuse.
   d) Is point P a point of the triangle?
   e) Is point R a point of the triangle?

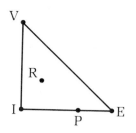

4. The following statements concern the sides and angles of a triangle. Classify each as true or false.
   a) Every triangle has three sides and three angles.
   b) Each pair of sides of a triangle intersect in one of its vertices.
   c) The vertices of a triangle are also the vertices of its angles.
   d) Every point on a side of a triangle is also a point on the triangle.
   e) Every point on an angle of a triangle is also a point on the triangle.

5. In the adjoining figure, $\overline{CB} \perp \overline{RO}$.
   a) The figure contains five triangles. Name them.
   b) How many of these triangles are right triangles?
   c) How many seem to be obtuse triangles?
   d) How many seem to be acute triangles?

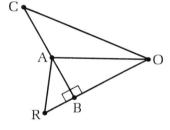

SET II

1. The two figures shown here represent the musical instrument called a "triangle" and a highway traffic sign.
   a) In what way is the musical instrument a poor illustration of a geometric triangle?
   b) In what way is the traffic sign a poor illustration of a geometric triangle?

2. The table below lists some different types of triangles. For example, "g" represents an obtuse scalene triangle.

	Scalene	Isosceles	Equilateral
Acute	a	b	c
Right	d	e	f
Obtuse	g	h	i

Use a straightedge to draw an example of each type of triangle that you think can exist. If you think the figure cannot exist, write "impossible."

3. The following questions refer to this statement:

   If a triangle is equilateral, then it is also isosceles.

   a) Is this statement true?
   b) Explain your answer.
   c) Is the converse of the statement true?

SET III

"Polyiamonds" are figures made by fitting together two or more equilateral triangles. The simplest polyiamond, a "diamond," consists of two equilateral triangles.

There are twelve different polyiamonds that contain six equilateral triangles each. They are shown on the next page. Trace and cut them out and see if you can figure out how to fit eight of them together to form the six-pointed star shown here.* It may be necessary to turn some of the pieces over.

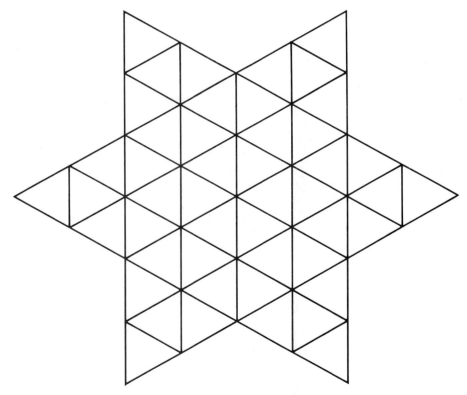

*From "Mathematical Games" by Martin Gardner. Copyright © 1964 by Scientific American, Inc. All rights reserved. This polyiamond puzzle and others appear in *Martin Gardner's Sixth Book of Mathematical Games from Scientific American* (W. H. Freeman and Company, 1971).

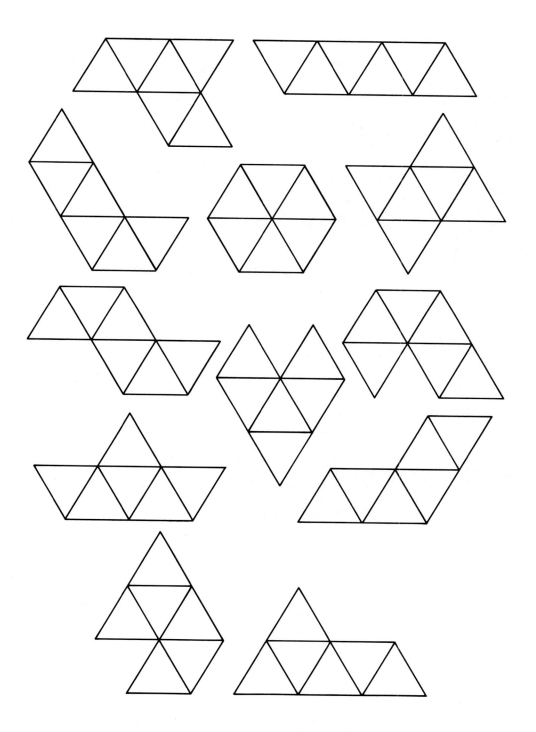

Lesson 2
Congruent Triangles

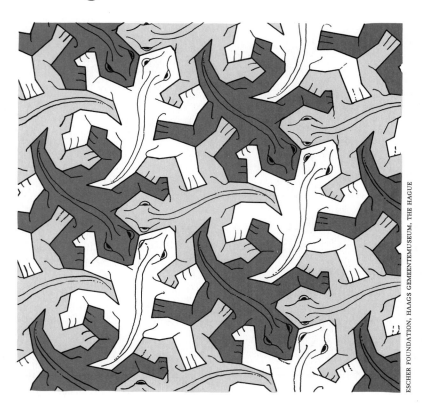

ESCHER FOUNDATION, HAAGS GEMEENTEMUSEUM, THE HAGUE

"Let us construct a two-dimensional universe out of an infinitely large number of identical but distinctly recognizable components."* This is what Maurits Escher has done with this drawing titled *Study of Regular Division of the Plane with Reptiles.* Although the picture does not contain an infinite number of reptiles, the repeating pattern shown can be extended indefinitely. Hence, it is truly what the artist has called it, "a fragment of infinity."

---

*M. C. Escher, in his essay "Approaches to Infinity," in *The World of M. C. Escher*, edited by J. L. Locher (Abrams, 1971).

All of the reptiles in the drawing are identical in size and shape. If a tracing were made of one of them, it could be placed so as to coincide exactly with any of the others. Such figures are said to be *congruent*.

The adjoining figure shows two congruent triangles. If △ABC were made to coincide with △DEF, A would fit on D, B on E, and C on F. We can represent this matching of the triangles' vertices symbolically as

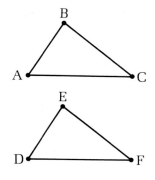

$$A \leftrightarrow D, \quad B \leftrightarrow E, \quad \text{and} \quad C \leftrightarrow F,$$

or, more briefly, as

$$ABC \leftrightarrow DEF.$$

This matching of the vertices of the two triangles is called a *one-to-one correspondence*: each vertex of one triangle corresponds to a vertex of the other triangle, and vice versa.

Setting up a one-to-one correspondence of vertices also establishes a one-to-one correspondence between sides and angles. From the correspondence

$$ABC \leftrightarrow DEF,$$

we have

$$\overline{AB} \leftrightarrow \overline{DE}, \quad \overline{BC} \leftrightarrow \overline{EF}, \quad \text{and} \quad \overline{AC} \leftrightarrow \overline{DF},$$

and

$$\angle A \leftrightarrow \angle D, \quad \angle B \leftrightarrow \angle E, \quad \text{and} \quad \angle C \leftrightarrow \angle F.$$

If the triangles can be made to coincide, then each of these pairs of corresponding sides must be equal and each pair of corresponding angles must be equal. This suggests a way of precisely defining congruent triangles without resorting to the use of the word "coincide."

► Definition
Two triangles are **congruent** iff there is a correspondence between their vertices such that the corresponding sides and corresponding angles of the triangles are equal.

Such a correspondence is called a *congruence* between the two triangles and is represented by the symbol ≅. To indicate that the two triangles shown here are congruent, we write

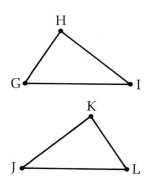

$$\triangle GHI \cong \triangle LKJ.$$

Notice that the way in which we have written this is based upon the congruence correspondence,

$$GHI \leftrightarrow LKJ,$$

between the vertices of the two triangles. A correspondence such as GHI ↔ JKL, on the other hand, is *not* a congruence, so we do not write △GHI ≅ △JKL.

Restating the definition of congruent triangles in terms of the second pair of triangles on page 149, we have

$$\triangle GHI \cong \triangle LKJ \quad \text{iff} \quad GHI \leftrightarrow LKJ$$
$$\text{such that} \quad GH = LK, \quad HI = KJ,$$
$$GI = LJ, \quad \angle G = \angle L, \quad \angle H = \angle K, \quad \text{and} \quad \angle I = \angle J.$$

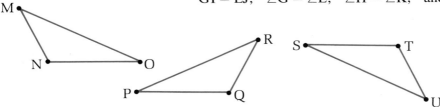

Look at the three triangles shown here. If $\triangle MNO \cong \triangle RQP$ and $\triangle RQP \cong \triangle UTS$, does it follow that $\triangle MNO \cong \triangle UTS$? In other words, is congruence of triangles transitive? It is easy to prove, on the basis of the definition of congruent triangles, that it is.

▶ **Theorem 9**

Two triangles congruent to a third triangle are congruent to each other.

Hypothesis: $\triangle MNO \cong \triangle RQP$ and $\triangle RQP \cong \triangle UTS$.
Conclusion: $\triangle MNO \cong \triangle UTS$.

Proof.

Statements	Reasons
1. $\triangle MNO \cong \triangle RQP$ and $\triangle RQP \cong \triangle UTS$.	Hypothesis.
2. $MN = RQ$, $\quad RQ = UT$, $NO = QP$, $\quad QP = TS$, $MO = RP$, $\quad RP = US$, $\angle M = \angle R$, $\quad \angle R = \angle U$, $\angle N = \angle Q$, $\quad \angle Q = \angle T$, $\angle O = \angle P$. $\quad \angle P = \angle S$.	If two triangles are congruent, then there is a correspondence between their vertices such that the corresponding sides and angles are equal.
3. $MN = UT$, $\quad \angle M = \angle U$, $NO = TS$, $\quad \angle N = \angle T$, $MO = US$, $\quad \angle O = \angle S$.	Transitive postulate of equality.
4. $\triangle MNO \cong \triangle UTS$.	If there is a correspondence between the vertices of two triangles such that their corresponding sides and angles are equal, then the triangles are congruent.

Exercises

1. The two triangles in the figure below are congruent. Imagine putting a tracing of △WAL over △NUT so that they coincide. Then

$$W \leftrightarrow N, \quad A \leftrightarrow U, \quad \text{and} \quad L \leftrightarrow T.$$

Which of the following correspondences convey the same idea?
a) WAL ↔ NUT.
b) LAW ↔ TUN.
c) WLA ↔ UTN.
d) UNT ↔ AWL.
e) NTU ↔ LWA.

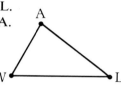

2. The two triangles shown here are congruent.
   a) Copy and complete the following three correspondences of the vertices to show this fact.

   $$A \leftrightarrow \text{▨}, \quad L \leftrightarrow \text{▨}, \quad \text{and} \quad M \leftrightarrow \text{▨}.$$

   Copy and complete the following congruence correspondences:
   b) ALM ↔ ▨.
   c) LAM ↔ ▨.
   d) ODN ↔ ▨.

3. The equal sides and angles of two congruent triangles can be read from a congruence correspondence between them. Using the fact that

   $$\triangle CAS \cong \triangle HEW,$$

   copy and complete each of the following equations.
   a) CA = ▨.
   b) EW = ▨.
   c) SC = ▨.
   d) ∠C = ▨.
   e) ∠W = ▨.

4. Name all of the pairs of equal segments and equal angles in the adjoining figure if
   a) △KOL ≅ △KAL.
   b) △KOL ≅ △AKL.

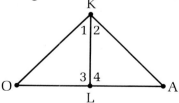

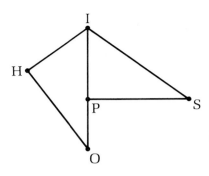

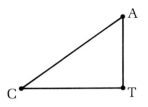

5. In the figure shown here, △PIS ≅ △TAC and △TAC ≅ △HIO.
   a) What conclusion can be drawn from this information?
   b) State the theorem that supports this conclusion.

6. We have proved that congruence of triangles is transitive. Use the figures shown here to answer the following questions.
   a) Is congruence of triangles reflexive? Explain.
   b) Is congruence of triangles symmetric? Explain.

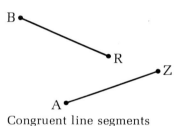

Congruent line segments

SET II

The term "congruent" may be applied to other geometric figures than triangles. In general, two figures are congruent if they have the same size and shape.

1. Since all line segments have the same shape, two line segments are congruent if they have the same size. In other words, *congruent line segments have equal lengths.* In terms of the figure shown here, $\overline{BR} \cong \overline{AZ}$ iff BR = AZ.
   a) Do you think all angles have the same shape?
   b) Do you think two angles that have the same size must have the same shape?
   c) Write a definition for *congruent angles* that does not use either of the words "size" or "shape."
   d) Express your definition in terms of the angles in the adjoining figure.

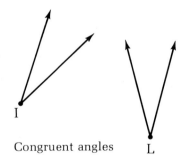

Congruent angles

2. Find and name each of the following in the figure shown here.
   a) Three pairs of line segments that appear to be congruent.
   b) Three pairs of numbered angles that appear to be congruent.
   c) Three pairs of triangles that appear to be congruent.

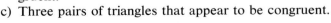

3. Suppose we know in the preceding figure that ∠HLE = ∠ALZ, HE = AZ, and HL = AL. On the basis of our definition of congruent triangles,

   a) what else would we have to know to prove that △HLE ≅ △ALZ?

   b) how many pairs of parts must we know to be equal to conclude that a pair of triangles are congruent in general?

4. Keeping in mind your answer to the previous question, write a complete proof for the following.

   Given: C is the midpoint of both $\overline{AE}$ and $\overline{NP}$; PE = AN; $\overline{PN} \perp \overline{PE}$ and $\overline{PN} \perp \overline{NA}$; ∡PCE and ∡ACN are vertical angles; ∠A = ∠E.

   Prove: △PEC ≅ △NAC.

   Par 10. (Each part of the given information separated by semicolons was counted as a step.)

SET III

The four reptiles shown here are congruent, yet one is different from the others in a very basic way. Which reptile is it and why?

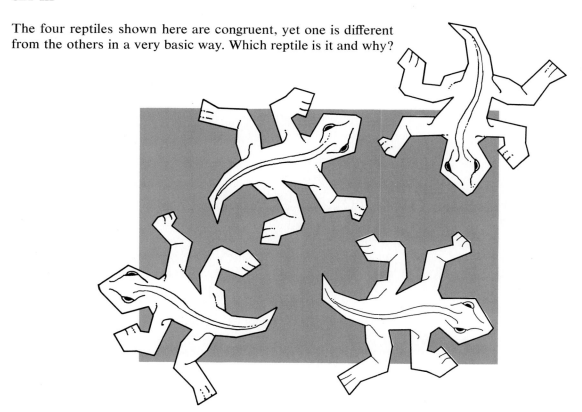

Lesson 3

Some Congruence Postulates

The largest geodesic dome in the world is in Montreal, Quebec.
Built as the United States Pavilion for Expo '67, it is 200 feet
high and has a spherical diameter of 250 feet. The dome is so
strong that it not only requires no interior support, but is capable
of supporting a tremendous weight.

The basic shape used in the construction of geodesic domes is
the triangle because it is a rigid form. If models of a triangle and a
quadrilateral (a four-sided figure) are made by threading drinking
straws together, the triangle will hold its shape but the quadrilateral will collapse.

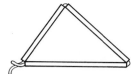  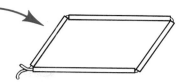

The shape of the triangle, then, is completely determined by the lengths of its sides. This means that if the sides of one triangle are equal to the sides of another, the triangles are congruent. Of the triangles in the figure below we are saying that if AB = DE, BC = EF, and AC = DF, then △ABC ≅ △DEF.

Although this seems like a reasonable assumption, we cannot prove it by means of our definition of congruent triangles and so we will adopt it as a postulate.

► **Postulate 10** (The S.S.S. Congruence Postulate)
If the three sides of one triangle are equal to the corresponding parts of another triangle, then the triangles are congruent.

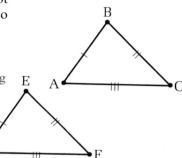

Remember that a triangle has six parts: three sides and three angles. Our postulate is based upon the idea that just three of these parts—the three sides—are sufficient to determine a triangle's shape. What about other sets of three parts?

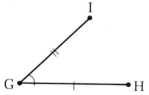

Let's try two sides and an angle and see what happens. Once the two sides and the angle of △GHI shown in the above figure have been drawn, is there any choice left about the measures of the other side and two angles? No, because the three vertices have already been located by the three parts drawn. Notice that the angle in the drawing is *included between the two sides* of the triangle. This suggests a second postulate about congruent triangles.

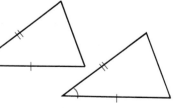

► **Postulate 11** (The S.A.S. Congruence Postulate)
If two sides and the included angle of one triangle are equal to the corresponding parts of another triangle, then the triangles are congruent.

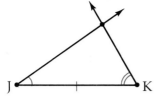

A third possibility of a set of three parts is illustrated by the adjoining figure. If we draw two angles of a triangle and their included side, the third vertex and hence the rest of the parts of the triangle are determined. A third postulate about congruent triangles states this.

▶ **Postulate 12**   (The A.S.A. Congruence Postulate)
If two angles and the included side of one triangle are equal to the corresponding parts of another triangle, then the triangles are congruent.

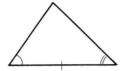 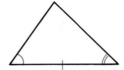

Exercises

SET I

1. Draw a triangle and label its vertices O, N, and E.
   a) Which side of the triangle is included by ∡O and ∡N?
   b) Which angle is included by sides NE̅ and OE̅?
   c) Which angles include OE̅?
   d) Which sides include ∡N?

2. *Without* drawing a figure, answer the following questions about △TWO.
   a) Which side is included by ∡W and ∡O?
   b) Which angle is included by sides TW̅ and WO̅?
   c) Which angles include TO̅?
   d) Which sides include ∡T?

3. Obtuse Ollie is trying to learn the congruence postulates.
   a) He says that two triangles are congruent if two sides and an angle of one triangle are equal to the corresponding parts of the other. Is this correct?
   b) He also says that two triangles are congruent if the three angles of one triangle are equal to the corresponding parts of the other. Is this correct?
   c) Ollie thinks that the two triangles shown here are congruent because of the A.S.A. Congruence Postulate. (The arcs and tick marks indicate the pairs of parts known to be equal.) Is this correct?

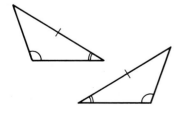

4. In each of the following figures, the tick marks and arcs identify equal parts. If we can conclude that the triangles are congruent on the basis of one of our congruence postulates, name it. Otherwise, write "no conclusion possible."

a)

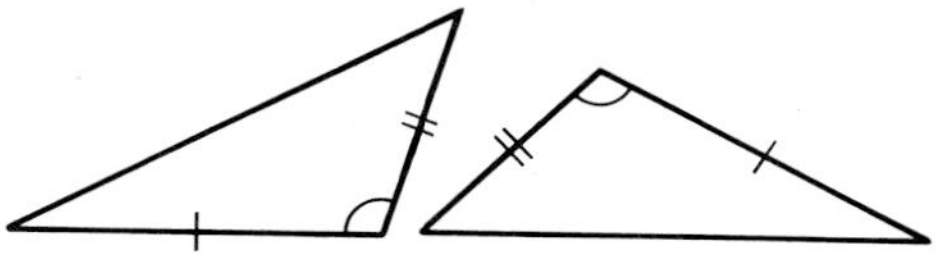

f)

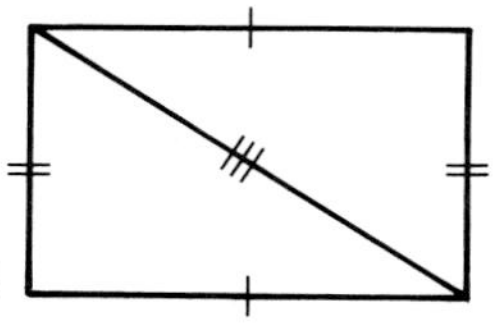

b)

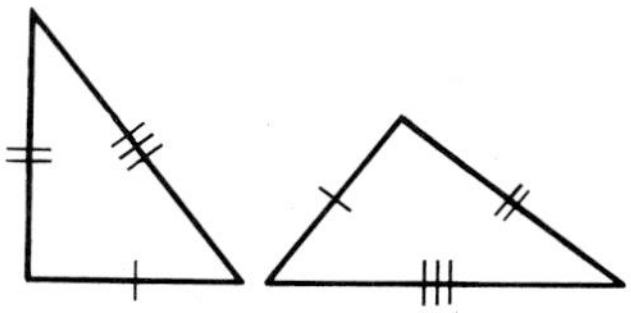

g)

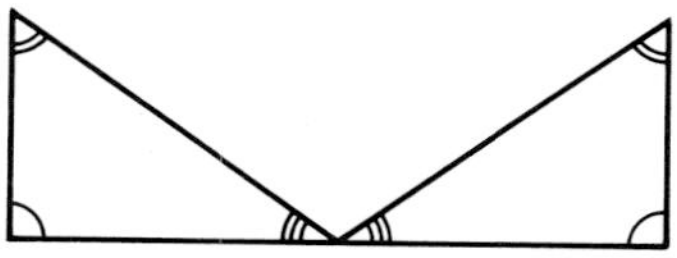

c)

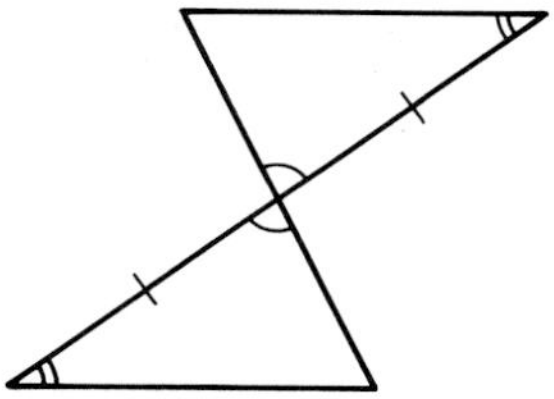

h)

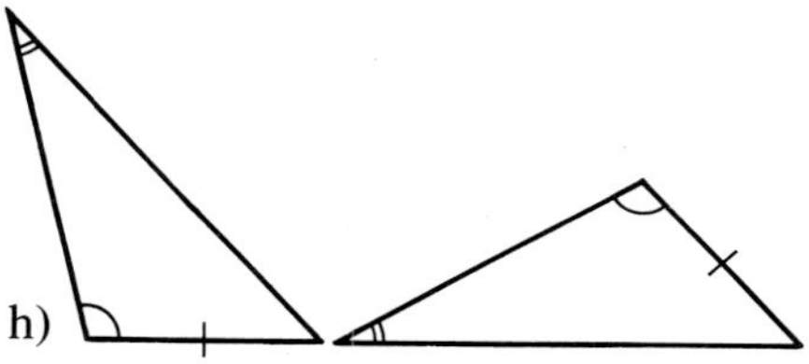

d)

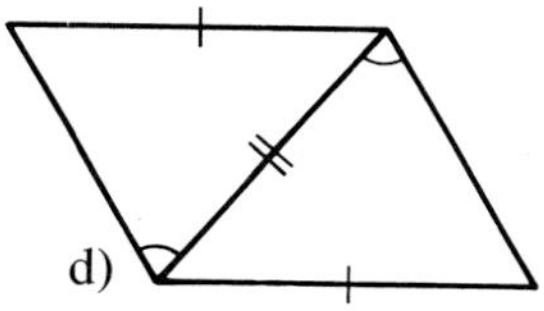

i)

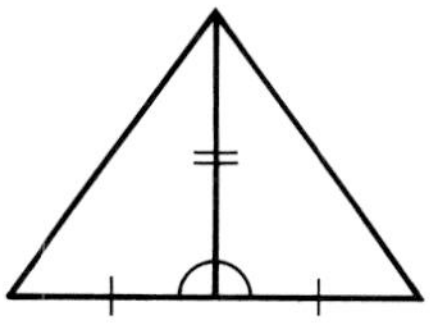

e)

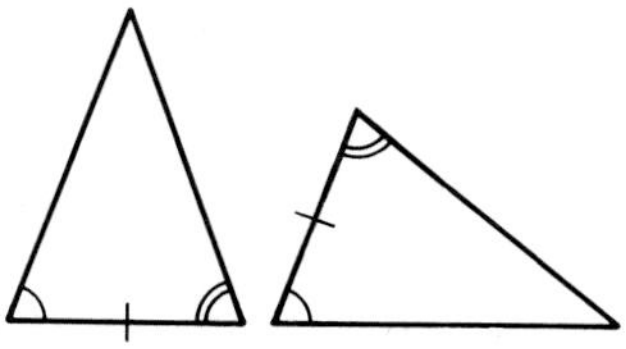

j)

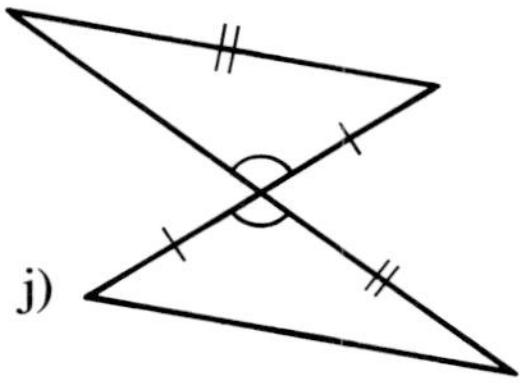

SET II

Use your straightedge to make a neat drawing for each of the following exercises.

1. Draw two isosceles right triangles in which the legs of one triangle are equal to the legs of the other. Why should your triangles be congruent to each other?

2. Draw two equilateral triangles in which a side of one is equal to a side of the other. Are the triangles necessarily congruent? Explain.

3. Draw two triangles in which all three angles and one side of one are equal to the corresponding parts of the other. Are the triangles necessarily congruent? Explain.

4. Draw two triangles that are obviously *not congruent*, but such that *two sides and an angle of one are equal to two sides and an angle of the other.*

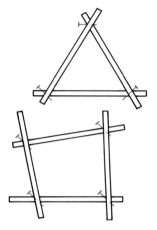

SET III

A geodesic dome owes its strength to the fact that the triangle is a rigid form. If three drinking straws are pinned together at their ends to form a triangle, the triangle will hold its shape, even if picked up and moved around.

Four straws pinned together to form a quadrilateral, however, result in a structure that is flexible. It can easily be changed into a variety of shapes.

Several frameworks are shown below. Which ones do you think are rigid and which are flexible? You might build some "straws and pins" models to check your answers.

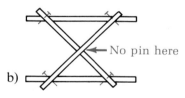

b)   ◄—No pin here

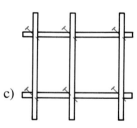

c)

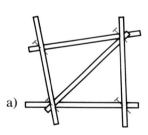

a)

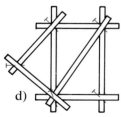

d)

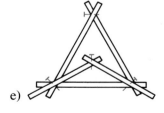

e)

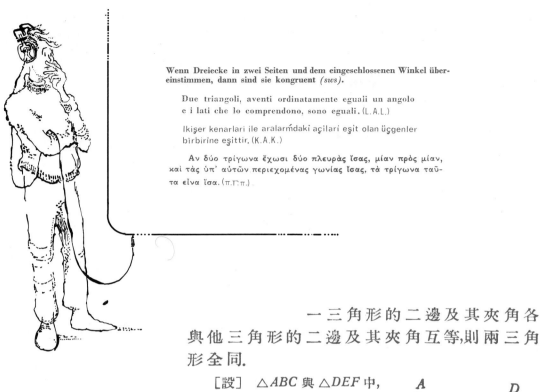

Wenn Dreiecke in zwei Seiten und dem eingeschlossenen Winkel über-
einstimmen, dann sind sie kongruent *(sws)*.

Due triangoli, aventi ordinatamente eguali un angolo
e i lati che lo comprendono, sono eguali. (L.A.L.)

Ikişer kenarlari ile aralarm̂daki açilari eşit olan üçgenler
birbirine eşittir. (K.A.K.)

Αν δύο τρίγωνα ἔχωσι δύο πλευρὰς ἴσας, μίαν πρὸς μίαν,
καὶ τὰς ὑπ' αὐτῶν περιεχομένας γωνίας ἴσας, τὰ τρίγωνα ταῦ-
τα εἶνα ἴσα. (π.Γ.π.)

一三角形的二邊及其夾角各
與他三角形的二邊及其夾角互等,則兩三角
形全同.

［設］ △*ABC* 與 △*DEF* 中,

$AB = DE,$

$BC = EF,$

$\angle B = \angle E.$

［求證］ △*ABC* ≅ △*DEF*.

Lesson 4

Proving Triangles Congruent

The postulate for proving triangles congruent that we call "S.A.S."
goes by different names in other countries. In Germany, it is the
"S.W.S." postulate because the German words for side and angle
are "seite" and "winkel." In Italy it is "L.A.L.," in Turkey it is
"K.A.K.," and in Greece the postulate is called "Π.Γ.Π." (pi-
gamma-pi). It is stated above in each of these languages and in
Chinese as well. In the Chinese version, the hypothesis and
conclusion of the postulate are restated in terms of the figure.

159

Although there are other ways by which triangles can be proved congruent, we will use at first just the three postulates introduced in Lesson 3. In Turkey, they are called the "K.A.K.," "A.K.A.," and "K.K.K." postulates. You should not only know our names for them, but also be able to state each one as a complete sentence.

Here are a couple of examples of proofs based upon them.

Example 1.

Given: AC = AD; $\overrightarrow{AB}$ bisects $\angle$CAD.
Prove: $\triangle$ABC $\cong$ $\triangle$ABD.

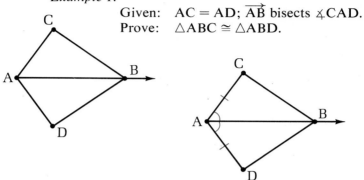

After marking the given information on the figure at the left above, it looks like the figure at the right. Two pairs of parts are not sufficient to prove that triangles are congruent. We need a third pair. In this case, it is the common side, $\overline{AB}$; since AB = AB, the triangles are congruent by the S.A.S. postulate.

Proof.

Statements	Reasons
1. AC = AD.	Given.
2. $\overrightarrow{AB}$ bisects $\angle$CAD.	Given.
3. $\angle$CAB = $\angle$BAD.	If an angle is bisected, it is divided into two equal angles.
4. AB = AB.	Reflexive postulate.
5. $\triangle$ABC $\cong$ $\triangle$ABD.	S.A.S. postulate.

Before studying the second example, recall that our definition of congruent triangles says:

Two triangles are congruent iff there is a correspondence between their vertices such that the corresponding sides and corresponding angles of the triangles are equal.

We will frequently use the following abbreviated statement of this definition as a reason in proofs:

Corresponding parts of congruent triangles are equal.

Example 2.

Given: ∡1 and ∡2 are vertical angles;
         G is the midpoint of $\overline{HF}$;
         ∠H = ∠F.

Prove: HI = FE.

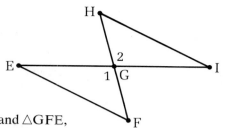

Since $\overline{HI}$ and $\overline{FE}$ are corresponding sides of △GHI and △GFE, we can prove them equal by proving the triangles congruent. The given information is sufficient to prove them congruent by the A.S.A. postulate.

Proof.

Statements	Reasons
1. ∡1 and ∡2 are vertical angles.	Given.
2. ∠1 = ∠2.	Vertical angles are equal.
3. G is the midpoint of $\overline{HF}$.	Given.
4. HG = GF.	The midpoint of a line segment divides it into two equal segments.
5. ∠H = ∠F.	Given.
6. △GHI ≅ △GFE.	A.S.A. postulate.
7. HI = FE.	Corresponding parts of congruent triangles are equal.

SET I

Exercises

1. In △GUI and △TAR shown here, GU = AR and ∠U = ∠A. Name the other pair(s) of parts you would need to prove equal in order to show that the triangles are congruent by the
    a) S.A.S. postulate.
    b) A.S.A. postulate.
    c) S.S.S. postulate.

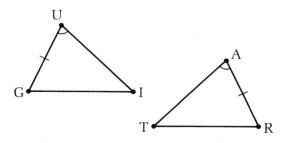

2. If △ZIT ≅ △HER, then is it true that
   a) ZI = HE? Why?
   b) ZT = ER? Why?
   c) ∠I = ∠E? Why?

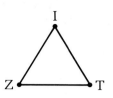

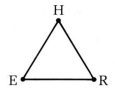

3. Supply a reason for each statement in the following proof.

Given: R is the midpoint of $\overline{AP}$; HA = HP;
        $\overrightarrow{RA}$ and $\overrightarrow{RP}$ are opposite rays.
Prove: $\overleftrightarrow{HR} \perp \overleftrightarrow{AP}$.

Proof.

Statements

a) R is the midpoint of $\overline{AP}$.
b) AR = RP.
c) HA = HP.
d) HR = HR
e) △HAR ≅ △HPR.
f) ∠1 = ∠2.
g) $\overrightarrow{RA}$ and $\overrightarrow{RP}$ are opposite rays.
h) ∡1 and ∡2 are a linear pair.
i) ∡1 and ∡2 are right angles.
j) $\overleftrightarrow{HR} \perp \overleftrightarrow{AP}$.

SET II

Write a complete proof for each of the following. Remember that this means copying the figure, given, and prove before writing the statements and reasons.

1. Given: FL = ET;
           ∡L and ∡T are right angles;
           ∠F = ∠E.
   Prove: △FLU ≅ △ETU.
   Par 5.

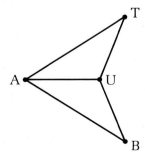

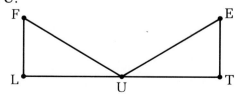

2. Given: TU = UB;
           TA = AB.
   Prove: ∠T = ∠B.
   Par 5.

162   *Chapter 4: CONGRUENT TRIANGLES*

3. Given: △IOA is equilateral;
∡1 and ∡2 are a linear pair;
∡2 and ∡3 are supplementary;
PI = AN.
   Prove: △PIO ≅ △NAO.
   Par 8.

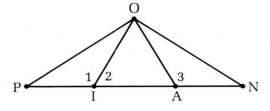

SET III

Obtuse Ollie decided to invent some of his own proofs. He likes
to draw his figures without bothering with a ruler. What do you
think of each of the following problems he made up?

1. Given: DM = RU.
   Prove: △RUM ≅ △MDR.
   Par 3.

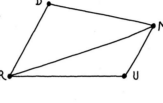

2. Given: $\overline{VO} \perp \overline{IL}$; VO = OL;
∡1 and ∡2 are right angles;
AO = IO; ∠V = ∠L; ∠1 = ∠2.
   Prove: △VIO ≅ △LAO.
   Par 7.

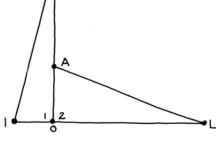

3. Given: ∡1 and ∡2 are a linear pair; OR = RN;
∡1 and ∡2 are complementary.
   Prove: HO = HN.
   Par 7.

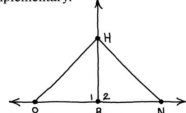

Lesson 4: Proving Triangles Congruent   163

"Very, very exclusive."

Lesson 5

More Congruence Proofs

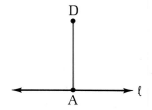

There doesn't seem to be any easy way for the two cowboys in this cartoon to get over to that drive-in. Yet there *is* an easy way for them to find out how far away it is! In fact, if they know the approximate length of their horses' strides, they can measure the distance without any special instruments or even getting off their horses!

The method uses congruent triangles. We will assume that the two men are directly across from the drive-in, so that in the adjoining figure point A represents their position, point D represents the drive-in, and line ℓ represents the edge of the cliff they are standing beside. The problem is to find the distance AD. Notice that $\overline{AD}$ is perpendicular to line ℓ.

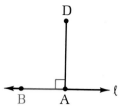

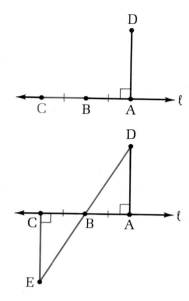

First, the two men can ride a short distance along the edge of the cliff to a point that we will call B, counting strides as they go along. (See the first figure above.) One man can then stop while the other rides on for an equal distance to a point we will call C. (See the second figure.)

Finally, the man at C can turn directly away from the cliff and ride until he comes to the point at which the man who stayed behind him is directly between him and the drive-in. (See the adjoining figure.) If we call this point E, the distance from C to E is equal to the distance from A to D. Do you see *why*?

It is because the two triangles in the figure are congruent, and since $\overline{CE}$ and $\overline{AD}$ are corresponding sides of these triangles, they must be equal. How do we know that the triangles are congruent? The parts that have been marked in the figure below suggest the reason. A proof for this is shown.

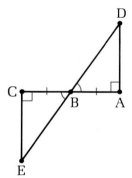

Given: $\overline{CA}$ and $\overline{DE}$ intersect at B; $\overline{AD} \perp \overline{CA}$; $\overline{CE} \perp \overline{CA}$; and CB = BA.

Prove: CE = AD.

Proof.

Statements	Reasons
1. $\overline{AD} \perp \overline{CA}$, $\overline{CE} \perp \overline{CA}$.	Given.
2. $\angle C$ and $\angle A$ are right angles.	Perpendicular lines form right angles.
3. $\angle C = \angle A$.	Any two right angles are equal.
4. CB = BA.	Given.
5. $\angle CBE = \angle ABD$.	Vertical angles are equal.
6. $\triangle CBE \cong \triangle ABD$.	A.S.A. postulate.
7. CE = AD.	Corresponding parts of congruent triangles are equal.

Exercises

Copy the figure for each of the following proofs and mark it to identify the equal parts. Then give a reason for each statement.

1. Given: ∠1=∠2; FI=IA.
   Prove: T̄I̅ bisects ∢FTA.

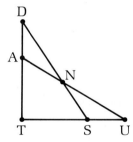

   Proof.

   Statements

   a) ∠1=∠2.
   b) ∢1 and ∢3 are supplementary;
      ∢2 and ∢4 are supplementary.
   c) ∠3=∠4.
   d) FI=IA.
   e) IT=IT.
   f) △FIT≅△AIT.
   g) ∠5=∠6.
   h) T̄I̅ bisects ∢FTA.

2. Given: ∠D=∠U; DT=TU.
   Prove: DA=SU.

   Proof.

   Statements

   a) ∠D=∠U.
   b) DT=TU.
   c) ∠T=∠T.
   d) △DTS≅△UTA.
   e) TS=TA.
   f) DA+AT=DT; TS+SU=TU.
   g) DA=DT−AT; SU=TU−TS.
   h) DA=TU−TS.
   i) DA=SU.

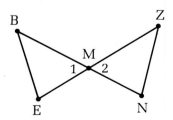

Write a complete proof for each of the following.

1. Given: ∢1 and ∢2 are vertical angles;
            BM = MZ, EM = MN.
   Prove: BE = ZN.
   Par 5.

2. Given: $\overline{AD} \perp \overline{UD}$, $\overline{AD} \perp \overline{DI}$; $\angle 1 = \angle 2$.
   Prove: $AU = AI$.
   Par 7.

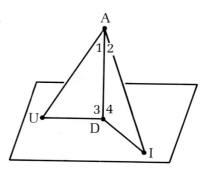

3. Given: O is the midpoint of $\overline{HN}$;
   $\angle HOD = \angle AON$; $OA = OD$.
   Prove: $\triangle AHO \cong \triangle DNO$.
   Par 8.

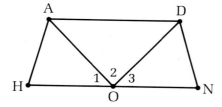

SET III

This road map is Dilcue's first attempt at map-making. It shows seven towns and the highways joining them. Assuming that the highways are straight, can you figure out what is wrong with the map other than the inaccuracies in scale?

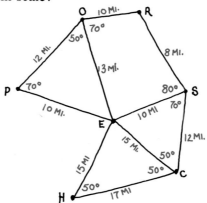

Lesson **6**
The Isosceles Triangle
Theorem

In Lewis Carroll's *Through The Looking Glass*, Alice wonders what it is like on the other side of the mirror over her fireplace. She climbs up on the mantel and passes through the glass into a room in which everything is reversed.

The figures at the top of this page and the next show what a couple of triangles would look like to Alice on the other side of the looking glass. The first triangle is scalene and its mirror image

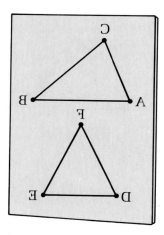

looks different from the triangle itself. The reflection of the second triangle, however, does not seem to be reversed (even though the letters naming its vertices are). We are saying that the following correspondence between the vertices of the triangle and its reflection seems to be a congruence:

$$DFE \leftrightarrow EFD.$$

It is easy to prove that it is and, since $\angle D$ corresponds to $\angle E$ in this congruence, these angles must be equal. In other words, the base angles of an isosceles triangle are equal.

► **Theorem 10** (The Isosceles Triangle Theorem)
If two sides of a triangle are equal, the angles opposite them
are equal.

Hypothesis: In △DEF, DF = EF.
Conclusion: ∠D = ∠E.

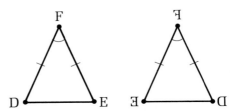

Our figure shows two views of the same isosceles triangle:
one from "in front of" and one from "behind the looking glass."

Proof.

Statements	Reasons
1. DF = EF.	Hypothesis.
2. ∠F = ∠F.	Reflexive postulate.
3. EF = DF.	Symmetric postulate.
4. △DFE ≅ △EFD.	S.A.S. postulate.
5. ∠D = ∠E.	Corresponding parts of congruent triangles are equal.

By means of this theorem, it is easy to prove that a triangle
having *three* equal sides must have *three* equal angles. We will
call this a corollary to the theorem.

► **Corollary**
If a triangle is equilateral, it is also equiangular.

Is the converse of the Isosceles Triangle Theorem true? In
other words, if two angles of a triangle are equal, must the sides
opposite them be equal? We can easily prove that this is so.

► **Theorem 11**
If two angles of a triangle are equal, the sides opposite them
are equal.

This theorem also has a corollary.

► **Corollary**
If a triangle is equiangular, it is also equilateral.

1. Complete the following proof of the corollary to Theorem 10 by supplying the reasons.

Exercises

If a triangle is equilateral, it is also equiangular.

Hypothesis: △MAY is equilateral.
Conclusion: △MAY is equiangular.

Proof.

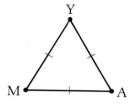

Statements

a) △MAY is equilateral.
b) MY = YA.
c) ∠M = ∠A.
d) MY = MA.
e) ∠A = ∠Y.
f) ∠M = ∠A = ∠Y.
g) △MAY is equiangular.

2. Complete the following proof of Theorem 11 by supplying the reasons.

If two angles of a triangle are equal, the sides opposite them are equal.

Hypothesis: In △APR, ∠A = ∠R.
Conclusion: AP = PR.

Proof.

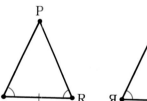
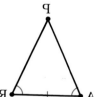

Statements

a) ∠A = ∠R.
b) AR = RA.
c) ∠R = ∠A.
d) △APR ≅ △RPA.
e) AP = PR.

3. The following questions refer to △OCT.
   a) If OT = CT, what can you conclude?
   b) If ∠O = ∠C = ∠T, what can you conclude?
   c) If OC and CT are not equal, can you conclude that ∠O and ∠T are not equal? Explain.

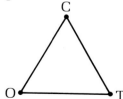

Write a complete proof for each of the following.

1. Given: $\angle J = \angle 1$; $\angle 2 = \angle L$.
   Prove: $JU = LY$.
   Par 5.

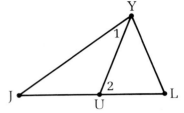

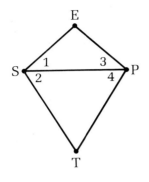

2. Given: $SE = EP$; $ST = TP$.
   Prove: $\angle EST = \angle EPT$.
   Par 7.

3. Given: $\angle 1$ and $\angle 2$ are a linear pair,
   $\angle 3$ and $\angle 4$ are a linear pair;
   $\angle 1 = \angle 4$.
   Prove: $\triangle FEB$ is isosceles.
   Par 6.

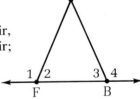

SET III

A Looking Glass Proof. Alice found a book in the room behind the mirror written in a language she didn't recognize. It started out like this:

JABBERWOCKY

'Twas brillig, and the slithy toves
Did gyre and gimble in the wabe:
All mimsy were the borogoves,
And the mome raths outgrabe.

If the book had been about geometry rather than poetry, it might have had the following problem. Can you prove it in "looking glass" style?

Given: $\angle D = \angle E$; $DE = EC$.
Prove: $\triangle DEC$ is equiangular.
Par 5.

Lesson 7

Overlapping Triangles

Do you think the lady in this picture is attractive? It depends upon which one you see!

In order to recognize both ladies in the picture, it is necessary to perceive its parts in two different relationships. The ability to see a variety of relationships in geometric figures is very important in developing proofs about them. If you see only the "obvious," you may be unable to think of your own proofs.

In the figure below, there are several pairs of triangles that appear to be congruent.

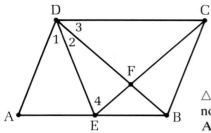

For example, it looks as if △DEF ≅ △CBF and △DEB ≅ △CBE. Guessing from the appearance of a figure, of course, does not constitute a proof. Suppose we know only the following facts: AD = ED, ∠1 = ∠3, and ∠A = ∠4. Is there a pair of triangles in the figure that can be proved congruent on the basis of this information?

Finding such a pair is somewhat like trying to see the "other lady" in the picture. Marking the equal parts on the figure is a good way to begin. Notice that three of these parts are in △ADE. If we knew that ∠1 = ∠2 instead of ∠1 = ∠3, then it would be obvious that △ADE ≅ △EDF by the A.S.A. postulate. The trouble is that we don't. What other possibilities are there?

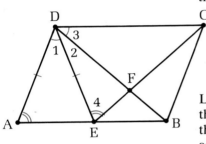

The known parts are concentrated on the left of the figure. Look at ⊿A and AD again. Are they parts of a triangle other than △ADE? Yes. The triangle is △ADB. Corresponding to these parts are ⊿4 and ED. In what triangles are they? △EDF and △EDC. It is the second of these triangles that we can prove congruent to △ADB. The figure is redrawn with these triangles shown in different colors so that they are easier to see. The third pair of parts that we can prove equal are ⊿ADB and ⊿EDC. Do you see how?

The proof is given below.

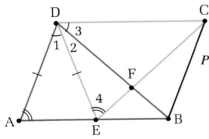

Given: AD = ED, ∠1 = ∠3, and ∠A = ∠4.
Prove: △ADB ≅ △EDC.

Proof.

Statements	Reasons
1. ∠1 = ∠3.	Given.
2. ∠1 + ∠2 = ∠3 + ∠2.	Addition postulate.
3. ∠ADB = ∠1 + ∠2; ∠EDC = ∠3 + ∠2.	Betweenness of rays.
4. ∠ADB = ∠EDC.	Substitution postulate.
5. AD = ED and ∠A = ∠4.	Given.
6. △ADB ≅ △EDC.	A.S.A. postulate.

As you can see from this problem, it is helpful in working with overlapping triangles to draw each one in a different color to make them stand out. It is also important to write a correct congruence correspondence between them. Then, by referring to it rather than to the figure, we can easily name the equal parts.

SET I

1. The following questions refer to the figure shown here. State a conclusion that can be proved by each of the following sets of facts. Also give the reason for each of your conclusions.
   a) $\angle 3 = \angle 4$.
   b) $\triangle OAG \cong \triangle OAB$ and $\triangle OAB \cong \triangle BAR$.
   c) $\angle G = \angle R$, $GA = RA$, and $\angle 1 = \angle 2$.
   d) $GO = RB$ and $\angle GOB = \angle RBO$.

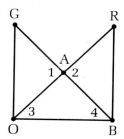

2. Copy the figure for the proof below, using one color for $\triangle TRY$ and another color for $\triangle ARY$. Then give a reason for each statement.

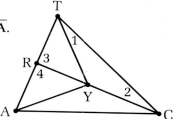

Given: $\angle 1 = \angle 2$; $AY = YC$;
R is the midpoint of $\overline{TA}$.

Prove: $\overline{RC} \perp \overline{TA}$.

Proof.

<u>Statements</u>

a) $\angle 1 = \angle 2$.
b) $TY = YC$.
c) $AY = YC$.
d) $TY = AY$.
e) $RY = RY$.
f) R is the midpoint of $\overline{TA}$.
g) $TR = RA$.
h) $\triangle TRY \cong \triangle ARY$.
i) $\angle 3 = \angle 4$.
j) $\angle 3$ and $\angle 4$ are right angles.
k) $\overline{RC} \perp \overline{TA}$.

Write a complete proof for each of the following.

1. Given: AE = AN; ∠WEN = ∠YNE.
   Prove: △WEN ≅ △YNE.
   Par 5.

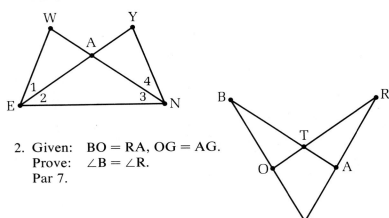

2. Given: BO = RA, OG = AG.
   Prove: ∠B = ∠R.
   Par 7.

3. Given: HU and NP bisect each other.
   Prove: △HBE ≅ △UBR.
   Par 7.

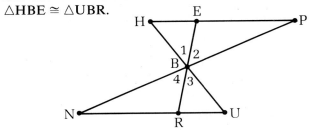

SET III

Euclid proved the Isosceles Triangle Theorem by means of a rather complicated method. The figure he used is on the next page. It looks something like a bridge and in the Middle Ages it became known as The Bridge of Fools. The idea was that dull students could not understand it and so could not "cross the bridge" in order to continue their study of geometry. It has been said that when the movie producers get around to filming Euclid's *Elements*, there will undoubtedly be a spectacular rescue over the burning Bridge of Fools!

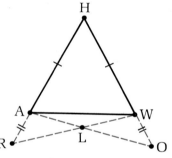

As you can see from the adjoining figure, Euclid added some extra line segments to the bottom of isosceles △HAW. They have been drawn so that H-A-R, H-W-O, and AR = WO. We want to show that if HA = HW in △HAW, then ∠HAW = ∠HWA.

1. First, Euclid proved △HRW ≅ △HOA. How do you think he did this?

2. Next, since $\overline{RW}$ and $\overline{OA}$ are corresponding sides of these triangles, he knew that RW = OA. He used this to help prove △RAW ≅ △OWA. How do you suppose he did this?

3. Finally, since ∡RAW and ∡OWA are corresponding angles of these triangles, he knew that ∠RAW = ∠OWA. He used this to prove that ∠HAW = ∠HWA. How could he do this?

Lesson **8**

Some Straightedge and Compass Constructions

Ever since the time of Euclid, two tools have been used in making geometric drawings: the straightedge and the compass. The straightedge is used to draw lines and the compass is used to draw circles. Drawings made with just these two instruments are called constructions to distinguish them from those made with other tools such as a ruler and a protractor.

In this lesson, we will learn how to make three basic constructions related to line segments and angles. The first is a method for bisecting a line segment.

► **Construction 1**

To bisect a line segment.

A •————————• B   A •————————• B

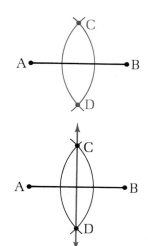

Method.

Let $\overline{AB}$ be the given line segment. With A and B as centers, draw two circles that have the same radius and that intersect each other in two points, C and D. (As you can see from the first adjoining figure, it is sufficient for the purpose of the construction to draw just a part of each circle.) Draw $\overleftrightarrow{CD}$.

This line not only bisects $\overline{AB}$ but is also perpendicular to it. As a result, it is called the *perpendicular bisector* of the segment. It is easy to prove that $\overleftrightarrow{CD}$ is the perpendicular bisector of $\overline{AB}$ by means of the following theorem.

A •————————• B

► **Theorem 12**

In a plane, two points equidistant from the endpoints of a line segment determine the perpendicular bisector of the line segment.

The figures below show three different relationships that the two points C and D might have to the line segment $\overline{AB}$. In all three cases, the theorem can be proved by means of congruent triangles. The proof below is based upon the first figure.

Hypothesis: $\overline{AB}$ with $CA = CB$ and $DA = DB$.
Conclusion: $\overleftrightarrow{CD}$ is the perpendicular bisector of $\overline{AB}$.

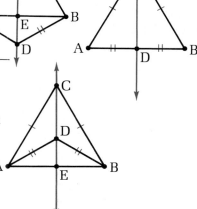

Proof.

Statements	Reasons
1. CA = CB, DA = DB.	Hypothesis.
2. CD = CD.	Reflexive postulate.
3. △ACD ≅ △BCD.	S.S.S. postulate.
4. ∠ACD = ∠BCD.	Corresponding parts of congruent triangles are equal.
5. CE = CE.	Same as step 2.
6. △ACE ≅ △BCE.	S.A.S. postulate.
7. ∠CEA = ∠CEB.	Same as step 4.
8. ⊾CEA and ⊾CEB are right angles.	If two angles in a linear pair are equal, they are right angles.
9. $\overleftrightarrow{CD} \perp \overline{AB}$.	If two lines form right angles, they are perpendicular.
10. AE = EB.	Same as step 4.
11. $\overleftrightarrow{CD}$ bisects $\overline{AB}$.	If a line segment is divided into two equal segments, it is bisected.

Lesson 8: Straightedge and Compass Constructions   179

► Construction 2
To bisect an angle.

Method.
Let ∡A be the given angle. With A as center, draw a circle that intersects the sides of the angle in points B and C. With B and C as centers, draw two circles that have the same radius and that intersect each other in point D. Draw $\overrightarrow{AD}$. This ray bisects the angle.

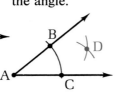

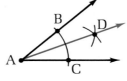

The proof that this method makes sense, as well as the one described in the following construction, are left as exercises.

► Construction 3
To construct an angle equal in measure to a given angle.

Method.
Let ∡A be the given angle. Draw a ray $\overrightarrow{BC}$ as one side of the angle to be constructed. With A as center, draw a circle that intersects the sides of ∡A in points D and E. Also draw a circle with the same radius and with B as center that intersects $\overrightarrow{BC}$ in point F.

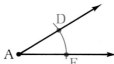

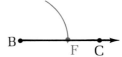

With F as center and the distance between D and E as radius, draw a circle intersecting the previous one in point G. Draw $\overrightarrow{BG}$. ∠B = ∠A.

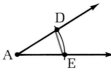

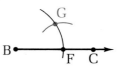

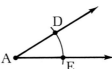

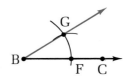

1. According to Theorem 12, in a plane, two points equidistant from the endpoints of a line segment determine the perpendicular bisector of the line segment. The two points can be on opposite sides of the segment, as the first figure for the theorem illustrates. One point can be on the segment itself, as shown by the second figure. Or both points can be on the same side of the segment, as shown by the third figure. Although the proof in this lesson is based upon the first figure, it also fits, with a few minor changes, the other two.
   a) Read the proof again and decide how it should be changed to fit the second figure.
   b) Does the proof have to be changed to fit the third figure?

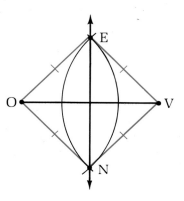

2. In the figure shown here, the perpendicular bisector of $\overline{OV}$ has been constructed as described in Construction 1. The reason the construction works is based upon the fact that the circles drawn with O and V as centers have the same radius. This means that EO = EV.
   a) What relationship does E therefore have to points O and V?
   b) Why is N equidistant from O and V?
   c) What do points E and N therefore determine?

3. In this figure, $\overrightarrow{KN}$ has been constructed as described in Construction 2 so that it bisects ∡IKL. Notice that, by the method used, KI = KL and IN = LN. Explain by means of congruent triangles why the construction works.

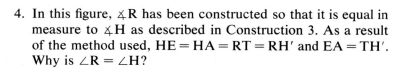

4. In this figure, ∡R has been constructed so that it is equal in measure to ∡H as described in Construction 3. As a result of the method used, HE = HA = RT = RH' and EA = TH'. Why is ∠R = ∠H?

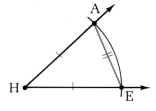 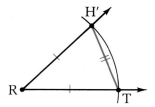

1. Draw two quadrilaterals of about the same shape but somewhat larger than the ones shown below.
   a) Use your straightedge and compass to bisect each side of each figure. Join the midpoints of the four sides of each figure in order, so that two new quadrilaterals are formed.
   b) What do you notice?

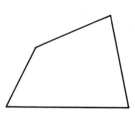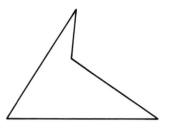

2. Make somewhat enlarged copies of the two triangles below.
   a) Use your straightedge and compass to bisect each of the three angles of each figure.
   b) What do you notice?

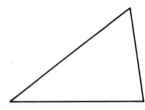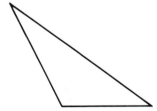

3. Since a line segment can be easily bisected by straightedge and compass, it can also be divided into four equal parts.
   a) Draw a line segment that is several inches long and do this.
   b) What other number of equal parts, less than ten, can a line segment be easily divided into?

4. It is actually possible, by just a straightedge and compass, to divide a line segment into *any* number of equal parts. There is a general method that will work in all cases. We will show how it can be used to trisect a line segment; that is, divide it into three equal parts.
   a) Draw a line segment several inches long and label its endpoints F and E.
   b) Draw a ray from F as shown in figure b.

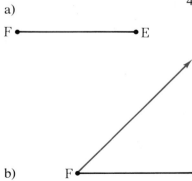

a)

F •————————————• E

b)

c)

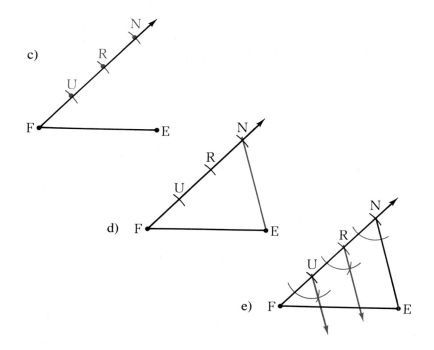

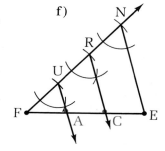

c) Use your compass to mark off three equal segments end-to-end on the ray, starting at F as shown in figure c.
d) Draw $\overline{NE}$.
e) Construct two angles with their vertices at R and U and one side along $\overleftrightarrow{FN}$ so that they are equal in measure to ∡N.
f) Label the points A and C in which the other sides of these angles intersect $\overline{FE}$. These points divide the segment into three equal parts.

The proof that this method does trisect the segment is based upon ideas more advanced than those we now know so we will not attempt to justify it.

SET III

People have been trying to figure out a way to trisect any given angle using just a straightedge and compass ever since the time of Euclid. Mathematicians, however, gave the problem up long ago because in 1837 it was proved by means of algebra to be impossible!

Nevertheless, this does not prevent amateurs from trying. One attempt is shown in the ad on the next page, which appeared in the *Los Angeles Times* several years ago.

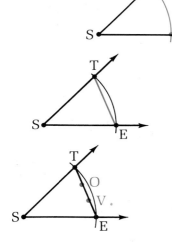

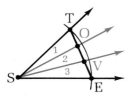

The method that people usually first think of is described as follows. Let ∢S be the given angle. With S as center, draw a circle that intersects the sides of the angle in points T and E. Draw $\overline{TE}$. Trisect $\overline{TE}$ (using the method described in Exercise 4 of Set II) and label the points of trisection O and V. Draw $\overrightarrow{SO}$ and $\overrightarrow{SV}$.

Now it looks as if $\overrightarrow{SO}$ and $\overrightarrow{SV}$ trisect ∢TSE. This is not the case, however, because ∢1, ∢2, and ∢3 do not all have the same measure.

Draw an obtuse angle and carry out the method described above. (You may want to guess the positions of the points that trisect the line segment to save time.) What do you observe?

If you would like to know more about the problem of angle trisection and how people have tried to solve it, you will want to read Martin Gardner's "Mathematical Games" in the June 1966 issue of *Scientific American*.

Chapter 4/Summary and Review

Basic Ideas

Postulates

10. *The S.S.S. Congruence Postulate.* If the three sides of one triangle are equal to the corresponding parts of another triangle, then the triangles are congruent.   155
11. *The S.A.S. Congruence Postulate.* If two sides and the included angle of one triangle are equal to the corresponding parts of another triangle, then the triangles are congruent.   155
12. *The A.S.A. Congruence Postulate.* If two angles and the included side of one triangle are equal to the corresponding parts of another triangle, then the triangles are congruent.   156

Theorems

Constructions

Exercises

SET I

1. The three triangles shown here are right triangles.
   a) With respect to the lengths of its sides, what kind of triangle does △COL appear to be?
   b) Which side is the hypotenuse of △UMB?
   c) Write a correspondence between △COL and △INE that seems to be a congruence.
   d) If △COL ≅ △UBM and △UBM ≅ △ENI, does it necessarily follow that △COL ≅ △ENI?

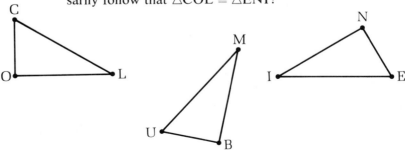

2. The following questions refer to the figure below, in which several pairs of equal parts have been marked.
   a) Name the pairs of triangles that seem to be congruent.
   b) Which one of these pairs of triangles can be proved congruent? Upon what postulate might the proof be based?
   c) Name another pair of parts that, if known to be equal, would be sufficient to prove △HOI ≅ △RCD.

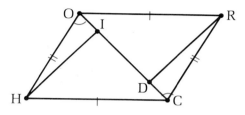

3. Obtuse Ollie says that if ∠1 = ∠2, then OS = SE because of the theorem that says, if two angles of a triangle are equal, the sides opposite them are equal. Is this correct?

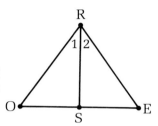

4. Do you think it is possible to do the following constructions by using just a straightedge and compass? Explain in each case.
   a) Divide any angle into four equal angles.
   b) Divide any angle into six equal angles.

___

SET II

1. Given: DA = DI; ∡1 and ∡2 are vertical angles.
   Prove: ∠A = ∠2.
   Par 5.

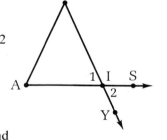

2. Given: $\overrightarrow{LT}$ bisects ∡ULP and $\overrightarrow{LP}$ bisects ∡TLI; ∠1 = ∠2; LU = LI.
   Prove: △TUL ≅ △PIL.
   Par 7.

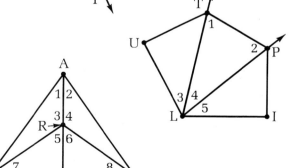

3. Given: ∠1 = ∠2; ∠5 = ∠6.
   Prove: ∠7 = ∠8.
   Par 8.

4. Given: $\angle 1 = \angle 2$; $\angle 3 = \angle 4$.
   Prove: $\triangle LAX$ is isosceles.
   Par 7.

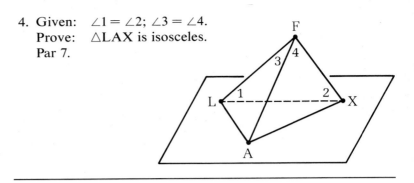

SET III

The following problem is from a popular puzzle book published in Russia.*

An equilateral triangle can be separated into four congruent triangles as shown in the figures below. Without the top triangle, the remaining three triangles form a four-sided figure called a trapezoid.

Can you figure out how to separate it into four congruent parts?

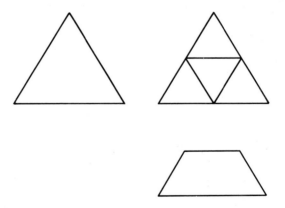

*An English language edition of *The Moscow Puzzles*, by Boris A. Kordemsky, was edited by Martin Gardner (Scribner's, 1972).

Chapter 5

TRANSFORMATIONS

Lesson **1**

The Reflection of a Point

This woodcut by the artist Maurits Escher is titled *Day and Night*. The interlocking black and white birds at the top of the picture are flying over two different views of the same landscape. The two views, except for the reversal of dark and light, are mirror images of each other.

Each point on one side of the picture corresponds to an "image point" on the other side. The first diagram on the facing page shows some of these points. The beak of the leftmost black bird, B, corresponds to the beak of the rightmost white bird, B′. The church steeple and windmill on the left, C and W, correspond to their images on the right, C′ and W′.

If a mirror were placed on line ℓ so that it faced toward the left, the reflections of points B, C, and W would appear in the mirror to be where B', C', and W' are. Because of this, we will refer to points B', C', and W' as the *images* of points B, C, and W *reflected through line* ℓ.

In the diagram at the right below, a segment has been drawn from each point to its image. The reflection line, ℓ, is related to each of these segments in two ways: it is *perpendicular* to each segment and it *bisects* each segment.

It is easy to see that as a point moves toward the reflection line, so does its image. In fact, if the point moves onto the reflection line, it and its image merge.

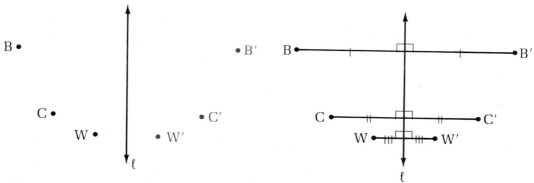

► Definition

One point is the **reflection (image) through line ℓ** of another point iff ℓ is the perpendicular bisector of the segment that joins the two points. If the point lies on ℓ, it is its own reflection.

Line ℓ is the *mirror* of the reflection.

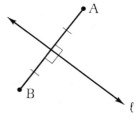

   In the adjoining figure, line ℓ is the perpendicular bisector of the segment joining points A and B. If ℓ is the mirror of reflection, is B the reflection of A or is A the reflection of B?
   It is obvious from the definition of reflection stated above that we may consider either point to be the reflection of the other. We will state this observation as a theorem.

► Theorem 13

If the reflection of point A through line ℓ is point B, then the reflection of point B through line ℓ is point A.

   Suppose a line is chosen at random in a plane. Does every point in the plane have a reflection through that line? Do any points in the plane have more than one reflection? Can different points have the same reflection? A little (mental) reflection on these matters makes the following postulate seem reasonable.

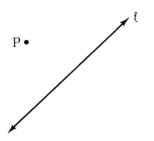

▶ **Postulate 13** (The Point Reflection Postulate)
There is a one-to-one correspondence between points in a plane and their reflections through a given line.

Where is the reflection of point P through line ℓ in the adjoining figure? One way to find it is by construction.

▶ **Construction 4**
To construct the reflection of a point through a line.

Let the point be P and the line be ℓ. First, with P as center, draw a circle that intersects ℓ in two points, A and B. Then, with A and B as centers, draw two more circles with the same radius as the first one. The circles with centers at A and B intersect in two points: point P and its reflection, point P'.

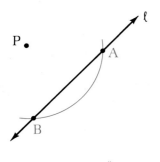

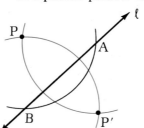

Exercises

For some of the exercises in this chapter, it is important that your copy of the figure used be especially accurate. In such cases, you may find it helpful to place your paper over the page and trace the figure. Such figures will be labeled *Trace*.

SET I

1. In the adjoining figure, ℓ is the mirror of reflection for the points shown.
   a) If R is the reflection of T, must T be the reflection of R?
   b) What point seems to be the reflection of point I?
   c) What point is the reflection of G?
   d) What relationship does line ℓ have to $\overline{TR}$ and $\overline{IE}$?

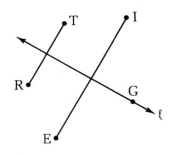

Copy the following figures. Then use a mirror to locate the reflection of each point through the given line. The mirror should be placed with its reflecting edge along the line and held so that it is perpendicular to your paper. Use the prime notation used in this lesson to name each image point. For example, the image of point L would be named point L′.

2.

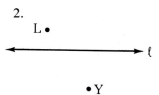

3. (*Trace.*)

N•

•X

4.

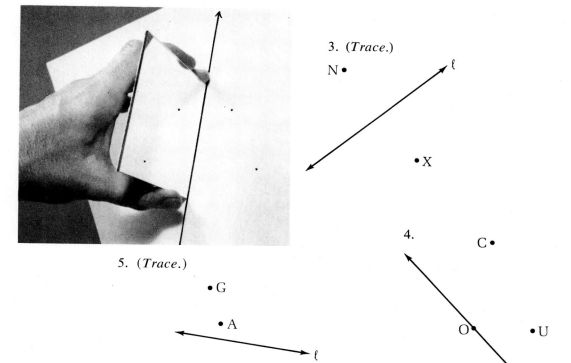

5. (*Trace.*)

•G

•A

•R

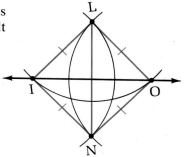

---

1. In this figure, N, the reflection of point L through $\overleftrightarrow{IO}$, has been constructed as described in Construction 4. As a result of the method used, IL = IN = OL = ON.
   a) What points are equidistant from L and N?
   b) What relationship does $\overleftrightarrow{IO}$ have to $\overline{LN}$?
   c) Why is N the reflection through $\overleftrightarrow{IO}$ of point L?

Lesson 1: The Reflection of a Point   193

Copy the following figures and then use your straightedge and compass to make the constructions described.

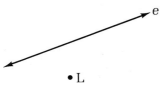

2. Construct the reflection of point L through line *e*.

3. Construct a line *o* so that P is the reflection of A through *o*.

• A

P •

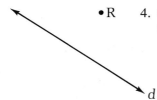

4. Construct a line that contains point R and is perpendicular to line *d*.

5. **Given:** M is the reflection of P through $\overleftrightarrow{UA}$.
   Prove: PA = MA.
   Par 8. (Hint: Use the definition of the reflection of a point through a line to prove the triangles congruent.)

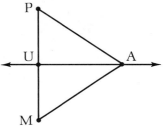

SET III

A Construction That Does Not Require a Compass. Trace the figure below, in which point T is the reflection of point A through line *ℓ*. Can you figure out a way to find the reflection of point C through *ℓ* that uses *only a straightedge*?

Hint: If a set of points is collinear, their reflections through a given line are also collinear.

C •

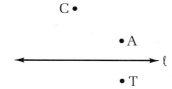

"Separate checks, please."

Lesson 2

More on Reflections

It looks as if the fellow in this cartoon doesn't want to pay for all eight drinks—especially since he has been somewhat irritated by observing that his friend copied everything he did during the meal. When he lined up the four glasses in front of him in a row, the other guy did, too. And when he moved each pair of glasses a certain distance apart, the other guy spaced his glasses the same distance apart as well. At least the other guy didn't put his glasses in the same order: he was original for a change and reversed them.

These observations suggest several ideas about the reflection of a set of points through a line. The adjoining figure shows three points, A, B, and C, and their reflections through line ℓ, A′, B′, and C′. It appears from this figure that if a set of collinear points is reflected through a line, their images are also collinear. We will assume that this is true and use it to prove some other reflection properties.

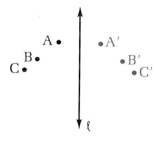

▶ **Postulate 14**
Reflection of a set of points through a line preserves collinearity.

Because reflection of a set of points through a line preserves collinearity, it follows that the reflection of a line through a line is also a line.

195

By means of Postulate 14, we can prove that several other relationships among a set of points are also preserved when the points are reflected through a line. Each relationship is illustrated by a figure and the proofs are left as exercises.

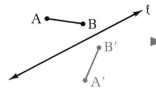

▶ **Theorem 14**

Reflection of a set of points through a line preserves distance.

If A′ and B′ are the reflections of A and B through ℓ, then A′B′ = AB.

▶ **Theorem 15**

Reflection of a set of points through a line preserves betweenness.

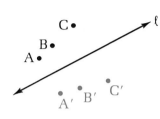

If A-B-C and A′, B′, and C′ are the reflections of A, B, and C through ℓ, then A′-B′-C′.

Since line segments and rays are defined in terms of betweenness, it follows from Theorem 15 that the reflection of a line segment through a line is also a line segment and that the reflection of a ray is also a ray. Furthermore, since an angle is a set of rays and a triangle is a set of line segments, the reflection of an angle through a line is also an angle and the reflection of a triangle is also a triangle.

▶ **Theorem 16**

Reflection of a set of points through a line preserves angle measure.

If ∡A′ is the reflection of ∡A through ℓ, then ∠A′ = ∠A.

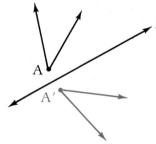

▶ **Theorem 17**

A triangle and its reflection through a line are congruent.

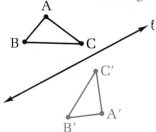

If △A′B′C′ is the reflection of △ABC through line ℓ, then △ABC ≅ △A′B′C′.

Although many properties of a set of points are preserved when the points are reflected through a line, there is one basic property that is changed. Imagine tracing △DEF in the figure shown here, from D to E to F and back to D. In doing this, we move *clockwise*. Tracing its image through line ℓ from D' to E' to F' and back to D', we move *counterclockwise*. If we trace their vertices in the orders named, the two triangles are said to have clockwise and counterclockwise *orientations*, respectively. Reflecting a triangle through a line *reverses* its orientation.

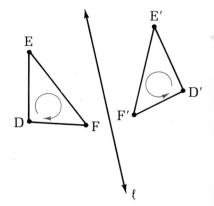

Exercises

1. In the figure shown here, △PMN is the reflection image of △AUT through line ℓ. State a theorem or postulate that supports each of the following conclusions.
   a) UL = ML.
   b) If A, L, and U are collinear, then P, L, and M are collinear.
   c) ∠N = ∠T.
   d) If U-I-T, then M-I-N.
   e) △PMN ≅ △AUT.

2. The orientation of a geometric figure depends upon the order in which we name its vertices. For example, consider the following names and orientations for the triangle shown here.

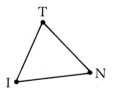

Name	Orientation
TIN	Counterclockwise
TNI	Clockwise
INT	Counterclockwise

   a) If the triangle were named △NTI, what would be its orientation?
   b) Give a name for the triangle other than the one listed above that corresponds to a clockwise orientation.
   c) Suppose △TIN were reflected through a line. What would be the orientation of △T'I'N'?

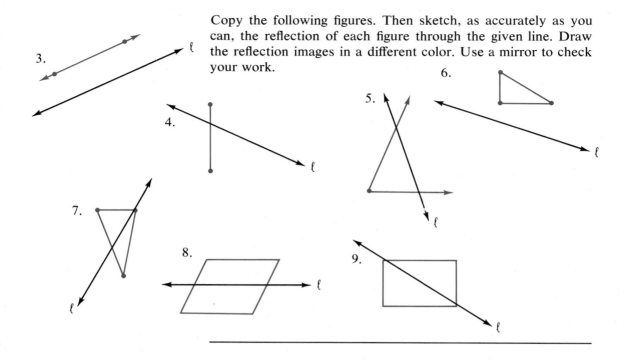

Copy the following figures. Then sketch, as accurately as you can, the reflection of each figure through the given line. Draw the reflection images in a different color. Use a mirror to check your work.

3.

4.

5.

6.

7.

8.

9.

SET II

1. Rather than writing a complete proof of Theorem 14, answer the following questions based upon the figure shown here. (It illustrates just one of several possible relationships of two points to a reflection line. A complete proof would require considering every one.)

Reflection of a set of points through a line preserves distance.

Hypothesis: O and N are the reflections of B and E through line ℓ.

Conclusion: BE = ON.

Draw $\overline{BO}$ and $\overline{EN}$, intersecting line ℓ in points R and Z, respectively. Also draw $\overline{BZ}$ and $\overline{OZ}$.
a) What relationship does line ℓ have to $\overline{BO}$ and $\overline{EN}$?
b) Why is $\triangle BRZ \cong \triangle ORZ$?
c) Why is BZ = OZ and $\angle 1 = \angle 2$?
d) What relationship do ∡3 and ∡4 have to ∡1 and ∡2, respectively?
e) Why is $\angle 3 = \angle 4$?
f) Why is $\triangle BZE \cong \triangle OZN$?
g) Why is BE = ON?

Complete the following proofs of Theorems 15 and 16 by supplying the reasons.

2. Reflection of a set of points through a line preserves betweenness.

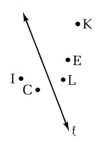

Hypothesis: N, I, and C are the reflections of K, E, and L through line ℓ, respectively, and K-E-L.

Conclusion: N-I-C.

Proof.

$$\text{Statements}$$

a) N, I, and C are the reflections of K, E, and L through line ℓ, respectively, and K-E-L.
b) K, E, and L are collinear and KE + EL = KL.
c) N, I, and C are collinear.
d) NI = KE, IC = EL, and NC = KL.
e) NI + IC = NC.
f) N-I-C.

3. Reflection of a set of points through a line preserves angle measure.

Hypothesis: C, O, and B are the reflections of A, L, and T through line ℓ, respectively.

Conclusion: ∠COB = ∠ALT.

Proof.

$$\text{Statements}$$

a) C, O, and B are the reflections of A, L, and T through line ℓ, respectively.
b) Draw $\overline{CB}$ and $\overline{AT}$.
c) CO = AL, OB = LT, and CB = AT.
d) △COB ≅ △ALT.
e) ∠COB = ∠ALT.

4. Use the figure shown here to prove Theorem 17.

A triangle and its reflection through a line are congruent.

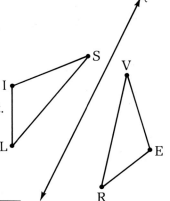

Hypothesis: △SIL is the reflection of △VER through line ℓ.

Conclusion: △SIL ≅ △VER.

Par 3.

Suppose the fellow in the cartoon at the beginning of this lesson had put a package of cigarettes on the table and noticed its reflection in the mirror. The lettering on the side of the package seems to say "choice quality," only it looks peculiar. Being a bit tipsy, the fellow decides that the other guy smokes Turkish cigarettes, the Turkish word for QUALITY being ᎤᏅⱯᎡᏞᎢᎩ. He is surprised, however, that the Turkish and English words for CHOICE are the same. Can you explain how this can be?

Lesson 3

Line Symmetry

The left side of a human face is close to being the mirror image of the right side. But not exactly. The first photograph above is of Edgar Allan Poe as he actually looked. The second and third photographs are composite pictures made by merging each side of the first picture with its corresponding reflection.

If a vertical line is drawn through the center of each of the composite pictures, each point of the picture in one half-plane has an image point in the other half-plane. Such a figure is said to be *symmetric* with respect to the line.

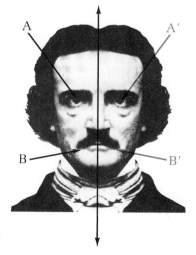

Photograph of Edgar Allan Poe reprinted with permission of the American Antiquarian Society. Composite photographs from *Symmetry* by I. Bernal, W. C. Hamilton, and J. S. Ricci. W. H. Freeman and Company. Copyright © 1972.

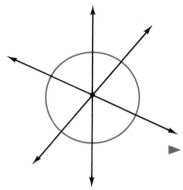

▶ Definition

A set of points has *line symmetry* iff there is a line ℓ such that the reflection through ℓ of each point of the set is also a point of the set.

The biologist's term for line symmetry is "bilateral symmetry." A simple test to determine whether a figure has it is to fold the figure along the supposed symmetry line and see if the two halves of the figure coincide.

The most symmetric of all geometric figures is the *circle* because, in the plane of the circle, *any* line through its center is a line of symmetry.

▶ Definition

A *circle* is the set of all points in a plane that are at a given distance from a given point in the plane.

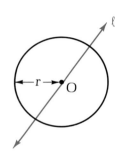

The given distance is called the *radius* of the circle and the given point is called its *center*. A circle is usually named by the point that is its center.

The adjoining figure represents a circle whose center is point O and whose radius is *r*. A point in the plane is on circle O iff its distance from O is *r*. It is easy to see why circle O is symmetric with respect to any line through its center, such as line ℓ. Since point O lies on ℓ, the reflection of point O through ℓ is itself. And since reflection of a set of points through a line preserves distance, the reflection through ℓ of every point on circle O is at a distance *r* from O. Hence, the reflection through line ℓ of every point on circle O is also a point on circle O. If you read our definition of line symmetry again, you will see that this conclusion proves that circle O is symmetric with respect to line ℓ. And the same argument can be applied to *any* line in the plane of the circle that passes through its center.

An example of a geometric figure that has only *one* line of symmetry is an isosceles triangle that has exactly two equal sides. In the adjoining figure, △ABC is isosceles with AC = BC. The triangle is symmetric with respect to the line through its vertex, C, and the midpoint of its base, M.

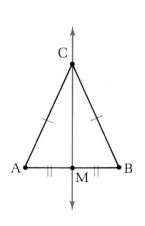

To see why, notice that both C and M are equidistant from the endpoints of $\overline{AB}$. Therefore, $\overleftrightarrow{CM}$ is the perpendicular bisector of $\overline{AB}$ because of the theorem that, in a plane, two points equidistant from the endpoints of a line segment determine the perpendicular bisector of the line segment. Hence, B is the reflection of A through $\overleftrightarrow{CM}$, and C is its own reflection through $\overleftrightarrow{CM}$ because it lies on it. Therefore, the reflections of $\overline{CA}$, $\overline{CB}$, and $\overline{AB}$ are $\overline{CB}$, $\overline{CA}$, and $\overline{BA}$, respectively. This means that the reflection through line $\overleftrightarrow{CM}$ of every point of △ABC is also a point of △ABC. So △ABC is symmetric with respect to $\overleftrightarrow{CM}$.

1. In which of the following figures does line ℓ look like a line of symmetry for the given line segment?

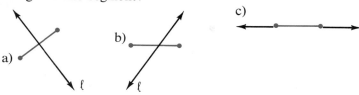

2. In which of the following figures does line ℓ look like a line of symmetry for the given angle?

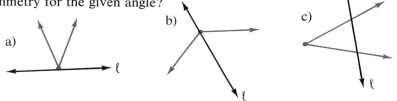

Many geometric figures other than the circle have more than one line of symmetry. For example, an equilateral triangle has three symmetry lines as shown below.

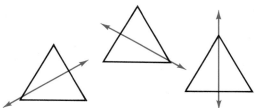

3. Trace the following figures and draw every line that you think is a symmetry line for each.

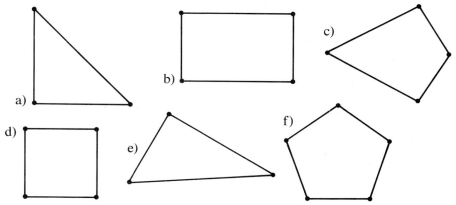

Lesson 3: Line Symmetry   203

4. How many lines of symmetry do you think each of these figures has?

a) A butterfly.

b) A snowflake.

c) A galaxy.

SET II

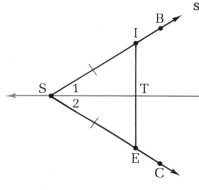

1. It is easy to prove that an angle is symmetric with respect to the line that bisects it. Answer the following questions about this proof.

Hypothesis: Line ℓ bisects $\angle$BSC.
Conclusion: $\angle$BSC is symmetric with respect to ℓ.

Of course, S is its own image through line ℓ. Let I be a point on $\overrightarrow{SB}$ at any given distance from S. Choose E on $\overrightarrow{SC}$ so that SE = SI. Draw $\overline{IE}$, intersecting line ℓ at T.

a) Why is $\angle 1 = \angle 2$?
b) Why is $\triangle$SIT $\cong$ $\triangle$SET?
c) Why is TI = TE?
d) What relationship do points S and T have to $\overline{IE}$?
e) Why is ℓ the perpendicular bisector of $\overline{IE}$?
f) Why is E the reflection of I through ℓ?

We have shown, in effect, that the reflection through ℓ of every point on $\angle$BSC is also on $\angle$BSC. Hence $\angle$BSC is symmetric with respect to ℓ.

2. (*Trace.*)

Copy the following figures and then add enough parts to each so that it is symmetric with respect to the lines in the figure.

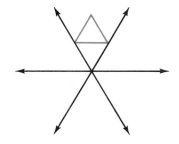

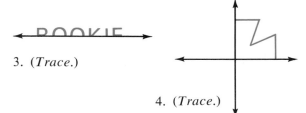

3. (*Trace.*)

4. (*Trace.*)

A stereoscope is an instrument that seems to change a pair of two-dimensional pictures into a single three-dimensional one. It does this by enabling the viewer to merge the two pictures, one of which is seen only with the left eye and the other only with the right, into one image.

Although a stereoscope does not use a mirror, you can get an idea of the kind of image it produces by placing a mirror along line ℓ in the figure below so that its reflecting surface faces to the right. Hold the mirror perpendicular to the page and put your head close enough to the mirror so that your left eye sees the left-hand picture and your right eye sees the image of the right-hand picture merge with it.

If you stare at the merged figure for a little while, it should begin to look three-dimensional. Can you explain how this works? Is line ℓ a symmetry line with respect to the two pictures?*

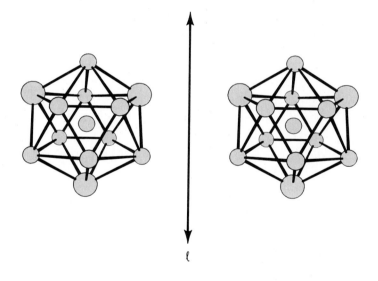

ℓ

---

*Adapted from Q. Johnson, G. S. Smith, O. H. Krikorian, and D. E. Sands, *Acta Crystallographica*, B26 (1970): 109.

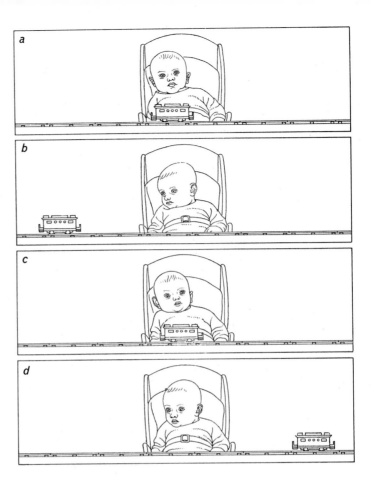

Lesson 4

Translations

How much does an infant understand about what he sees? Some recent experiments with very small children have had rather surprising results. One of them is illustrated by the sequence of diagrams shown here.

A toy train with flashing lights was placed in front of a 12-week-old baby. After a short period the train moved to the left along the track and stopped, as shown in diagram *b*. Then the train moved back to the center and repeated the trip from center to left and back again ten times. The eleventh time, the train slowly moved from the center to the right, as shown in diagram *d*. Instead of looking at it in its new position, however, the child looked again to the left and bewilderedly stared at the place where the train had stopped before. At the same time the train was in full view in its new place, flashing lights and all! This experiment was performed with a large number of twelve-week-old babies and every one of them made the same mistake!*

*From "The Object in the World of the Infant" by T. G. R. Bower. Copyright © 1971 by Scientific American, Inc. All rights reserved.

The change of position of the train on the track is a simple example of a *translation*. A translation can be thought of as the result of *sliding, without turning,* a geometric figure from one position to another. Translations and reflections are special types of geometric *transformations*.

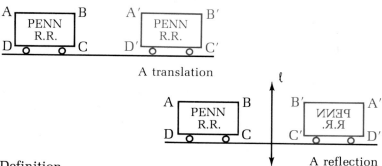

A translation

A reflection

► Definition

A *transformation* is a one-to-one correspondence between the points of a plane and the same or other points of the plane.

For example, in the first figure above, A′B′C′D′ is a translation image of ABCD. To each point of one figure there corresponds a point of the other. In the second figure, A′B′C′D′ is a reflection image of ABCD (through line ℓ). Notice that a translation *preserves orientation*, whereas a reflection *reverses* it.

There is an interesting connection between these two transformations. Look at the figure below in which A′B′C′D′ is the reflection of ABCD through line ℓ_1, and A″B″C″D″ is the reflection of A′B′C′D′ through line ℓ_2. We might also say that *A″B″C″D″ is a translation of ABCD.* In other words, a translation can be considered the result of two successive reflections. We will call the result of two or more successive transformations their *composite*.

Notice that in the figure below the reflection mirrors, ℓ_1 and ℓ_2, appear to be *parallel*.

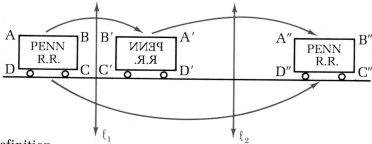

► Definition

Two lines are *parallel* iff they lie in the same plane and do not intersect.

The symbol for "parallel" is ‖; to indicate that lines ℓ_1 and ℓ_2 are parallel, we write $\ell_1 \parallel \ell_2$. Parallel lines play an important role in Euclidean geometry and later we will study them more extensively. For the present, they are useful in defining the word "translation."

► Definition
A transformation is a *translation* iff it is the composite of two reflections through parallel lines.

Another example of a translation is shown below. The reflection of $\triangle ABC$ through ℓ_1 is $\triangle DEF$ and the reflection of $\triangle DEF$ through ℓ_2 is $\triangle GHI$. If $\ell_1 \parallel \ell_2$, then $\triangle GHI$ is a translation image of $\triangle ABC$. If a tracing of $\triangle ABC$ were made, it could slide onto $\triangle GHI$ without being turned.

Since translations are composites of reflections, it follows that those properties preserved by reflections are also preserved by translations: collinearity, distance, betweenness of points, and angle measure.

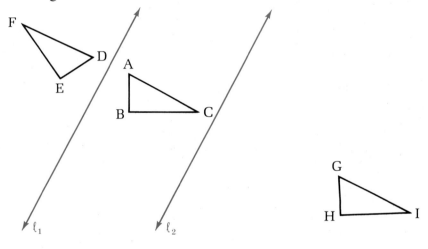

Exercises

SET I

1. In the figure at the top of the facing page, $\triangle HAR$ is the reflection of $\triangle BUC$ through ℓ_1 and $\triangle EST$ is the reflection of $\triangle HAR$ through ℓ_2; $\ell_1 \parallel \ell_2$.
   a) Through what transformation is $\triangle EST$ the image of $\triangle BUC$?
   b) Why is $\triangle BUC \cong \triangle HAR$ and $\triangle HAR \cong \triangle EST$?
   c) Why is $\triangle BUC \cong \triangle EST$?

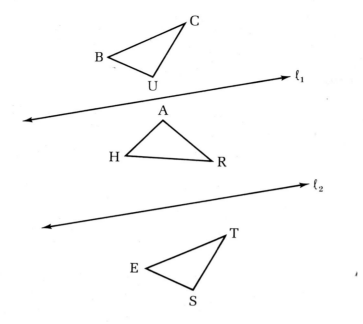

2. In the figure below, $\ell_1 \parallel \ell_2$. △DEF, △GHI, △JKL, and △MNO are reflection images of △ABC through either or both of the lines.

Which triangle is the reflection image of
a) △ABC through ℓ_1?
b) △GHI through ℓ_2?
c) △ABC through ℓ_2?
d) △MNO through ℓ_1?

Which triangle is the translation image of △ABC as a result of successive reflections through
e) ℓ_1 and ℓ_2?
f) ℓ_2 and ℓ_1?

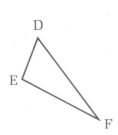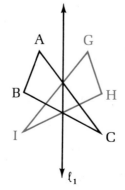

3. You know that the composite of two successive reflections through parallel lines is a translation. In the cartoon, this is illustrated by the man in the barber chair and those of his images that are looking in the same direction. Using the cartoon as a clue, what do you think is the composite of
   a) *three* successive reflections through parallel lines?
   b) *four* successive reflections through parallel lines?

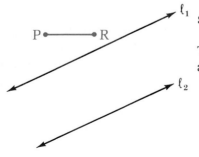

SET II

Trace the following figures on your paper. Then sketch, as accurately as you can, the images described.

1. a) Reflect $\overline{PR}$ through ℓ_1 and name its image $\overline{AG}$.
   b) Reflect $\overline{AG}$ through ℓ_2 and name its image $\overline{UE}$.
   c) Explain how you know that $PR = UE$.
   d) What other relationship do $\overline{PR}$ and $\overline{UE}$ seem to have?

2. a) Reflect △IST through ℓ_1 and name its reflection △ASN so that △ASN ≅ △IST.
   b) Reflect △ASN through ℓ_2 and name its reflection △BUL so that △BUL ≅ △ASN.
   c) Draw $\overline{IB}$, $\overline{SU}$, and $\overline{TL}$.
   d) If a tracing of △IST were made and slid onto △BUL, its vertices would move along the three "tracks" you have just drawn. Name two relationships that these tracks seem to have.

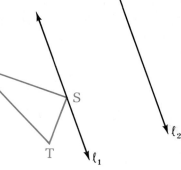

3. Trace this figure so that ℓ_1 and ℓ_2 are centered on your paper.
   a) Measure what you think is the distance between ℓ_1 and ℓ_2.
   b) Translate $\overline{ZA}$ by reflecting it through ℓ_1 and its image through ℓ_2. Label the translation image $\overline{GR}$.
   c) Measure the distance from Z to G and the distance from A to R.
   d) Translate $\overline{ZA}$ by reflecting it through ℓ_2 and its image through ℓ_1. Label the translation image $\overline{EB}$.
   e) Measure the distance from Z to E and the distance from A to B.
   f) If two parallel lines are x units apart, what do you think is the distance between a point and its translation image through the two lines?

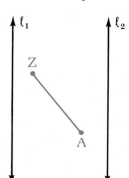

4. a) Translate OZ by reflecting it through ℓ_1 and its image through ℓ_2.
   b) If OZ were translated by reflecting it through ℓ_2 and its image through ℓ_1, where do you think its translation image would be?

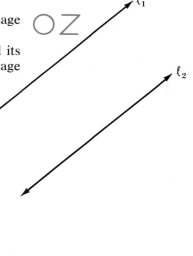

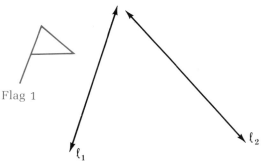

Flag 1

5. a) Reflect the "flag" through ℓ_1 and name its image Flag 2.
   b) Reflect Flag 2 through ℓ_2 and name its image Flag 3.
   c) Is Flag 3 a translation of Flag 1?

SET III

Make two identical drawings of Woodstock traveling on his "elevator" path so that one is a translation of the other. They should each be about an inch tall and should be spaced several inches apart.

Can you find a line through which to reflect one of your drawings of Woodstock and a second line through which to reflect its image so that the result is the translation you have drawn?

Lesson 5

Rotations

You have become acquainted with two fundamental geometric transformations, called *reflections* and *translations*. A third transformation is illustrated by the two drawings of the tortoise shown below. There is a one-to-one correspondence between the points of the rightside-up tortoise and the upside-down tortoise, yet it is neither a reflection nor a translation.

One way to illustrate this correspondence would be to put a sheet of tracing paper over the left-hand drawing and pin it at point P. If a tracing were made of the top tortoise and the paper were then turned about the pin, the tracing would eventually coincide with the drawing of the bottom tortoise. For this reason, the bottom tortoise is called a *rotation image* of the top one.

Another way to illustrate the correspondence is by reflections. You know that two successive reflections through parallel lines results in a translation. If the lines intersect, then two successive reflections through them results in a rotation instead. The right-hand drawing shows that the tortoise rotation is the composite of two such reflections.

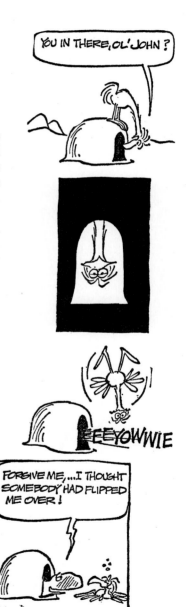

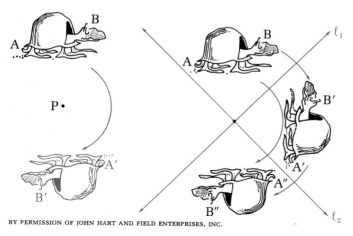

BY PERMISSION OF JOHN HART AND FIELD ENTERPRISES, INC.

213

Because of this, we can define "rotation" in the same way that we defined "translation," with the exception of just one word.

► Definition
A transformation is a *rotation* iff it is the composite of two reflections through intersecting lines.

Another example of a rotation is shown below. The reflection of △ABC through ℓ_1 is △DEF and the reflection of △DEF through ℓ_2 is △GHI. Since ℓ_1 and ℓ_2 intersect in point P, △GHI is a rotation image of △ABC. Point P is called the *center* of the rotation because it is about this point that a tracing of △ABC could be turned to coincide with △GHI.

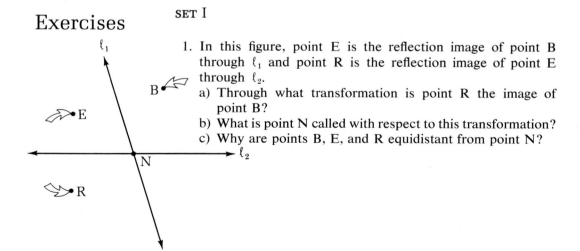

Exercises

SET I

1. In this figure, point E is the reflection image of point B through ℓ_1 and point R is the reflection image of point E through ℓ_2.
   a) Through what transformation is point R the image of point B?
   b) What is point N called with respect to this transformation?
   c) Why are points B, E, and R equidistant from point N?

2. Place two mirrors along lines ℓ_1 and ℓ_2 so that they meet at P and their reflecting surfaces face toward each other.
   a) How many fish do you see, including both the original and all of the images?
   b) How many of these are rotation images of the original fish?
   c) How many of these are reflection images of the original fish?

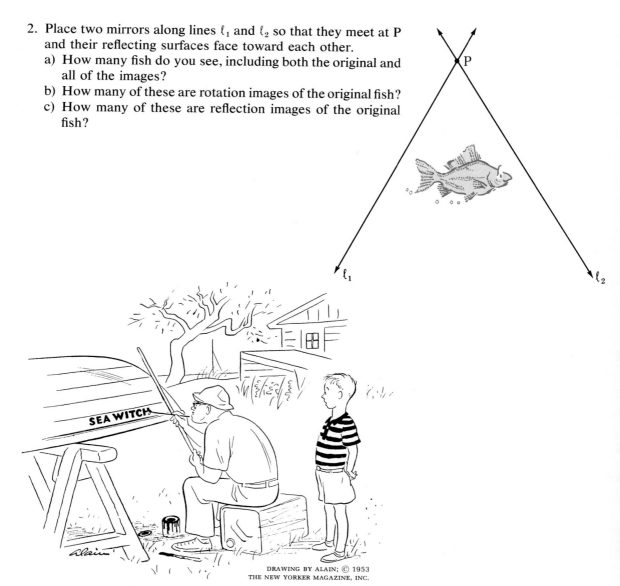

DRAWING BY ALAIN; © 1953
THE NEW YORKER MAGAZINE, INC.

"Want to know something, Dad?"

3. The most obvious way to get the words at the right to appear rightside up would be to turn this page upside down: in other words, rotate it 180°.

   Another way would be to use two mirrors to form a rotation image of the words. Can you discover and describe a way to place the mirrors so that the words in the image appear rightside up?

SEA WITCH

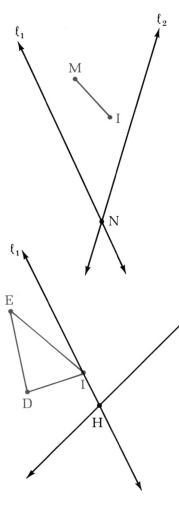

Trace the following figures on your paper. Then sketch, as accurately as you can, the images described.

1. a) Measure one of the acute angles formed by ℓ_1 and ℓ_2.
   b) Rotate $\overline{MI}$ by reflecting it through ℓ_1 and its image through ℓ_2. Label the rotation image $\overline{LA}$ so that L is the image of M.
   c) Draw $\measuredangle$MNL. This angle is formed by rays from the center of the rotation through a point of the figure and through the rotation image of the point; its measure is called the *magnitude* of the rotation.
   d) Measure $\measuredangle$MNL. How does the magnitude of the rotation seem to compare with your answer to part *a*?

2. a) Reflect $\triangle$EDI through ℓ_1 and name its reflection $\triangle$NBI so that $\triangle$NBI $\cong$ $\triangle$EDI.
   b) Reflect $\triangle$NBI through ℓ_2 and name its reflection $\triangle$URG so that $\triangle$URG $\cong$ $\triangle$NBI.
   c) Circles that have the same center are called *concentric*. Draw three concentric circles centered at H so that one contains E, one contains D, and one contains I.
   d) What do you notice?
   e) If a tracing of $\triangle$EDI were made, about what point could it be rotated so that it would coincide with $\triangle$URG?
   f) What is this point called?

3. a) Reflect NO through line ℓ_1 and reflect the resulting image through ℓ_2.
   b) What seems to be the measure of the angles formed by ℓ_1 and ℓ_2?
   c) What seems to be the magnitude of the rotation of NO?

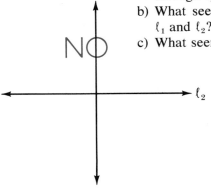

4. a) Reflect the "flag" through ℓ_1 and name its image Flag 2.
   b) Reflect Flag 2 through ℓ_2 and name its image Flag 3.
   c) Find the magnitude of the rotation.

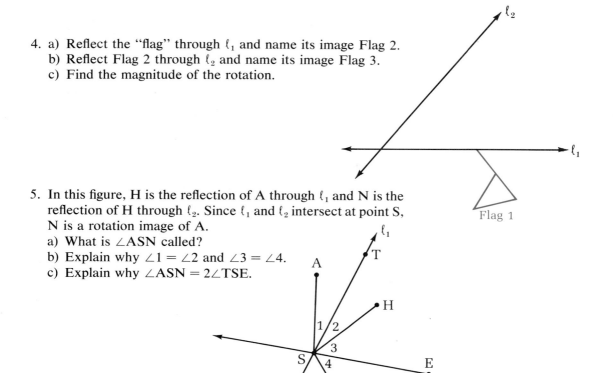

Flag 1

5. In this figure, H is the reflection of A through ℓ_1 and N is the reflection of H through ℓ_2. Since ℓ_1 and ℓ_2 intersect at point S, N is a rotation image of A.
   a) What is $\angle$ASN called?
   b) Explain why $\angle 1 = \angle 2$ and $\angle 3 = \angle 4$.
   c) Explain why $\angle$ASN $= 2\angle$TSE.

SET III

Little children like being rotated in a swing.

Trace the two drawings below. Can you find a line through which to reflect one of them and a second line through which to reflect its image so that the result is the same as the rotation shown?

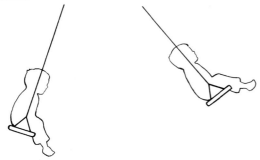

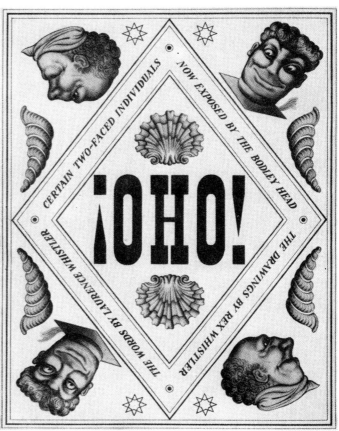

Lesson 6

Point Symmetry

The title page of a book of drawings by an English artist shown here is most unusual. Except for the words within the diamond-shaped border, the page looks exactly the same if turned upside-down. This is due to the fact that it has a special kind of symmetry called "point symmetry."

If we turn the first exclamation mark rightside-up, the title of the book has line symmetry, as shown in the left-hand figure at the top of the facing page.

Since the name "¡OHO!" expresses surprise, perhaps an equally appropriate name would have been "¡ONO!"

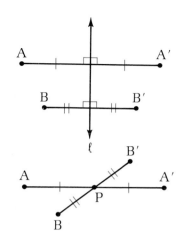

Although this title does not have line symmetry, it does possess point symmetry. To understand what we mean by this, compare the two adjoining figures. The first shows two points and their reflections through a *line*. The second shows two points and their reflections through a *point*.

▶ Definition

One point is the ***reflection (image) through point P*** of another point iff P is the midpoint of the segment that joins the two points. Point P is its own reflection.

It is very easy to locate the reflection of a point through a point by construction.

▶ **Construction 5**

To construct the reflection of a point through a point.

Let P be the point to be reflected through point O. Draw $\overleftrightarrow{OP}$. With O as center and OP as radius, draw a circle. The circle intersects $\overleftrightarrow{OP}$ in two points: point P and its reflection, point P′.

Now we are ready to define what we mean by "point symmetry."

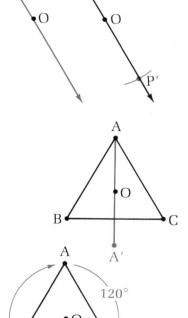

▶ Definition

A set of points has ***point symmetry*** iff there is a point P such that the reflection through P of each point of the set is also a point of the set.

A simple test to determine whether a figure has point symmetry is to turn it upside-down and see if it looks the same. In other words, a figure that has point symmetry is unchanged in appearance by a 180° rotation.

Point symmetry is a special case of a more general type of symmetry called *rotational symmetry*. If a figure can be rotated through less than 360° so that it coincides with itself, the figure has rotational symmetry. For example, an equilateral triangle does *not* have point symmetry. If vertex A of the first triangle shown here, for instance, is reflected through point O, its image is not a point on the triangle. Neither is there any other point through which every point of the triangle can be reflected onto the triangle. Nevertheless, the triangle *does* have rotational symmetry because it can be rotated through 120° about point O so that it coincides with itself.

Exercises

• H

• O

• I

• P

1. Trace this figure and then construct the reflections of points H, O, and P through point I. Label them H′, O′, and P′, respectively.
   a) What relationship does I have to $\overline{HH'}$, $\overline{OO'}$, and $\overline{PP'}$?
   b) H, O, and P are collinear. Do their images appear to be collinear?
   c) On the basis of your drawing, what other properties of a set of points seem to be preserved when it is reflected through a point?

2. In the figure shown here, △I′N′C′ is the reflection image of △INC through point A.
   a) What other transformation of △INC do you think would produce △I′N′C′ as its image?
   b) We have proved that a triangle and its reflection through a line are congruent. Do a triangle and its reflection through a point seem to be congruent?
   c) You know that reflection through a line reverses orientation. Does reflection through a point reverse orientation?

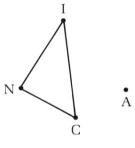

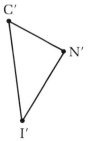

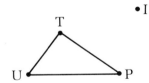

3. Copy the figure shown here.
   a) Use a straightedge and compass to construct the reflection image of △TUP through point I.
   b) Does the completed figure have point symmetry?
   c) Does it have line symmetry?

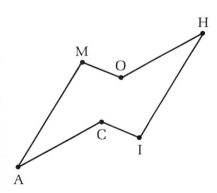

4. The adjoining figure is symmetric with respect to a point.
   a) Trace it and use your straightedge and compass to find the point. Name the point N.
   b) Draw three segments determined by corners of the figure that are bisected by N.
   c) Does the figure have rotational symmetry?

5. The figure shown here has rotational symmetry.
   a) Trace it and draw the angle through which point P would have to be rotated in order to coincide with point U.
   b) Measure the angle with your protractor.
   c) Give the measure of a larger angle through which the figure can be rotated so that it will also coincide with itself.
   d) Does the figure have point symmetry?

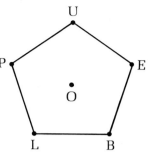

SET II

1. A kaleidoscope is a simple optical instrument that uses two mirrors to produce symmetrical patterns. This photograph shows the face of a toy monkey and five of its images in a kaleidoscope with mirrors that meet at a 60° angle.

   The pattern of the monkey faces appears to be very symmetrical.
   a) How many lines of symmetry does it seem to have?
   b) Does it seem to have point symmetry?

   Place two mirrors along lines ℓ_1 and ℓ_2 of the fish diagram on page 215 again so that they meet at P and their reflecting surfaces face toward each other. You should see a kaleidoscopic pattern of six fish.
   c) How many lines of symmetry does it have?
   d) Does it have point symmetry?

   Although the mirror arrangements for the monkeys and the fish are the same, the resulting kaleidoscope patterns are different in their symmetry.
   e) Can you explain why?

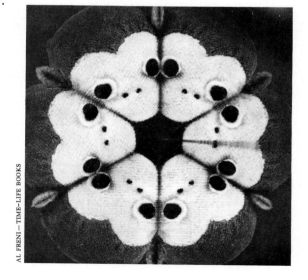

AL FRENI—TIME-LIFE BOOKS

Lesson 6: Point Symmetry   221

2. Crossword puzzles are usually designed so that they are symmetrical. The top four rows have been completed in the diagram below. Copy it and fill in the last three rows so that the diagram has point symmetry.

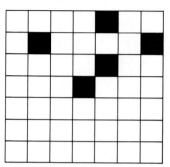

Copy the following figures and sketch, as accurately as you can, the reflection of each figure through the point named P.

3.

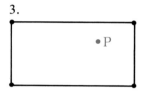

• P

4.

• P

5. (*Trace.*)

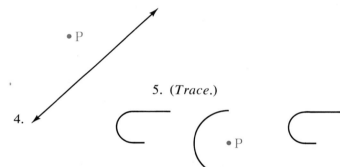

• P

6. Given: The reflections of points S and I through O are points X and U, respectively.
   Prove: ∠S = ∠X.
   Par 6.

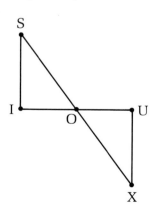

The swimming pool in this cartoon is closed on Mondays. There is something rather remarkable in the picture other than the fact that there are mice swimming in the pool. What is it?

COURTESY OF JOHN MC CLELLAN

Chapter 5 / Summary and Review

Basic Ideas

Constructions

Postulates

13. *The Point Reflection Postulate.*   There is a one-to-one correspondence between points in a plane and their reflections
through a given line.   192
14. Reflection of a set of points through a line preserves collinearity.   195

Theorems

13. If the reflection of point A through line ℓ is point B, then the reflection of point B through line ℓ is point A.   191
14. Reflection of a set of points through a line preserves distance. 196
15. Reflection of a set of points through a line preserves between-ness.   196
16. Reflection of a set of points through a line preserves angle measure.   196
17. A triangle and its reflection through a line are congruent.   196

SET I

Exercises

1. What points remain fixed in each of the following trans-formations?
   a) Reflection through a line.
   b) Translation.
   c) Rotation.

2. Copy this figure and use your straightedge and compass to construct the reflection image of point P
   a) through line ℓ.
   b) through point O.

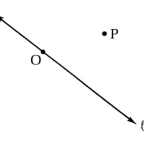

Trace the figures below and sketch the indicated reflection images.

3. Reflect through line ℓ.

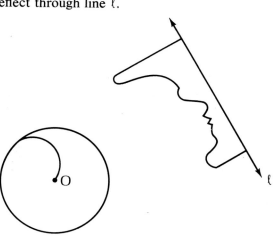

4. Reflect through point O.

5. In this figure, $\ell_1 \parallel \ell_2$ and ℓ_3 intersects ℓ_1 and ℓ_2.
   a) Reflect $\triangle ABC$ through ℓ_1 and label its image $\triangle DEC$.
   b) Reflect $\triangle DEC$ through ℓ_2 and label its image $\triangle FGH$.
   c) Reflect $\triangle FGH$ through ℓ_3 and label its image $\triangle JIK$.
   d) What transformation relates $\triangle ABC$ to $\triangle FGH$? Explain.
   e) What transformation relates $\triangle DEC$ to $\triangle JIK$? Explain.

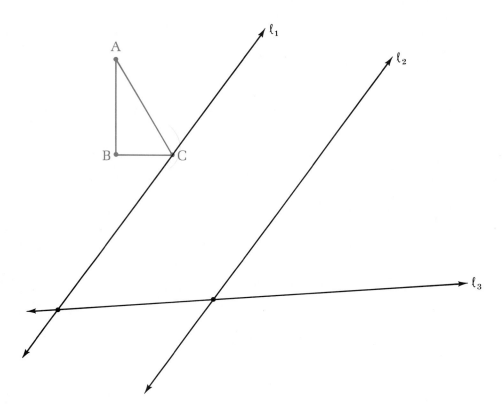

6. An equilateral triangle has both line symmetry and rotational symmetry.
   a) Construct two equilateral triangles with your straightedge and compass.
   b) Shade in part of the interior of one so that the shaded figure possesses line symmetry but not rotational symmetry.
   c) Shade in part of the interior of the other triangle so that it possesses rotational symmetry but not line symmetry.

Escher has created many mosaics based upon animal shapes. In fact, an entire book has been written about them.* You saw one mosaic based upon reptiles on page 148 of this book. Four more are shown here.

These mosaics illustrate both the various types of geometric transformations and the types of symmetry we have studied. Identify these in each case. (Ignore the different shadings and imagine that each pattern continues indefinitely without any borders.)

1. Beetles.

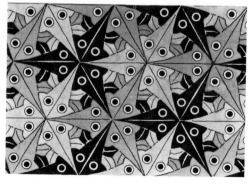

2. Birds.

3. Flatfish.

4. Bulldogs.

*Caroline H. MacGillavry, *Symmetry Aspects of M. C. Escher's Periodic Drawings* (published for the International Union of Crystallography by A. Oosthoek's Uitgeversmaatschappij NV, Utrecht, 1965). Mosaics courtesy of Escher Foundation, Haags Gemeentemuseum, The Hague.

Some playing cards look exactly the same if turned upside down. For example, the two of spades does, but the five of clubs does not.

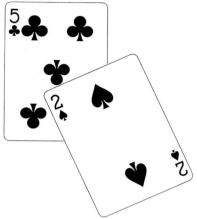

1. Can you figure out which cards in a deck of 52 cards possess point symmetry? The basic patterns on the numbered cards are shown here.

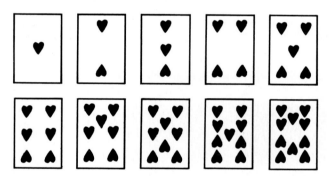

2. What playing cards have line symmetry?

Chapter 6

INEQUALITIES

Lesson 1

Postulates of Inequality

The bartender in this cartoon likes to keep everything in order and one of his customers is in the wrong place. In the diagram, the heights of the little fellow and the two men standing next to him have been represented by a, b, and c. To show that a is greater than b, we write $a > b$ and to show that b is less than c, we write $b < c$. Since $b < c$, we also know that $c > b$, which is the order the bartender would prefer since he wants the taller fellow to be at the left.

"I wonder, sir, if you would indulge me in a rather unusual request?"

230

If $a > b$ and $c > b$, can we conclude anything about the relative sizes of a and c? The first two adjoining figures reveal that the answer to this is no. What if $a > b$ and $b > c$? The third figure suggests that it follows that $a > c$. We will assume this conclusion as a postulate about inequality. It is stated below along with some other basic assumptions. Like those in the postulates of equality, the letters represent real numbers.

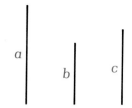

▶ **Postulate 1**   (The Three Possibilities Postulate)
Either $a > b$, $a = b$,   or $a < b$.

▶ **Postulate 2**   (The Transitive Postulate)
If $a > b$ and $b > c$,   then $a > c$.

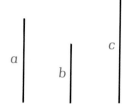

▶ **Postulate 3**   (The Addition Postulate)
If $a > b$,   then $a + c > b + c$.

▶ **Postulate 4**   (The Subtraction Postulate)
If $a > b$,   then $a - c > b - c$.

▶ **Postulate 5**   (The Multiplication Postulate)
If $a > b$ and $c > 0$,   then $ac > bc$.

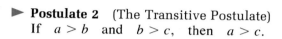

▶ **Postulate 6**   (The Division Postulate)
If $a > b$ and $c > 0$,   then $\dfrac{a}{c} > \dfrac{b}{c}$.

Notice that Postulates 2–6 are expressed in terms of the symbol $>$. They are equally valid in terms of the symbol $<$.
    We will use these postulates to prove two theorems about inequalities.

The Addition Theorem of Inequality.
If $a > b$ and $c > d$,   then $a + c > b + d$.

The "Whole Greater than Its Part" Theorem.
If $a = b + c$ and $c > 0$,   then $a > b$.

    Although the words "greater than" and "less than" concern comparisons of *numbers*, we will also use them in referring to *line segments* and *angles*. If, for example, AB > CD, it is natural to say that $\overline{AB}$ is *longer* than $\overline{CD}$. If ∠E > ∠F, we say that ∡E is *larger* than ∡F.

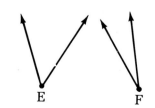

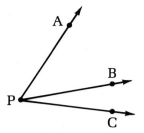

The "whole greater than its part" theorem gets its name from the way that it is applied to geometric figures. For example, in the figure at the left, $\overrightarrow{PB}$ is between $\overrightarrow{PA}$ and $\overrightarrow{PC}$. Angle APC is evidently larger than ∡APB, but can you prove it?

Since $\overrightarrow{PB}$ is between $\overrightarrow{PA}$ and $\overrightarrow{PC}$, we know that $\angle APC = \angle APB + \angle BPC$. Also, $\angle BPC > 0$ because, by definition, the measure of every angle is greater than zero. It follows from the "whole greater than its part" theorem that $\angle APC > \angle APB$. The proof of this useful theorem is left as an exercise.

SET I

Complete the following proofs by stating the reasons.

1. *The addition theorem of inequality.*
   If $a > b$ and $c > d$, then $a + c > b + d$.
   Proof.

   Statements

   a) $a > b$.
   b) $a + c > b + c$.
   c) $c > d$.
   d) $b + c > b + d$.
   e) $a + c > b + d$.

2. *The "whole greater than its part" theorem.*
   If $a = b + c$ and $c > 0$, then $a > b$.
   Proof.

   Statements

   a) $a = b + c$.
   b) $a - b = c$.
   c) $c > 0$.
   d) $a - b > 0$.
   e) $a > b$.

3. Translate the following statements into inequalities:
   a) ∡S is larger than ∡T.
   b) $\overline{EA}$ is shorter than $\overline{AL}$.

Name the postulate or theorem that justifies each of the conclusions about the following geometric figures.

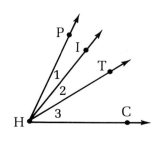

4. a) If $\angle IHC > \angle 1$, then $2\angle IHC > 2\angle 1$.
   b) If $\angle 1 < \angle 2$ and $\angle 2 < \angle 3$, then $\angle 1 < \angle 3$.
   c) If $\angle IHC = \angle 2 + \angle 3$, then $\angle IHC > \angle 3$.
   d) If $\angle IHC > \angle 3$ and $\angle PHT = \angle 3$, then $\angle IHC > \angle PHT$.

5. a) If neither SE < EI nor SE > EI, then SE = EI.
   b) If SK > EI, then SK − EK > EI − EK.
   c) If ST < SE and TR < EI, then ST + TR < SE + EI.
   d) If SI > SR and RI > SI, then RI > SR.

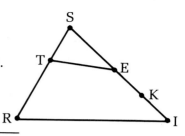

SET II

1. Given: F-L-Y.
   Prove: FY > LY.
   Par 3.

2. Given: CR < UR and RE < RV.
   Prove: CE < UV.
   Par 4.

3. Given: BT = BU.
   Prove: ∠BTN > ∠TUB.
   Par 5.

SET III

A bricklayer has eight bricks. Seven of the bricks weigh the same amount and one is a little heavier than the others. If the man has a balance scale, how can he find the heaviest brick in only two weighings?*

*From *Litton's Problematical Recreations*, edited by James F. Hurley (Van Nostrand Reinhold, 1971).

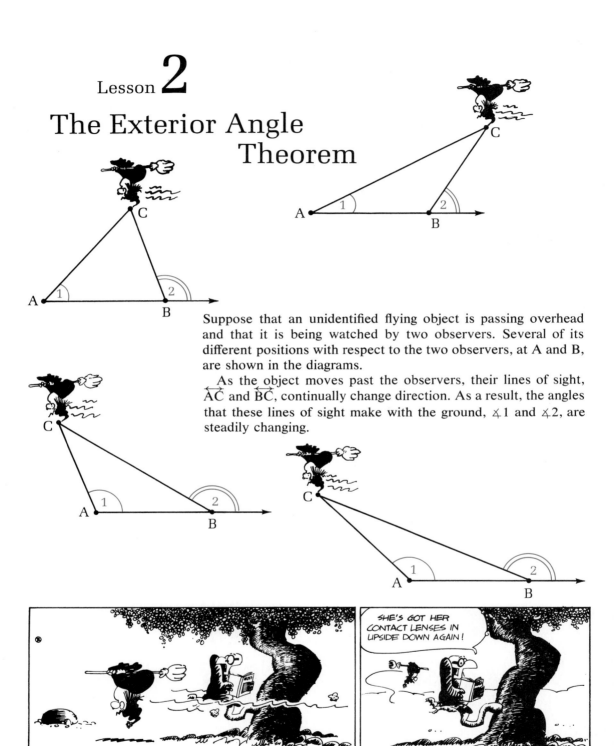

Lesson 2

The Exterior Angle Theorem

Suppose that an unidentified flying object is passing overhead and that it is being watched by two observers. Several of its different positions with respect to the two observers, at A and B, are shown in the diagrams.

As the object moves past the observers, their lines of sight, $\overleftrightarrow{AC}$ and $\overleftrightarrow{BC}$, continually change direction. As a result, the angles that these lines of sight make with the ground, $\angle 1$ and $\angle 2$, are steadily changing.

SHE'S GOT HER CONTACT LENSES IN UPSIDE DOWN AGAIN!

RUSSELL MYERS

BROOM-HILDA © 1973 THE CHICAGO TRIBUNE

234

The angle for observer B seems in every diagram to be larger than that for observer A. In symbols, $\angle 2 > \angle 1$. Do you think that this would continue to be true even after the object had flown several miles past the observers?

Angle 1 is an angle of the triangle ABC. It is sometimes called an interior angle of the triangle to distinguish it from the exterior angles of the triangle, of which ∡2 is an example. Notice that ∡2 forms a linear pair with the interior angle of the triangle at B, ∡ABC.

▶ Definition
An **exterior angle of a triangle** is an angle that forms a linear pair with one of the angles of the triangle.

The other two angles of the triangle are called *remote interior angles* with respect to the exterior angle under consideration. In the flying-object diagrams, ∡A and ∡C are the remote interior angles with respect to the exterior angle at B, ∡2.

Look again at the diagrams and you will notice that not only does it always seem that $\angle 2 > \angle A$, but also that $\angle 2 > \angle C$. We can prove that these inequalities are always true; in other words, that an exterior angle of a triangle is always larger than either remote interior angle of the triangle. At this point in your study of geometry, you would not be expected to think of such a proof yourself.

▶ **Theorem 18**   (The Exterior Angle Theorem)
An exterior angle of a triangle is greater than either remote interior angle.

Hypothesis: ∡CBD is an exterior angle of △ABC.
Conclusion: $\angle CBD > \angle A$; also, $\angle CBD > \angle C$.

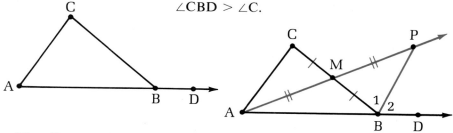

We will prove the theorem for just one of the two conclusions, since the proof of the other is similar. We begin by marking the midpoint of $\overline{CB}$, M, drawing a line through A and M such that $MP = AM$, and then drawing $\overline{BP}$. It is easy to prove that $\triangle ACM \cong \triangle PBM$ so that $\angle C = \angle 1$. Since $\angle CBD > \angle 1$, we can conclude that $\angle CBD > \angle C$.

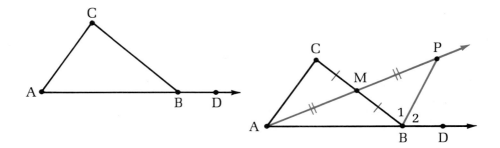

Proof.

Statements	Reasons
1. ∡CBD is an exterior angle of △ABC.	Hypothesis.
2. Let **M** be the midpoint of $\overline{CB}$.	A line segment has exactly one midpoint.
3. CM = MB.	A midpoint divides a segment into two equal segments.
4. Draw $\overleftrightarrow{AM}$.	Two points determine a line.
5. Choose **P** on the ray opposite $\overrightarrow{MA}$ so that MP = AM.	The Unique Point Corollary.
6. Draw $\overline{BP}$.	Same reason as step 4.
7. ∠AMC = ∠BMP.	Vertical angles are equal.
8. △ACM ≅ △PBM.	S.A.S. postulate.
9. ∠C = ∠1.	Corresponding parts of congruent triangles are equal.
10. ∠CBD = ∠1 + ∠2.	Betweenness of rays.
11. ∠CBD > ∠1.	The "whole greater than its part" theorem.
12. ∠CBD > ∠C.	Substitution.

Exercises

1. In the adjoining figure, $\overrightarrow{UT}$ and $\overrightarrow{UA}$ are opposite rays.
   a) Name the exterior angle of the triangle shown.
   b) Name the two remote interior angles with respect to it.
   c) Write two inequalities for the figure that follow from the Exterior Angle Theorem.

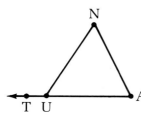

2. In the adjoining figure, ∡WAH and ∡LAE are exterior angles of △WAL.
   a) Name two relationships that exist between ∡WAH and ∡WAL.
   b) Name two relationships that exist between ∡WAH and ∡LAE.
   c) Is ∡HAE an exterior angle of the triangle? Explain.
   d) How many exterior angles does a triangle have at one vertex?
   e) How many exterior angles does a triangle have in all?

3. Use the three figures shown here to help in determining whether the following statements are true or false.
   a) All six exterior angles of a triangle may be obtuse.
   b) All six exterior angles of a triangle may have different measures.
   c) An exterior angle of a triangle may be smaller than one of the interior angles of the triangle.
   d) An exterior angle of a triangle may be equal to one of the interior angles of the triangle.
   e) Four of the exterior angles of an obtuse triangle are obtuse angles.

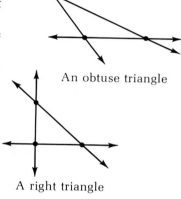

An obtuse triangle

An acute triangle

A right triangle

4. The following questions refer to the figure below. In answering them, you may assume that the points that look collinear are collinear.

   According to the Exterior Angle Theorem, which angles must be
   a) smaller than ∡7?
   b) larger than ∡5?
   c) smaller than ∡4?

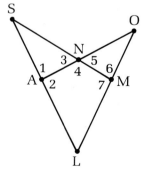

Write complete proofs for each of the following.

1. Given: ∡1 and ∡4 are exterior angles
   of △COD; ∠1 > ∠4.
   Prove: ∠1 > ∠2.
   Par 4.

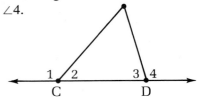

2. Given: ∡PKE is an exterior angle
   of △PIK; $\overline{PI} \perp \overline{IK}$.
   Prove: ∡PKE is obtuse.
   Par 7.

3. Given: ∡1 and ∡2 are a linear pair,
   ∡3 and ∡4 are a linear pair.
   Prove: ∠1 > ∠S.
   Par 4.

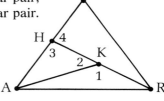

Dilcue decides to call an angle such as ∡ PEH in the figure at the left an "inferior" angle of △ERC.

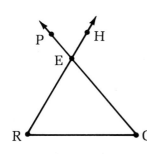

1. On the basis of this example, how would you define an "inferior" angle of a triangle?

2. How many "inferior" angles does a triangle have?

3. In an attempt to invent an "Inferior Angle Theorem," Dilcue says that "each inferior angle of a triangle is greater than either of its remote interior angles." Can you give a logical argument proving that there is *no* triangle for which this is true?

Lesson 3

Triangle Side and Angle Inequalities

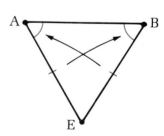

This strange looking picture is a sixteenth-century portrait of one of the kings of England, Edward VI. It is drawn with a peculiar perspective, so that it is almost unrecognizable when viewed from the front. When viewed from the edge, however, the perspective is such that the distortion disappears and the portrait looks normal!

The diagrams at the right show the two ways of looking at the picture as seen from overhead. The top edge of the picture is represented by $\overline{AB}$ and the eyes by point E. In the first diagram, if our eyes are centered with respect to the edges of the picture, EA = EB. We can conclude by means of the Isosceles Triangle Theorem that $\angle A = \angle B$. So the angles formed by the picture and our lines of sight to its left and right edges are equal.

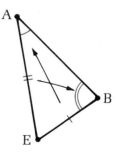

In the second diagram, on the other hand, we are looking at the picture from a point much closer to one edge than the other: EA > EB. How do the angles at A and B compare in measure now? Evidently, $\angle B > \angle A$. Notice that of the two angles the larger angle, $\angle B$, is opposite the longer side, $\overline{EA}$.

This relationship of unequal sides and unequal angles holds true for all triangles.

► **Theorem 19**

If two sides of a triangle are unequal, the angles opposite them
are unequal and the larger angle is opposite the longer side.

Hypothesis: △ABC with BC > AC.
Conclusion: ∠A > ∠B.

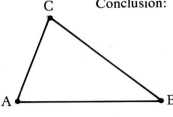
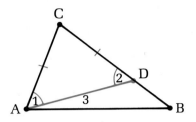

To prove this theorem, we will construct an isosceles triangle
within the given triangle as shown in the figure at the right above.
Since ∠CAB > ∠1 and ∠1 = ∠2, we can conclude that ∠CAB
> ∠2. Notice that ∠2 > ∠B, since ∡2 is an exterior angle of
△ADB. Therefore, ∠CAB > ∠B.

Proof.

Statements	Reasons
1. △ABC with BC > AC.	Hypothesis.
2. Choose point D on $\overrightarrow{CB}$ such that CD = CA.	The Unique Point Corollary.
3. Draw $\overline{AD}$.	Two points determine a line.
4. ∠1 = ∠2.	If two sides of a triangle are equal, the angles opposite them are equal.
5. ∠CAB = ∠1 + ∠3.	Betweenness of rays.
6. ∠CAB > ∠1.	The "whole greater than its part" theorem.
7. ∠CAB > ∠2.	Substitution.
8. ∠2 > ∠B.	An exterior angle of a triangle is greater than either remote interior angle.
9. ∠CAB > ∠B.	Transitive postulate of inequality.

The converse of this theorem is also true:

► **Theorem 20**

If two angles of a triangle are unequal, the sides opposite them
are unequal and the longer side is opposite the larger angle.

1. It is easy to prove Theorem 20 indirectly. Remember that the indirect method consists of considering all of the possibilities and then eliminating all but one by showing that they lead to logical contradictions.

Exercises

If two angles of a triangle are unequal, the sides opposite them are unequal and the longer side is opposite the larger angle.

Hypothesis: △RUN with ∠R > ∠U.

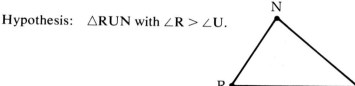

Instead of writing a two-column proof for this theorem, answer the following questions about it.

a) What, in terms of the figure, is the conclusion of the theorem?
b) In order to prove this conclusion, two other possible relationships between these sides must be eliminated. What are they? (Hint: What does the three possibilities postulate say?)
c) Suppose NU < NR. Then it follows that ∠R < ∠U. Why?
d) What does this contradict?
e) Suppose NU = NR. Then it follows that ∠R = ∠U. Why?
f) What does this contradict?
g) What is the only possibility remaining?

2. What conclusion can you draw from each of the following assumptions about △FLY?
a) FL = FY.
b) ∠L < ∠F.
c) LY > LF.

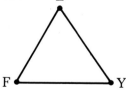

3. The adjoining figure is not accurately drawn. On the basis of the measures of the angles indicated, answer the following questions.

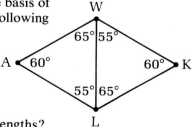

a) Are the two triangles congruent?
b) Which side of △WAL is the longest?
c) Which side of △WLK is the longest?
d) Do these two segments necessarily have equal lengths?

Lesson 3: Triangle Side and Angle Inequalities   241

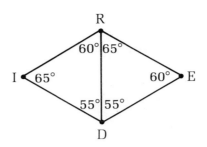

4. The figure above is not accurately drawn. On the basis of the measures of the angles indicated, answer the following questions.
   a) Which side of △RID is the longest?
   b) Which side of △RDE is the longest?
   c) Can you draw any conclusion about the relative lengths of the two segments you named in parts a and b?
   d) Are the triangles congruent?

SET II

Write complete proofs for each of the following.

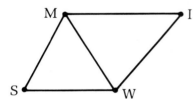

1. Given: MI > MW; ∠S > ∠SMW.
   Prove: MI > SW.
   Par 4.

2. Given: △HIK with IE = IK.
   Prove: ∠1 > ∠2.
   Par 4.

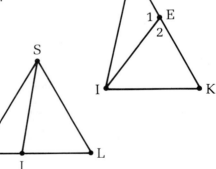

3. Given: △SAL is equilateral.
   Prove: SI > IL.
   Par 7.

If this accordion were accurately drawn, which segment would be
the shortest? (Hint: The answer is *not* $\overline{CD}$.)

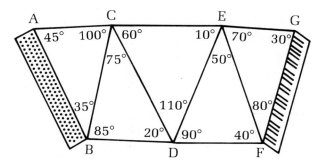

Lesson 4

The Triangle Inequality Theorem

An anteater is thirsty and decides to get a drink from a nearby stream. Then he plans to walk over to an anthill for lunch.

Suppose that the stream is perfectly straight and that the anteater and anthill are 45 feet and 30 feet from it, respectively, as shown in the diagram below. (The anteater is at point A, $\overleftrightarrow{BC}$ represents the stream, and the anthill is at point H.) To what point on the stream should the anteater walk in order to make his path from A to $\overleftrightarrow{BC}$ to H *as short as possible*?

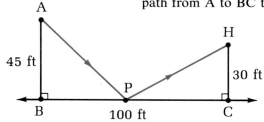

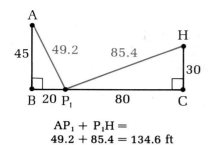

$$AP_1 + P_1H =$$
$$49.2 + 85.4 = 134.6 \text{ ft}$$

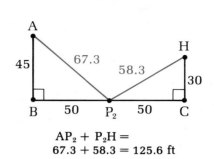

$$AP_2 + P_2H =$$
$$67.3 + 58.3 = 125.6 \text{ ft}$$

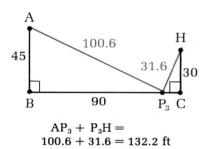

$$AP_3 + P_3H =$$
$$100.6 + 31.6 = 132.2 \text{ ft}$$

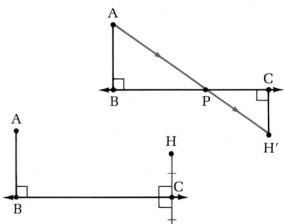

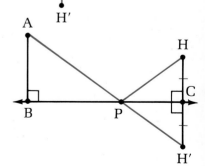

The diagrams above show three possibilities. Of these three paths, the second one is the shortest. Is there some other path that is even shorter? If so, what is it?

If the anthill were on the other side of the stream, the shortest path would be obvious. It would be along the segment joining A and H. This suggests a way to solve the problem. Reflect point H through $\overleftrightarrow{BC}$ and label its image H'. Draw $\overline{AH'}$. The point in which $\overleftrightarrow{AH'}$ intersects $\overleftrightarrow{BC}$ is the point on the stream to which the anteater should walk.

It is easy to show that AP + PH = AH'. Since A-P-H', AH' = AP + PH'; PH = PH' because reflection of a set of points through a line preserves distance. Therefore, AH' = AP + PH by substitution.

Lesson 4: The Triangle Inequality Theorem   245

There is one little hole in our argument about the shortest path. We have assumed from the appearance of the figure that, if A-P-H', AP + PH' is less than any other path AQ + QH' where Q is not between A and H'. This is equivalent to assuming that, in △AQH', the sum of the lengths of two sides, AQ + QH', is greater than the length of the third side, AH'.

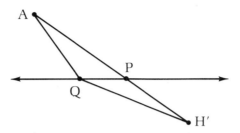

We can prove that this relationship is true of *every* triangle and will call it the Triangle Inequality Theorem.

▶ **Theorem 21** (The Triangle Inequality Theorem)
The sum of the lengths of any two sides of a triangle is greater than the length of the third side.

Exercises   SET I

1. Complete the following proof of the Triangle Inequality Theorem by supplying the reasons.

   The sum of the lengths of any two sides of a triangle is greater than the length of the third side.

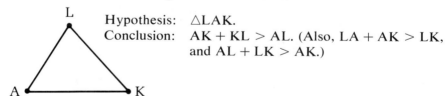

Hypothesis: △LAK.
Conclusion: AK + KL > AL. (Also, LA + AK > LK, and AL + LK > AK.)

   The proof of this theorem, like those of the other inequality theorems we have proved, requires the addition of some extra parts to the figure. This time we will construct an isosceles triangle next to the given triangle and then relate its equal angles to other angles in the figure, in somewhat the same way that we proved Theorem 19.

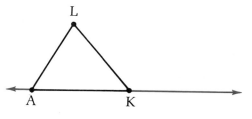

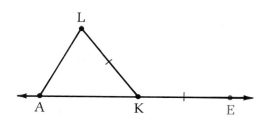

Proof.

<u>Statements</u>

a) Draw $\overleftrightarrow{AK}$.
b) Choose point E on the ray opposite $\overrightarrow{KA}$ so that KE = KL.
c) Draw $\overline{LE}$.
d) $\angle 1 = \angle 2$.
e) $\angle ALE = \angle 3 + \angle 1$.
f) $\angle ALE > \angle 1$.
g) $\angle ALE > \angle 2$.
h) In $\triangle ALE$, AE > AL.
i) AE = AK + KE.
j) AK + KE > AL.
k) AK + KL > AL.

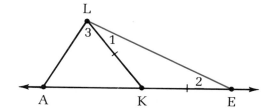

2. The figure below contradicts the Triangle Inequality Theorem. This means that it would be impossible to draw a triangle that actually had sides of the indicated lengths.
   a) What is the contradiction?
   b) Could $\triangle$SEA exist if SE = 7 in. instead of 10 in.?
   c) Could $\triangle$SEA exist if SE = 8 in. instead of 10 in.?

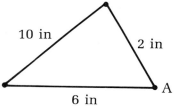

3. Triangle BAY has been drawn to scale.
   a) Even though the lengths of the sides have not been written on the figure, can you conclude that BA < BY + YA?
   b) Is it true that $\angle B + \angle A > Y$?
   c) Does the following statement seem like a reasonable theorem?

   The sum of the measures of any two angles of a triangle is greater than the measure of the third angle.

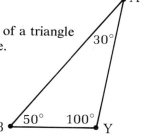

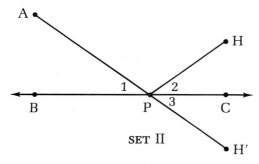

SET II

1. We have found that the shortest path the anteater could take to the stream and then to the anthill can be determined by reflecting, in effect, the anthill through the stream. In the figure above, A represents the anteater's starting point, $\overleftrightarrow{BC}$ represents the stream, and H′ represents the reflection of H, the anthill, through $\overleftrightarrow{BC}$.

   Can you explain why, if the anteater travels from A to P to H, the angle at which it approaches the stream is equal to the angle at which it leaves it? In other words, why $\angle 1 = \angle 2$?

Write complete proofs for each of the following.

2. Given: $\angle ODN = \angle N$.
   Prove: $PO + ON > PD$.
   Par 4.

3. Given: $\triangle OCE \cong \triangle ANE$.
   Prove: $CA > CO$.
   Par 6.

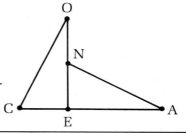

SET III

R •

A roadrunner at point R wants to run to Road 1, then to Road 2, and then to point C to visit a coyote. Can you figure out the shortest path he could take? Copy the figure shown here and make a neat drawing to show it.

C •

Road 2

Chapter 6/Summary and Review

Basic Ideas

Exterior angle of a triangle   235

Postulates of Inequality   231

1. Three Possibilities
2. Transitive
3. Addition
4. Subtraction
5. Multiplication
6. Division

Theorems of Inequality   231

Addition
Whole Greater than Its Part

Theorems

18. *The Exterior Angle Theorem.*   An exterior angle of a triangle is greater than either remote interior angle.   235
19. If two sides of a triangle are unequal, the angles opposite them are unequal and the larger angle is opposite the longer side. 240
20. If two angles of a triangle are unequal, the sides opposite them are unequal and the longer side is opposite the larger angle. 240
21. *The Triangle Inequality Theorem.*   The sum of the lengths of any two sides of a triangle is greater than the length of the third side.   246

Exercises

SET I

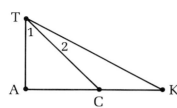

1. State the postulate or theorem that justifies each of the following conclusions about the adjoining figure.
   a) AK > AC, since AK = AC + CK.
   b) If TA < TC and TC < TK, then TA < TK.
   c) ∠TCK > ∠A.
   d) If ∠K > ∠2, then TC > CK.
   e) TA + AK > TK.
   f) If TA < TC and AC < CK, then TA + AC < TC + CK.
   g) If neither ∠1 < ∠K nor ∠1 = ∠K, then ∠1 > ∠K.

2. Complete the following proof by supplying the reasons.

   Given: △PIN with exterior ∡2;
            IP = IN.
   Prove: ∡P is acute.

   Proof.

   Statements

   a) △PIN with exterior ∡2.
   b) ∠2 > ∠P.
   c) IP = IN.
   d) ∠P = ∠1.
   e) ∠2 > ∠1.
   f) ∡1 and ∡2 are a linear pair.
   g) ∡1 and ∡2 are supplementary.
   h) ∠1 + ∠2 = 180°.
   i) ∠2 = 180° − ∠1.
   j) 180° − ∠1 > ∠1.
   k) 180° > 2∠1.
   l) 90° > ∠1 (so ∠1 < 90°).
   m) ∠P < 90°.
   n) ∡P is acute.

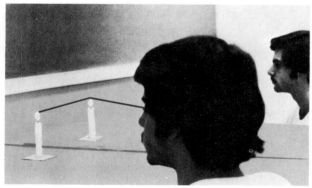

PHOTOGRAPHY BY ROB NOLAN

3. This photograph shows someone looking at the reflection of a candle in a mirror. The pair of line segments that have been added to the picture represent the path of one ray of light from the tip of the candle flame to the observer's eye. Of course, the flame sends out rays of light in many directions. How many of these rays are reflected to the same point?

By means of the Exterior Angle Theorem and one fact about light reflection, we can prove that there is only one. The figure below represents an overhead view of the mirror, $\overline{AB}$, the candle, C, and the eye of the observer, E. Whenever a ray of light is reflected from a mirror, the angle of incidence, ∢1, is equal to the angle of reflection, ∢2.

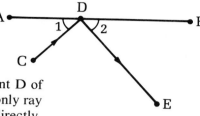

The figure shows a ray from C reflected from point D of the mirror to point E. We will prove that this is the only ray reflected by the mirror from C to E by reasoning indirectly. Suppose there is another ray from C reflected from another point, say F, of the mirror to point E.

a) Why is ∠1 > ∠3?

By the reflection property of light, ∠3 = ∠4.

b) Why, then, is ∠1 > ∠4?

c) Why is ∠4 > ∠2?

d) Why, then, is ∠1 > ∠2?

e) What does this contradict?

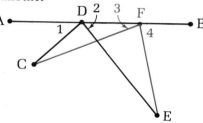

We have based our argument on the assumption that point F is to the right of point D. It can be shown that the assumption that it is to the left instead leads to the same contradiction. It follows that the ray reflected at D is the only ray from C that reaches E.

SET II

Write complete proofs for each of the following.

1. Given: BT > BO.
   Prove: ∠1 > ∠L.
   Par 4.

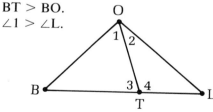

2. Given: Figure NAIL.
   Prove: NA + AI + IL > NL.
   Par 5.

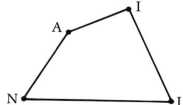

3. This figure represents a billiard table with three billiard balls, A, B, and C, such that ball B is between balls A and C. When a billiard ball strikes a cushion of the table, it behaves in the same way that light does when it is reflected by a mirror: the angle of incidence is equal to the angle of reflection.

On the assumption that there are no other balls on the table, can you figure out how someone could hit ball A so that it would hit ball C?

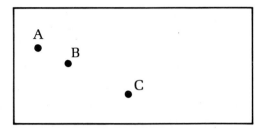

SET III

The distances from Hobbiton to Bucklebury, Whitfurrows, and Frogmorton are 3, 4, and 5 kilometers, respectively. Frodo begins at Whitfurrows and drives along the outside roads to Bucklebury, Frogmorton, and back to Whitfurrows. He figures at the end of the trip that he has traveled 25 kilometers altogether. Is this possible?

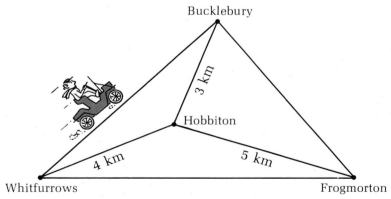

Chapter 7

PARALLEL LINES

BY PERMISSION OF JOHN HART AND FIELD ENTERPRISES, INC.

Lesson 1

Parallel Lines

Even though the comely young lady in this cartoon couldn't care less about the idea of parallel lines, they play an important role in Euclidean geometry. Peter has drawn an illustration of them on the ground, with the comment that they never meet. This, as you may recall, is the basis for our definition of such lines.

▶ Definition
Two lines are *parallel* iff they lie in the same plane and do not intersect.

We might describe Peter's drawing more precisely as an illustration of two parallel line *segments*. Rays and segments are parallel iff the lines that contain them are parallel.

254

In practice, our definition of parallel lines is difficult to use. Since lines are infinite in extent, how can we prove that there is no point somewhere in space where they intersect? The adjoining figure suggests a way of taking care of this. Lines ℓ_1 and ℓ_2 appear to be parallel. A third line, t, has been drawn across them, forming a number of angles. If some of these angles are equal, such as the pair that has been numbered, then it is easy to prove that the lines must be parallel.

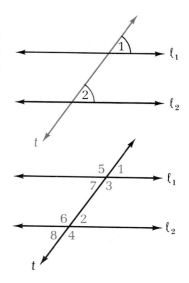

Before doing this, we will illustrate a few of the terms that we will be using. A *transversal* is a line that intersects two or more coplanar lines in different points. In the two figures shown here, line t is a transversal with respect to lines ℓ_1 and ℓ_2. A transversal that intersects two lines forms eight angles; certain pairs of these angles are given special names. In the figure at the right, $\angle 1$ and $\angle 2$ are called *corresponding angles*. Other pairs of corresponding angles are $\angle 3$ and $\angle 4$, $\angle 5$ and $\angle 6$, and $\angle 7$ and $\angle 8$.

Angles 7 and 2 are called *alternate interior angles*. The other pair of alternate interior angles is $\angle 3$ and $\angle 6$. Angles 7 and 6 are called *interior angles on the same side of the transversal*. The other pair of such angles is $\angle 3$ and $\angle 2$.

We have claimed that if ℓ_1 and ℓ_2 form equal corresponding angles with transversal t, then $\ell_1 \parallel \ell_2$. We can prove this by means of the indirect method.

▶ **Theorem 22**

If two lines form equal corresponding angles with a transversal, then the lines are parallel.

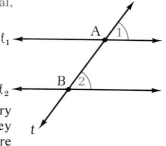

Hypothesis: Lines ℓ_1 and ℓ_2
with transversal t,
$\angle 1 = \angle 2$.
Conclusion: $\ell_1 \parallel \ell_2$.

Remember that an indirect proof consists of considering every possibility and then eliminating all but one by showing that they lead to logical contradictions. We want to prove that the lines are parallel. The other possibility is that they are not parallel.

Suppose the lines are not parallel. Then they must intersect (since they are coplanar), and the adjoining figure illustrates this. If ℓ_1 and ℓ_2 intersect in some point C, then they form a triangle, $\triangle ABC$. Since $\angle 1$ is an exterior angle of this triangle and $\angle 2$ is a remote interior angle, $\angle 1 > \angle 2$. (We know from the Exterior Angle Theorem that an exterior angle of a triangle is greater than either remote interior angle.) But this contradicts the hypothesis that $\angle 1 = \angle 2$. Therefore our assumption that the lines are not parallel is false and so $\ell_1 \parallel \ell_2$.

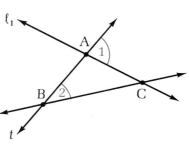

Using the angles formed by two lines and a transversal is such a convenient way to prove the lines parallel that we will prove three more theorems of this type.

► **Corollary 1**

If two lines form equal alternate interior angles with a transversal, then the lines are parallel.

► **Corollary 2**

If two lines form supplementary interior angles on the same side of a transversal, then the lines are parallel.

► **Corollary 3**

In a plane, two lines perpendicular to a third line are parallel to each other.

Exercises

SET I

1. The following questions refer to the adjoining figure.
   a) Which line in the figure is the transversal?
   b) One pair of corresponding angles is ∡1 and ∡5. Name three other pairs.
   c) One pair of alternate interior angles is ∡2 and ∡8. Name the other pair.
   d) One pair of interior angles on the same side of the transversal is ∡2 and ∡5. Name the other pair.

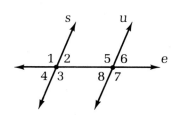

2. The following questions refer to the figure below.
   a) If $\angle 1 = \angle 2$, which lines in the figure must be parallel? Why?
   b) If $\angle 2 = \angle 3$, which lines in the figure must be parallel? Why?
   c) If you know only that $\angle 1 = \angle 3$, can you conclude that some lines are parallel?

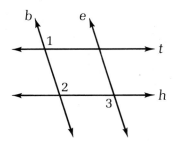

Complete the following proofs of the corollaries to Theorem 22 by supplying the reasons.

3. *Corollary 1.*   If two lines form equal alternate interior angles with a transversal, then the lines are parallel.

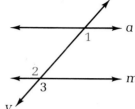

> Hypothesis: Lines *a* and *m*
> with transversal *y*,
> $\angle 1 = \angle 2$.
> Conclusion: $a \parallel m$.

Proof.

Statements

a) Lines *a* and *m* with transversal *y*, $\angle 1 = \angle 2$.
b) $\angle 2 = \angle 3$.
c) $\angle 1 = \angle 3$.
d) $a \parallel m$.

4. *Corollary 2.*   If two lines form supplementary interior angles on the same side of a transversal, then the lines are parallel.

> Hypothesis: Lines *a* and *t* with transversal *p*,
> $\angle 1$ and $\angle 2$ are supplementary.
> Conclusion: $a \parallel t$.

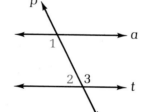

Proof.

Statements

a) Lines *a* and *t* with transversal *p*,
   $\angle 1$ and $\angle 2$ are supplementary.
b) $\angle 2$ and $\angle 3$ are supplementary.
c) $\angle 1 = \angle 3$.
d) $a \parallel t$.

5. *Corollary 3.*   In a plane, two lines perpendicular to a third line are parallel to each other.

> Hypothesis: Lines *b*, *e*, and *a* are coplanar,
> $a \perp b, a \perp e$.
> Conclusion: $b \parallel e$.

Proof.

Statements

a) Lines *b*, *e*, and *a* are coplanar, $a \perp b, a \perp e$.
b) $\angle 1$ and $\angle 2$ are right angles.
c) $\angle 1 = \angle 2$.
d) $b \parallel e$.

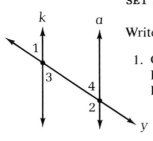

SET II

Write complete proofs for each of the following.

1. Given: $\angle 1 + \angle 2 = 180°$.
    Prove: $k \parallel a$.
    Par 5.

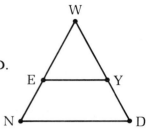

2. Given: WN = WD; $\angle$WEY = $\angle$D.
    Prove: $\overline{EY} \parallel \overline{ND}$.
    Par 5.

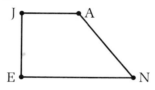

3. Given: $\overline{JE} \perp \overline{EN}$; $\angle$J is a right angle.
    Prove: $\overline{JA} \parallel \overline{EN}$.
    Par 4.

SET III

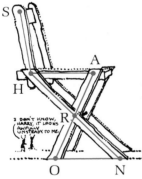

Dilcue has built himself a folding chair in which the legs, $\overline{HN}$ and $\overline{AO}$, are attached at their midpoints as shown in this picture. Even though the legs turned out to be unequal in length, the seat of the chair, $\overline{HA}$, is parallel to the floor, $\overleftrightarrow{ON}$. Can you explain why this should be?

Lesson 2

Perpendicular Lines

The Leaning Tower of Pisa has been standing for eight hundred years and every year it leans a little more. Built with a shallow foundation on sandy ground, it started to tip even before it was finished and, if nothing is done to prevent it, the tower will eventually collapse. Many people have come up with ideas on how to save it. One man, perhaps in the same condition as the fellow in this cartoon, decided to straighten it by putting a rope around the tower and pulling it with his car. When he drove off, he left not only the tower behind but also his bumper!

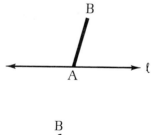

The problem, stated geometrically, is that the tower is not perpendicular to the ground. The first diagram at the right shows the view of the tower shown in the cartoon, where $\overline{AB}$ represents the tower and line ℓ the ground. If its admirer walked around to the opposite side, he would see the view shown in the second diagram.

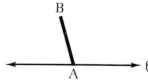

In neither case is $\overline{AB} \perp \ell$. In how many directions could the tower have been built so that from *every* view it would look perpendicular to the ground? Obviously, just one: directly up. In terms of our two-dimensional figures, there is exactly *one perpendicular* to a line at a given point on the line. It is also true that there is exactly one perpendicular to a line through a given point *not* on the line. We will prove both of these statements as theorems.

259

▶ **Theorem 23**
In a plane, through a point on a line there is exactly one line perpendicular to the line.

Hypothesis: Line ℓ and point P on it.
Conclusion: There is exactly one line
through P perpendicular to ℓ.

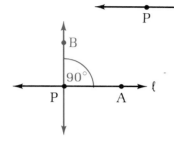

Proof.
We can choose a second point A on line ℓ since a line contains at least two points. Because of the Unique Ray Corollary, we know that there is exactly one ray, $\overrightarrow{PB}$, in the half-plane above ℓ such that ∠APB = 90°. Since ⊾APB has a measure of 90°, it is a right angle. So $\overrightarrow{PB}$ (and hence, $\overleftrightarrow{PB}$) is perpendicular to ℓ.

If the point is not on the line, the situation is the same.

▶ **Theorem 24**
Through a point not on a line there is exactly one line perpendicular to the line.

Hypothesis: Line ℓ and point P not on it.
Conclusion: There is exactly one line through P perpendicular to ℓ.

To prove this theorem, we will do two things. First, we will show that there is *at least* one perpendicular through P to the line. Second, we will show that there is *no more* than one perpendicular to the line through P.

Proof that there is at least one perpendicular.
Let P′ be the reflection of point P through line ℓ. (According to the Point Reflection Postulate, there is a one-to-one correspondence between points in a plane and their reflections through a given line.) Draw $\overline{PP'}$. Now ℓ ⊥ $\overline{PP'}$ because if one point is the reflection through a line of another point, the line is the perpendicular bisector of the segment that joins the two points. Hence $\overline{PP'}$ ⊥ ℓ, and this establishes the fact that through P there is at least one line perpendicular to ℓ.

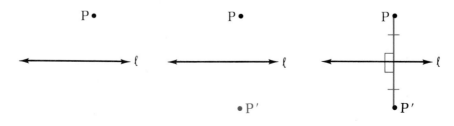

Proof that there is no more than one perpendicular.
We will use the indirect method. Suppose that there is more than one line through P perpendicular to ℓ: say $\overleftrightarrow{PA} \perp \ell$ and $\overleftrightarrow{PB} \perp \ell$. Then $\overleftrightarrow{PA} \parallel \overleftrightarrow{PB}$ because, in a plane, two lines perpendicular to a third line are parallel to each other. But this contradicts the fact that $\overleftrightarrow{PA}$ and $\overleftrightarrow{PB}$ intersect at point P. So the assumption that there is more than one perpendicular is false and the conclusion is established.

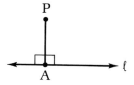

Now that we have proved that there is a unique perpendicular from a point to a line, we will show why the distance from the point to the line is measured along this perpendicular.

▶ **Theorem 25**

The perpendicular segment from a point to a line is the shortest segment joining them.

According to this theorem, if $\overline{PA} \perp \ell$ in the figure at the right, then $\overline{PA}$ is shorter than any other segment joining P to ℓ. This is easy to prove and is left as an exercise.

This theorem is the basis for defining the distance from a point to a line, something we took for granted in the problem of the surfer in the introduction to this course.

▶ **Definition**

The *distance from a point to a line* is the length of the perpendicular segment from the point to the line.

SET I

Exercises

1. No triangle can contain two right angles because if such a triangle existed, it would contradict Theorem 24. Use the adjoining figure to explain why.

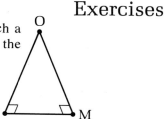

2. Acute Alice says that through a point on a line there are infinitely many lines perpendicular to the line. Obtuse Ollie is positive that she is wrong. What do you think?

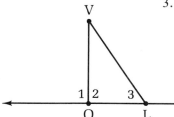

3. Supply the reasons for the following proof of Theorem 25.

The perpendicular segment from a point to a line is the shortest segment joining them.

Hypothesis: $\overline{VO} \perp t$ and $\overline{VL}$ is any other segment joining V to t.

Conclusion: $VO < VL$.

Proof.

Statements

a) $\overline{VO} \perp t$.
b) ∡1 and ∡2 are right angles.
c) $\angle 1 = \angle 2$.
d) $\angle 1 > \angle 3$.
e) $\angle 2 > \angle 3$.
f) $VL > VO$, so $VO < VL$.

SET II

In the following exercises, two more constructions are presented. You should know not only how to do each one but also why it works.

▶ **Construction 6**
To construct the line perpendicular to a given line through a point on it.

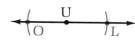

1. Carry out the construction as indicated in the following steps.
   a) Draw a line and choose a point U on it.
   b) With U as center, draw a circle intersecting the line at O and L.
   c) Finish the construction by bisecting $\overline{OL}$ by the method you learned in Construction 1.

2. The following questions refer to the construction you have just done.
   a) Just before you drew the final line, how many points in the figure were equidistant from O and L?
   b) Were they all needed to determine the final line?
   c) Why is the final line perpendicular to the original one?

3. For additional practice, copy the figure at the right.
   a) Construct two lines perpendicular to *a* at M and P.
   b) What relationship do the two lines at M and P have to
      each other?
   c) Why?

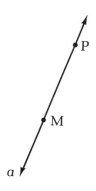

▶**Construction 7**
To construct the line perpendicular to a given line through
a point not on it.

• M

←————————→ *h*

4. An easy way to do this is to reflect the point through the
   line. Copy the above figure and then construct O so that it is
   the reflection of M through line *h*. (You may need to review
   Construction 4 on page 192.) Draw $\overleftrightarrow{MO}$; $\overleftrightarrow{MO}$ is perpendicular to line *h*.

5. For additional practice, copy the triangle at the right.
   a) Construct lines through the three vertices so that each is
      perpendicular to the line of the opposite side.
   b) What do you notice?

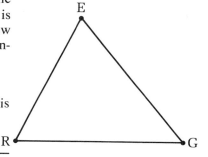

SET III

Dilcue rented a rowboat and took off from the dock as shown
in the figure below. He decided to row the boat on a course such
that his lines of sight back to the dock and to his favorite fishing
spot on the shore were always perpendicular to each other.
   In what kind of a path on the lake do you think he traveled?

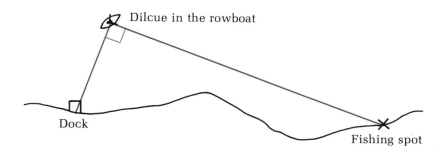

PHOTOGRAPH BY RENÉ MALTÊTE IN LAUGHING CAMERA 2, HANNS REICH VERLAG

Lesson 3

The Parallel Postulate

In the previous lesson we proved the following two theorems about perpendicular lines:

Theorem 23. In a plane, through a point on a line there is exactly one line perpendicular to the line.

Theorem 24. Through a point not on a line there is exactly one line perpendicular to the line.

Suppose that in each of these theorems the word "perpendicular" were changed to "parallel." Would they still be true?

$$P_1 \qquad \ell_1$$

Look at the figure above. Is there exactly one line through P_1 that is parallel to ℓ_1? No, because even one such line would violate our definition of parallel lines.

264

How about the adjoining figure? Is there exactly one line through P_2 that is parallel to ℓ_2? The answer in this case seems to be yes. Recall that in proving the corresponding theorem about perpendiculars, we proved two things: that there is *at least* one such line and *no more* than one such line.

It is easy to prove the first of these statements for parallels.

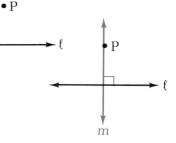

► **Theorem 26**

Through a point not on a line, there is at least one line parallel to the line.

        Hypothesis: Line ℓ and point P not on it.
        Conclusion: There is at least one line
                    through P parallel to ℓ.

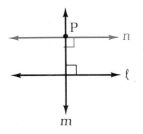

Proof.

Through point P, draw line $m \perp \ell$ as shown in the adjoining figure. We can do this because, through a point not on a line, there is exactly one line perpendicular to the line. Then through point P draw line $n \perp m$. We can do this because, in a plane, through a point on a line there is exactly one line perpendicular to the line. So $n \parallel \ell$, since in a plane two lines perpendicular to a third line are parallel to each other. Therefore, there is at least one line through P parallel to ℓ.

It now seems sensible to try to prove that, through a point not on a line, there is no more than one line parallel to the line. We might assume that there are *two* such lines parallel to the line and show that this assumption leads to a contradiction with something we already know.

The trouble with this is that *we won't be able to come up with a contradiction!* That there is only one line cannot be proved by means of the postulates and theorems we already have.* So if we wish to use this idea in future work, we must assume it without proof. We will call it the Parallel Postulate.

► **Postulate 15.** (The Parallel Postulate)

Through a point not on a line, there is no more than one line parallel to the line.

From this postulate, it follows that

► **Theorem 27**

In a plane, two lines parallel to a third line are parallel to each other.

---

*If this surprises you, you will be interested to know that it was not until the nineteenth century that mathematicians realized this! The assumption that there can be two lines through a point both parallel to a third line led to one of the so-called non-Euclidean geometries that we will study later in the course.

Exercises

1. *A Puzzle.* This figure represents a zebra beach blanket. Just the stripes show, so you see the top and bottom edges but not the side edges. Is the blanket longer from top to bottom, or from side to side?

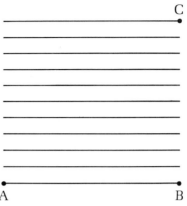

2. It is convenient to quote the following statement as a reason in proofs: "Through a point not on a line, there is exactly one line parallel to the line." This statement says two things, one of which we proved in this lesson and the other of which we assumed. Try to state each of them without looking them up.

3. With the statement of the Parallel Postulate, the list of reasons that we will need for adding extra points and lines to geometric figures is complete. This list contains both postulates and theorems. To see if you know how to use them, try to state or name the reason that justifies each of the following assertions about the adjoining figure.

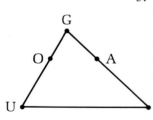

    It is possible to
a) draw $\overleftrightarrow{OA}$.
b) draw, through A, a line perpendicular to $\overline{GD}$.
c) draw, through O, a line perpendicular to $\overleftrightarrow{UD}$.
d) draw a ray that bisects $\angle U$.
e) choose a point X on $\overrightarrow{UD}$ so that $UX = UO$.
f) draw, through D, a line parallel to $\overline{GU}$.
g) choose a point Y so that it is the midpoint of $\overline{GD}$.
h) choose a ray $\overrightarrow{UZ}$ in the half-plane below $\overleftrightarrow{UD}$ so that $\angle DUZ = \angle DUG$.

4. Theorem 26 can be proved indirectly. Answer the questions included in the following proof.

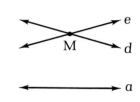

In a plane, two lines parallel to a third line are parallel to each other.

> Hypothesis: Lines *e*, *d*, and *a* are coplanar;
> $e \parallel a, d \parallel a$.
> Conclusion: $e \parallel d$.

Proof.
Suppose that *e* and *d* are not parallel.
a) Since they are coplanar, what can we then conclude?

This means that two lines passing through the same point (M in the figure at the right) have the same relationship to line *a*.
b) What is it?
c) What fact does this contradict?

Because of this contradiction, we know that our assumption is false and that $e \parallel d$.

SET II

1. You already know how to construct perpendicular lines. One way to construct parallel lines is illustrated below.

▶ **Construction 8**
To construct the line parallel to a given line through a point not on it.

First, any point Q on line ℓ is chosen and $\overleftrightarrow{PQ}$ is drawn.
a) What is being done in Step 2?
b) When line *m* is drawn in Step 3, why is it parallel to line ℓ?
c) Copy the original figure and construct a line parallel to ℓ through point P without using corresponding angles.

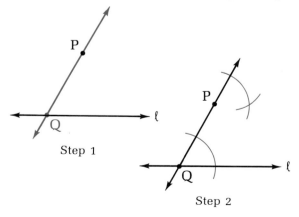

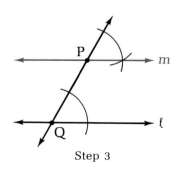

2. Given: $b \perp u$, $e \perp u$; $b \parallel \ell$.
   Prove: $\ell \parallel e$.
   Par 4.

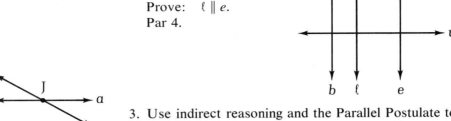

3. Use indirect reasoning and the Parallel Postulate to explain why, in a plane, a line that intersects one of two parallel lines must also intersect the other. Write your explanation in terms of the adjoining figure.

SET III

Many attempts have been made to prove the Parallel Postulate. One of them was by a brilliant, eighteenth-century mathematician, Joseph Louis Lagrange, whom Frederick the Great once called "the greatest mathematician of Europe." Just after he had begun a lecture to the French Academy on his proof, however, Lagrange suddenly realized that there was a flaw in it. He announced to his audience, "Il faut que j'y songe encore!" (I shall have to think it over again!) and left the room.

The mistake in Lagrange's proof must have been rather subtle if he himself had been fooled by it. The error in the following proof is not so subtle. Can you find it?

Through a point not on a line, there is no more than one line parallel to the line.

Hypothesis: Line ℓ and point P not on it.
Conclusion: There is no more than one line through P parallel to ℓ.

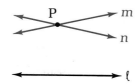

Proof.

Suppose that there is more than one line through P parallel to ℓ: say *m* ∥ ℓ and *n* ∥ ℓ. Choose any point Q on ℓ and draw transversal $\overleftrightarrow{PQ}$.

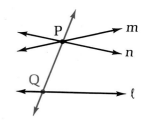

In the figure, ∠1 = ∠2 because if two parallel lines are cut by a transversal the corresponding angles are equal. For the same reason, ∠3 = ∠2. Therefore, ∠1 = ∠3. Now we have two different rays from point P that form with $\overrightarrow{PQ}$ angles that have the same measure in the half-plane to the right of $\overleftrightarrow{PQ}$. This contradicts the Unique Ray Corollary. Therefore, our assumption that there is more than one parallel is false. It follows that there is no more than one line through P parallel to ℓ.

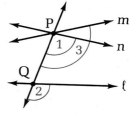

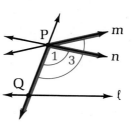

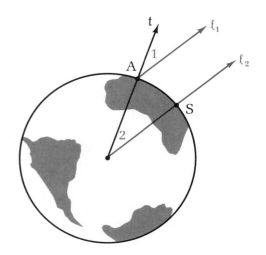

Lesson 4

Some Consequences of the Parallel Postulate

The fact that the world is round was known to the ancient Greeks. One of them, Eratosthenes, who lived in the third century B.C., determined the earth's circumference without sailing around it or, in fact, traveling more than 500 miles. How did he do it?

The diagram above illustrates the method that Eratosthenes used. He knew that the city of Alexandria in Egypt is about 500 miles due north of the city of Syene (now called Aswan). By comparing the apparent directions of the sun in the sky as observed at the same time from each of these cities, he was able to figure out the distance around the earth.

BY PERMISSION OF JOHN HART AND FIELD ENTERPRISES, INC.

The points A and S in the diagram represent Alexandria and Syene, respectively, and the lines ℓ_1 and ℓ_2 represent the direction of the sun as seen from each city. At noon on a certain day of the year, the sun was directly overhead in Syene, as shown by ℓ_2. In Alexandria at the same time, the direction of the sun was along ℓ_1, in contrast with the overhead direction, t. Eratosthenes measured the angle between these two directions and found that it was about 7.5°. This, together with the distance between Alexandria and Syene (500 miles), is sufficient to determine the earth's circumference. Since the sun is so far away from the earth, the lines ℓ_1 and ℓ_2 are very close to being parallel. For simplicity, Eratosthenes assumed that they were. Angles 1 and 2 are corresponding angles formed by the parallel lines ℓ_1 and ℓ_2 and the transversal t. If $\angle 1 = \angle 2$, then $\angle 2 = 7.5°$. Since 7.5° is 1/48 of 360°, 500 miles (the distance from A to S along the circle) is apparently 1/48 of the earth's circumference (the entire distance around the circle).

$$48 \cdot 500 = 24,000 \text{ miles.}$$

The modern value for the circumference of the earth is, to two significant figures, 25,000 miles, so Eratosthenes' estimate was remarkably accurate. His method depends on the following fact: If two parallel lines are cut by a transversal, the corresponding angles are equal. Although it seems as if we already know this, it is the *converse* of this statement that we have proved. If we want to use it, we must prove it as a new theorem.

▶ **Theorem 28**

If two parallel lines are cut by a transversal, the corresponding angles are equal.

Hypothesis: Lines ℓ_1 and ℓ_2
with transversal t;
$\ell_1 \parallel \ell_2$.
Conclusion: $\angle 1 = \angle 2$.

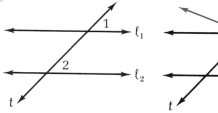

Proof.
Suppose that $\angle 1 \neq \angle 2$. Then we can draw a line ℓ_3 through P such that $\angle 3 = \angle 2$ (the Unique Ray Corollary). Now $\ell_3 \parallel \ell_2$ (because two lines that form equal corresponding angles with a transversal are parallel) and $\ell_1 \parallel \ell_2$ (by hypothesis). So through point P we have two lines parallel to ℓ_2. Since this contradicts the Parallel Postulate, our assumption that $\angle 1 \neq \angle 2$ is false. Therefore, $\angle 1 = \angle 2$.

Now that we have established that parallel lines form equal corresponding angles with a transversal, it is easy to prove some more theorems about the angles formed by parallel lines. You will recognize the first two as the converses of theorems we have already proved.

▶ **Corollary 1**
If two parallel lines are cut by a transversal, then the alternate interior angles are equal.

▶ **Corollary 2**
If two parallel lines are cut by a transversal, then the interior angles on the same side of the transversal are supplementary.

▶ **Corollary 3**
In a plane, if a line is perpendicular to one of two parallel lines, it is also perpendicular to the other.

Exercises

SET I

1. Prove Corollary 1 by using the figure below.

If two parallel lines are cut by a transversal, then the alternate interior angles are equal.

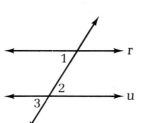

Hypothesis: Lines r and u are cut by transversal m, $r \parallel u$.
Conclusion: $\angle 1 = \angle 2$.
Par 4.

Complete the following proofs of Corollaries 2 and 3 by supplying the reasons.

2. *Corollary 2.* If two parallel lines are cut by a transversal, then the interior angles on the same side of the transversal are supplementary.

Hypothesis: Lines g and i are cut by transversal n, $g \parallel i$.
Conclusion: $\angle 1$ and $\angle 2$ are supplementary.

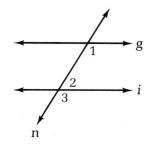

Proof.

Statements

a) $g \parallel i$.
b) $\angle 1 = \angle 3$.
c) $\angle 2$ and $\angle 3$ are supplementary.
d) $\angle 2 + \angle 3 = 180°$.
e) $\angle 2 + \angle 1 = 180°$.
f) $\angle 1$ and $\angle 2$ are supplementary.

3. *Corollary 3.* In a plane, if a line is perpendicular to one of two parallel lines, it is also perpendicular to the other.

Hypothesis: $r \parallel y$ and $e \perp r$.
Conclusion: $e \perp y$.

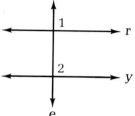

Proof.

Statements

a) $r \parallel y$.
b) $\angle 1 = \angle 2$.
c) $e \perp r$.
d) $\angle 1$ is a right angle.
e) $\angle 1 = 90°$.
f) $\angle 2 = 90°$.
g) $\angle 2$ is a right angle.
h) $e \perp y$.

SET II

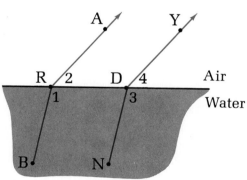

1. When a spoon is put in a glass of water, it appears to be bent. This is because light rays are bent when they go from water into air. In fact, parallel light rays are always bent by the same amount. In terms of the diagram shown here, this means that if $\overline{BR} \parallel \overline{ND}$, then $\angle BRA = \angle NDY$.

   Explain, geometrically, why the rays are still parallel when they emerge into the air; that is, why $\overrightarrow{RA} \parallel \overrightarrow{DY}$.

Lesson 4: Consequences of the Parallel Postulate   273

Write complete proofs for each of the following.

2. Given: $\overline{WI} \parallel \overline{EN}$; $WI = EN$.
   Prove: $\triangle WIN \cong \triangle NEW$.
   Par 5.

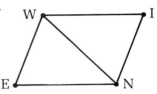

3. Given: $\overline{PO} \parallel \overline{TR}$; $\angle P$ and $\angle R$ are supplementary.
   Prove: $\angle T = \angle R$.
   Par 4.

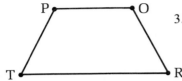

SET III

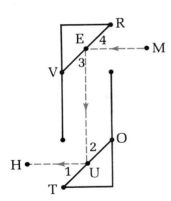

A periscope is an optical instrument that enables its user to look at objects that are not in a direct line of sight. In its simplest form it consists of a tube with a mirror at each end, as shown in the adjoining figure. A ray of light is shown entering the top, being reflected twice, and leaving at the bottom.

Since the angle at which a light ray hits a mirror is equal to the angle at which it is reflected, $\angle 1 = \angle 2$ and $\angle 3 = \angle 4$. Also, the two mirrors in the periscope are parallel: $\overline{VR} \parallel \overline{TO}$.

Can you explain why light rays leave the periscope in the same direction at which they enter? In other words, why $\overline{ME}$ and $\overline{UH}$ are always parallel?

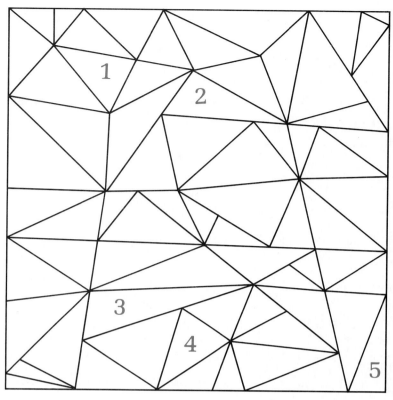

Lesson 5

The Angles of a Triangle

Triangle 2
54°
28°
+ 98°
180°

To determine the relative positions of different points on a piece of land, surveyors use a method called "triangulation" — so named because the points are used as the vertices of a network of triangles. This is illustrated in the diagram shown here.

Triangle 3
20°
61°
+101°
182°

A basic procedure in triangulation is to measure the angles of these triangles. Data for the five numbered triangles in the diagram are given at the right. Each angle has been measured to the nearest degree.

Triangle 4
41°
73°
+ 65°
179°

In each case, the sum of the measures of the angles of the triangle seems to be about the same number. Is it actually exactly the same, so that the small discrepancies in the sums are due solely to inaccuracies in measurement? It is easy to prove, by means of the Parallel Postulate, that this is the case.

Triangle 5
90°
71°
+ 19°
180°

275

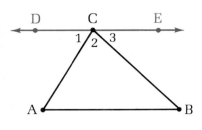

► **Theorem 29**

The sum of the measures of the angles of a triangle is 180°.

Hypothesis: △ABC.
Conclusion: ∠A + ∠B + ∠C = 180°.

Proof.

Statements	Reasons
1. △ABC.	Hypothesis.
2. Through point C, draw $\overleftrightarrow{DE} \parallel \overline{AB}$.	Through a point not on a line, there is exactly one line parallel to the line.
3. ∠1 = ∠A and ∠3 = ∠B.	If two parallel lines are cut by a transversal, the alternate interior angles are equal.
4. ∡DCB and ∡3 are supplementary.	If two angles are a linear pair, they are supplementary.
5. ∠DCB + ∠3 = 180°.	If two angles are supplementary, the sum of their measures is 180°.
6. ∠DCB = ∠1 + ∠2.	Betweenness of rays.
7. ∠1 + ∠2 + ∠3 = 180°.	Substitution.
8. ∠A + ∠C + ∠B = 180°.	Substitution.

This theorem has some useful corollaries:

► **Corollary 1**

If two angles of one triangle are equal to two angles of another, then the third pair of angles are equal.

► **Corollary 2**

The acute angles of a right triangle are complementary.

► **Corollary 3**

Each angle of an equilateral triangle has a measure of 60°

Another theorem that is easy to prove by means of the "triangle angle-sum" theorem concerns the measure of an exterior angle of a triangle. You will recall that we have already proved that an exterior angle of a triangle is greater than either remote interior angle. We can now prove that it is *equal to the sum of their measures.*

► **Theorem 30**

An exterior angle of a triangle is equal in measure to the sum of the measures of the remote interior angles.

Complete the following proofs by supplying the reasons.

Exercises

1. *Corollary 1 to Theorem 29.* If two angles of one triangle are equal to two angles of another, then the third pair of angles are equal.

Hypothesis: △PLI and △ERS
with ∠P = ∠E
and ∠L = ∠R.
Conclusion: ∠I = ∠S.

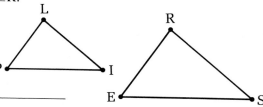

Proof.

Statements

a) △PLI and △ERS.
b) ∠P + ∠L + ∠I = 180°, ∠E + ∠R + ∠S = 180°.
c) ∠P + ∠L + ∠I = ∠E + ∠R + ∠S.
d) ∠P = ∠E and ∠L = ∠R.
e) ∠P + ∠L + ∠I = ∠P + ∠L + ∠S.
f) ∠I = ∠S.

2. *Theorem 30.* An exterior angle of a triangle is equal in measure to the sum of the measures of the remote interior angles.

Hypothesis: △FIL with exterior ∡ILE.
Conclusion: ∠ILE = ∠F + ∠I.

Proof.

Statements

a) △FIL with exterior ∡ILE.
b) ∠F + ∠I + ∠ILF = 180°.
c) ∡ILE and ∡ILF are a linear pair.
d) ∡ILE and ∡ILF are supplementary.
e) ∠ILE + ∠ILF = 180°.
f) ∠ILE + ∠ILF = ∠F + ∠I + ∠ILF.
g) ∠ILE = ∠F + ∠I.

3. Briefly explain why each of the following corollaries follows from the "triangle angle-sum" theorem.
   a) *Corollary 2.* The acute angles of a right triangle are complementary.
   b) *Corollary 3.* Each angle of an equilateral triangle has a measure of 60°.

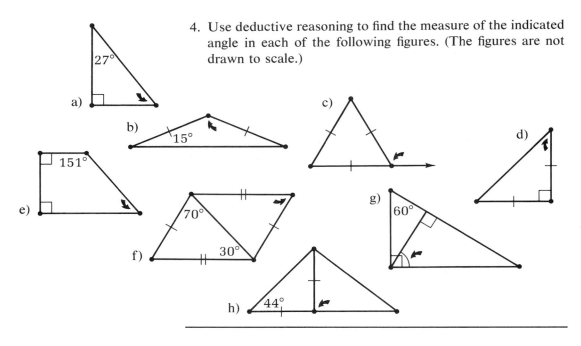

4. Use deductive reasoning to find the measure of the indicated angle in each of the following figures. (The figures are not drawn to scale.)

a)

b)   15°

c)

d)

e)   151°

f)   70°   30°

g)   60°

h)   44°

SET II

Write complete proofs for each of the following.

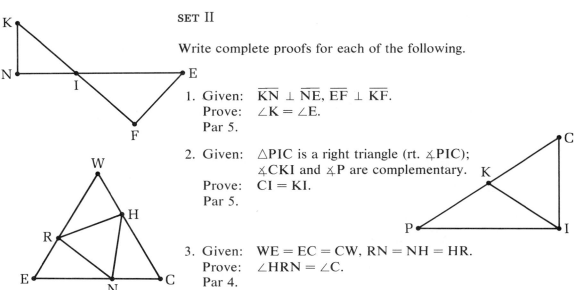

1. Given: $\overline{KN} \perp \overline{NE}$, $\overline{EF} \perp \overline{KF}$.
   Prove: $\angle K = \angle E$.
   Par 5.

2. Given: $\triangle PIC$ is a right triangle (rt. $\angle PIC$); $\angle CKI$ and $\angle P$ are complementary.
   Prove: $CI = KI$.
   Par 5.

3. Given: $WE = EC = CW$, $RN = NH = HR$.
   Prove: $\angle HRN = \angle C$.
   Par 4.

SET III

There is a symmetrical five-pointed star hidden somewhere in the diagram at the beginning of this lesson. Can you find it?

Lesson **6**

Two More Ways
to Prove Triangles Congruent

Imagine two identical ladders leaning against the opposite sides of a high wall. If the two ladders reach to the same height on the wall, are their lower ends necessarily equidistant from it?

In the figure below, $\overline{AB}$ represents the wall, $\overline{AC}$ and $\overline{AD}$ the ladders, and $\overleftrightarrow{CD}$ the ground. We will assume that the wall is perpendicular to the ground, so that $\overline{AB} \perp \overleftrightarrow{CD}$. If AC = AD, does it follow that CB = DB?

Triangles ABC and ABD seem to be congruent. They are right triangles that have equal hypotenuses and a common leg. Furthermore, $\overline{AC}$ and $\overline{AD}$ are also the legs of an isosceles triangle, $\triangle ACD$, so $\angle 1 = \angle 2$. Since $\angle 3 = \angle 4$, it follows that $\angle 5 = \angle 6$ (if two angles of one triangle are equal to two angles of another, then the third pair of angles are equal.) We now know that $\triangle ABC \cong \triangle ABD$ by the A.S.A. Congruence Postulate ($\angle 1 = \angle 2$, AC = AD, and $\angle 5 = \angle 6$). Therefore, CB = DB because they are corresponding sides of the triangles.

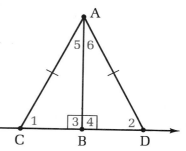

Note, at the start, that we knew these two right triangles had equal hypotenuses and a common leg. It seems evident that *any* two right triangles in which the hypotenuse and a leg of one are equal to the hypotenuse and a leg of the other must be congruent. Reasoning in the same way that we just have about the ladder problem, we will prove this and call it the H.L. Theorem.

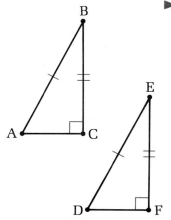

▶ **Theorem 31** (The H.L. Congruence Theorem)
If the hypotenuse and a leg of one right triangle are equal to the corresponding parts of another right triangle, then the triangles are congruent.

Hypothesis: △ABC and △DEF are right triangles (with rt. ∢s C and F); AB = DE, BC = EF.
Conclusion: △ABC ≅ △DEF.

Notice that the triangles have been drawn separately this time. This makes things more difficult but has the advantage of making the theorem more general, and hence, more useful. In order to use the properties of an isosceles triangle as we did before, we will copy, in effect, the reflection of one triangle so that it shares one leg with the other.

Proof.
Draw $\overleftrightarrow{DF}$ and choose point G on the ray opposite $\overrightarrow{FD}$ so that FG = AC (the Unique Point Corollary); also draw $\overline{EG}$. In △ABC and △GEF,

$$AC = FG,$$
$$\angle C = \angle 4 \text{ (they are both right angles),}$$
and
$$BC = EF \text{ (hypothesis),}$$

so △ABC ≅ △GEF by the S.A.S. Congruence Postulate. Now

$$GE = AB \text{ (they are corresponding sides}$$
$$\text{of the congruent triangles),}$$
and
$$AB = DE \text{ (hypothesis),}$$

so GE = DE (transitive). This means that △DEG is an isosceles triangle, and we can reason in the same fashion as we did for the first drawing representing the ladders to show that △DEF ≅ △GEF. Since we have already shown that △ABC ≅ △GEF, we know that △ABC ≅ △DEF because two triangles congruent to a third triangle are congruent to each other.

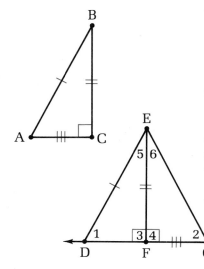

We now have four ways to prove triangles congruent (S.S.S., S.A.S., A.S.A., and H.L.). Since triangle congruence is such a convenient method for proving parts of geometric figures equal, we'll add one more. It is suggested by the figures shown here.

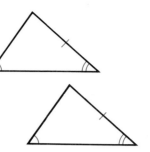

▶ **Theorem 32** (The A.A.S. Congruence Theorem)
If two angles and the side opposite one of them in one triangle are equal to the corresponding parts of another triangle, then the triangles are congruent.

This theorem is easy to prove and its proof is left to you.

SET I

Exercises

1. Name the third pair of parts in △NUT and △MEG needed to prove the triangles congruent by
   a) the H.L. theorem.
   b) the A.A.S. theorem.

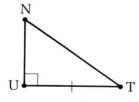 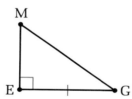

2. Use the figures below to prove Theorem 32.

   If two angles and the side opposite one of them in one triangle are equal to the corresponding parts of another triangle, then the triangles are congruent.

   Hypothesis: △GAR and △LIC with ∠G = ∠L,
   ∠R = ∠C, and AR = IC.
   Conclusion: △GAR ≅ △LIC.

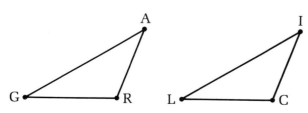

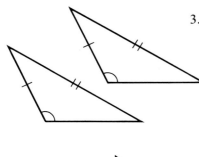

3. Each pair of triangles shown here has been drawn to scale.
   a) The appearance of the pair at the left makes another theorem for proving triangles congruent seem plausible. What would you name it?
   b) How does the pair at the left below support this idea?
   c) How does the pair at the right below contradict it?

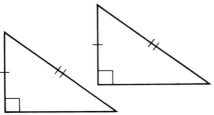

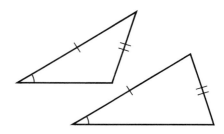

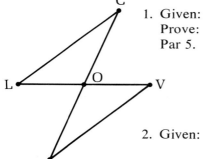

SET II

Write complete proofs for each of the following.

1. Given: O is the midpoint of $\overline{LV}$; ∠C = ∠E.
   Prove: △CLO ≅ △EVO.
   Par 5.

2. Given: ∢A and ∢E are right angles; SA = EG.
   Prove: $\overline{SE} \parallel \overline{AG}$.
   Par 7.

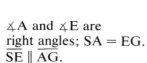

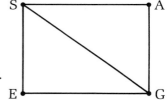

3. Given: △TYM with $\overline{MH} \perp \overline{TY}$ and $\overline{YE} \perp \overline{TM}$; MH = YE.
   Prove: △TYM is isosceles.
   Par 9.

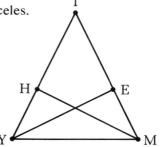

Three pairs of equal parts, if they have the right relationship, are sufficient to prove two triangles congruent. Nevertheless, it is a rather remarkable fact that two triangles can have *five* pairs of equal parts and yet *not* be congruent!

Can you sketch a triangle that is not congruent to the triangle shown here but that seems to have five parts equal to those of it? A small amount of deductive reasoning should tell you how to do it.

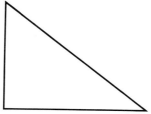

Chapter 7/Summary and Review

Basic Ideas

Angles formed by a transversal   255
Distance from a point to a line   261
Parallel lines   254
Transversal   255

Constructions

6. To construct the line perpendicular to a given line through a point on it.   262
7. To construct the line perpendicular to a given line through a point not on it.   263
8. To construct the line parallel to a given line through a point not on it.   267

Postulate

15. *The Parallel Postulate.*   Through a point not on a line, there is no more than one line parallel to the line.   265

Theorems

22. If two lines form equal corresponding angles with a transversal, then the lines are parallel.   255
   Corollary 1.   If two lines form equal alternate interior angles with a transversal, then the lines are parallel.   256
   Corollary 2.   If two lines form supplementary interior angles on the same side of a transversal, then the lines are parallel.   256
   Corollary 3.   In a plane, two lines perpendicular to a third line are parallel to each other.   256
23. In a plane, through a point on a line there is exactly one line perpendicular to the line.   260
24. Through a point not on a line, there is exactly one line perpendicular to the line.   260
25. The perpendicular segment from a point to a line is the shortest segment joining them.   261
26. Through a point not on a line, there is at least one line parallel to the line.   265

27. In a plane, two lines parallel to a third line are parallel to each other. 265
28. If two parallel lines are cut by a transversal, then the corresponding angles are equal. 271
   Corollary 1. If two parallel lines are cut by a transversal, then the alternate interior angles are equal. 272
   Corollary 2. If two parallel lines are cut by a transversal, then the interior angles on the same side of the transversal are supplementary. 272
   Corollary 3. In a plane, if a line is perpendicular to one of two parallel lines, it is also perpendicular to the other. 272
29. The sum of the measures of the angles of a triangle is 180°. 276
   Corollary 1. If two angles of one triangle are equal to two angles of another, then the third pair of angles are equal. 276
   Corollary 2. The acute angles of a right triangle are complementary. 276
   Corollary 3. Each angle of an equilateral triangle has a measure of 60°. 276
30. An exterior angle of a triangle is equal in measure to the sum of the measures of the remote interior angles. 276
31. *The H.L. Theorem.* If the hypotenuse and a leg of one right triangle are equal to the corresponding parts of another right triangle, then the triangles are congruent. 280
32. *The A.A.S. Theorem.* If two angles and the side opposite one of them in one triangle are equal to the corresponding parts of another triangle, then the triangles are congruent. 281

SET I

Exercises

1. Here is another proof that an exterior angle of a triangle is equal to the sum of the measures of the remote interior angles. Supply a reason for each statement.

   Hypothesis: △CED with exterior ∡CDA.
   Conclusion: ∠CDA = ∠C + ∠E.

Proof.   Statements

   a) △CED with exterior ∡CDA.
   b) Through D, draw $\overleftrightarrow{DR}$ ∥ $\overline{EC}$.
   c) ∠C = ∠1.
   d) ∠E = ∠2.
   e) ∠CDA = ∠1 + ∠2.
   f) ∠CDA = ∠C + ∠E.

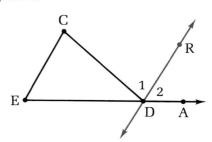

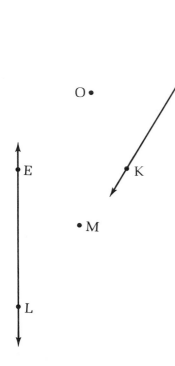

2. Copy this figure and construct each of the following.
   a) Through A, a line perpendicular to $\overleftrightarrow{AK}$.
   b) Through O, a line perpendicular to $\overleftrightarrow{AK}$.
   c) Through O, a line parallel to $\overleftrightarrow{AK}$.

3. Copy this figure and construct the reflection of $\overleftrightarrow{EL}$ through point M by finding E′ and L′—the reflections of E and L through M, respectively—and by drawing $\overleftrightarrow{E'L'}$.
   a) What seems to be the relationship between $\overleftrightarrow{EL}$ and $\overleftrightarrow{E'L'}$?
   b) Explain why this relationship exists.

4. We have proved the following statement as a theorem:

   In a plane, two lines perpendicular to a third line are parallel to each other.

   Would this statement be true if
   a) the first three words were omitted?
   b) the words "perpendicular" and "parallel" were inter-changed?
   c) the word "perpendicular" were changed to "parallel"?

5. Find the measure of the indicated angle in each of the figures below.

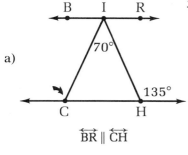

a)

$\overleftrightarrow{BR} \parallel \overleftrightarrow{CH}$

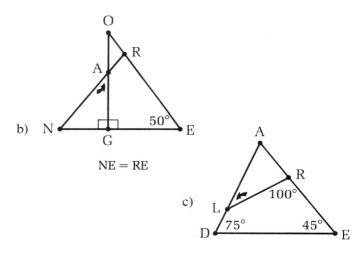

b)

NE = RE

c)

BY PERMISSION OF JOHN HART AND FIELD ENTERPRISES, INC.

SET II

1. The sun is so far away that its light to the earth travels in rays that are very close to being parallel. In the following problem, we will assume that they actually are.

   If the lens of a magnifying glass is held so that it is perpendicular to the light rays of the sun, the ray passing through its center is not bent. Those close to the edge of the lens, however, are bent a great deal. As a result, the sun's rays are focused into a single point.

   Answer the following questions about the figure shown here, which represents three rays of light being focused by the lens of the glass. Assume that $\overline{ER} \perp \overline{PU}$.

   a) If $\overleftrightarrow{SP} \parallel \overline{ER}$, what can you conclude about $\overleftrightarrow{SP}$ and $\overline{PU}$?
   b) Why?
   c) What can you conclude about $\angle 1$ and $\angle 2$?
   d) Why?
   e) If C is the midpoint of $\overline{PU}$, U is the reflection of P through $\overleftrightarrow{RE}$. Why?
   f) On the basis of part e, what can you conclude about $\angle 2$ and $\angle 3$?
   g) Why?

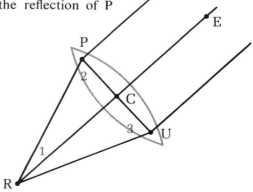

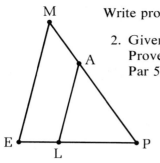

Write proofs for each of the following.

2. Given: AP > AL; $\overline{AL} \parallel \overline{ME}$.
   Prove: ∠E > ∠P.
   Par 5.

3. Given: $\overrightarrow{MA}$ bisects ∢PML;
   PM = PA.
   Prove: $\overline{PA} \parallel \overline{ML}$.
   Par 6.

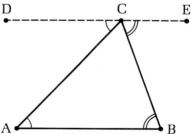

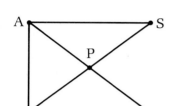

4. Given: $\overline{AN} \perp \overline{NE}$, $\overline{AS} \parallel \overline{NE}$;
   ∠S = ∠E.
   Prove: △SAN ≅ △ENA.
   Par 7.

SET III

Can you translate this passage from an Italian geometry book?

Teorema. *La somma degli angoli di un qualsiasi triangolo è eguale a un angolo piatto.*

Dato un triangolo ABC e considerata per un vertice, ad esempio per C, la parallela DE al lato opposto AB, si ha che l'angolo $\widehat{ACD}$ risulta eguale al suo alterno $\widehat{CAB}$ (rispetto alle parallele DE, AB e alla lore trasversale AC) e, similmente, l'angolo $\widehat{BCE}$ risulta eguale al suo alterno $\widehat{ABC}$ (rispetto alle stesse parallele e alla trasversale BC). Perciò la somma dei tre angoli $\widehat{A}, \widehat{B}, \widehat{C}$ del triangolo è eguale a

$$\widehat{DCA} + \widehat{ACB} + \widehat{BCE,}$$

cioè appunto a un angolo piatto.

Chapter **8**

QUADRILATERALS

Lesson 1

Quadrilaterals

This picture of a partly open window is a painting by a Belgian surrealist artist, René Magritte. At first glance, it seems that we are looking through the window at the sky. The partly open pane on the right, however, reveals that behind the window is darkness. Perhaps the "sky" is a picture fastened to the glass, or even a reflection of actual sky on our side of the window. But the top of the right-hand pane seems to rule out both of these possibilities. The painting represents an unreal window in an unreal room.

The frames of the window panes are physical models of geometric figures called *quadrilaterals*. A quadrilateral is a figure having four sides, such as:

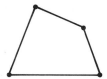 

290

but *not* such as:

Because we don't want to think of such figures as the two above as being quadrilaterals, we make the following definition:

▶ Definition
A *quadrilateral* is the union of four line segments that join four coplanar points, no three of which are collinear; each segment intersects exactly two others, one at each endpoint.

As they are in a triangle, the points are called the *vertices* of a quadrilateral and the line segments are called its *sides*. Two sides that intersect each other, such as $\overline{AB}$ and $\overline{BC}$ in quadrilateral ABCD shown at the right are called *consecutive*. A pair of sides that do *not* intersect, such as $\overline{AB}$ and $\overline{DC}$, are called *opposite*. The same terms are applied to the vertices and angles of a quadrilateral: ∡A and ∡D, for example, are *consecutive angles*; points A and C are *opposite vertices*.

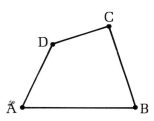

The four vertices of a quadrilateral actually determine *six* line segments, as the figure at the right shows. Four of the line segments are the sides of the quadrilateral; the other two are called its *diagonals*.

▶ Definition
A *diagonal* of a quadrilateral is a line segment that joins two opposite vertices.

Look at the six quadrilaterals below. The first three are different from the last three in a fundamental way. The first three are *convex* and the last three are *concave*.

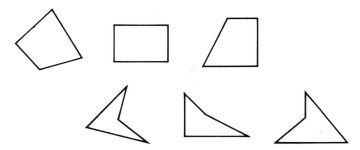

To distinguish between them, we can use the lines that contain their sides. If a line is drawn that contains any side of a convex quadrilateral, the rest of the quadrilateral lies in one of the half-planes that has the line for its edge. This is not true of concave quadrilaterals.

A convex quadrilateral

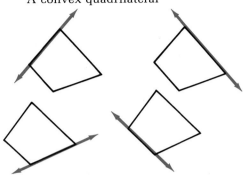

A concave quadrilateral

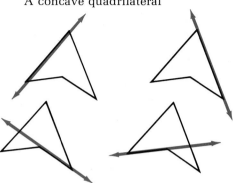

All of the quadrilaterals that have special names, such as "rectangle" and "square," are convex. Because of this, we will assume when referring to quadrilaterals in the future that they are convex unless we mention otherwise. It is easy to prove, no matter what its shape, that the sum of the measures of the angles of every such quadrilateral is the same number.

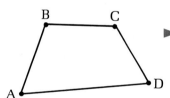

► **Theorem 33**

The sum of the measures of the angles of a quadrilateral is 360°.

Hypothesis: Quadrilateral ABCD.
Conclusion: ∠A + ∠B + ∠C + ∠D = 360°.

To prove this theorem, we will draw a diagonal of the quadrilateral to form two triangles.

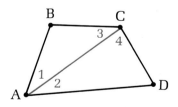

Proof.

Statements	Reasons
1. Quadrilateral ABCD.	Hypothesis.
2. Draw $\overline{AC}$.	Two points determine a line.
3. ∠1 + ∠B + ∠3 = 180°, ∠2 + ∠D + ∠4 = 180°.	The sum of the measures of the angles of a triangle is 180°.
4. ∠1 + ∠B + ∠3 + ∠2 + ∠D + ∠4 = 360°.	Addition.
5. ∠1 + ∠2 = ∠BAD, ∠3 + ∠4 = ∠BCD.	Betweenness of rays.
6. ∠BAD + ∠B + ∠BCD + ∠D = 360°.	Substitution.

1. According to our definition of a quadrilateral, the figures below are not quadrilaterals, even though each consists of four line segments determined by four points. In each case, tell why not.

Exercises

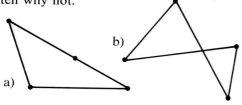

a)

b)

c)

2. Find the measure of the indicated angle in each of the following figures.

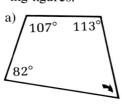

a)

107° 113°

82°

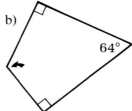

b)

64°

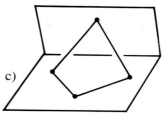

76° 145°

c) 102°

3. Some of the quadrilaterals shown here seem to possess the following properties.
   a) Two pairs of parallel sides.
   b) Exactly two equal sides.
   c) Four equal sides.
   d) Two sides that are parallel but not equal.
   e) Exactly two right angles.

   Indicate which ones in each case.

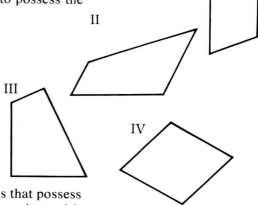

I

II

III

IV

4. Try to make accurate drawings of quadrilaterals that possess each of the following properties. If you think a certain quadrilateral cannot exist, write "impossible."
   a) Exactly one right angle.
   b) Exactly three right angles.
   c) Exactly three equal sides.
   d) Exactly three equal angles.
   e) Two sides parallel and the other two sides equal.
   f) Two pairs of parallel sides and no sides equal.
   g) Two pairs of equal sides and no sides parallel.

1. In quadrilateral ARGO, ∡A is a right angle and ∡R and ∡O are supplementary. Explain why sides $\overline{GR}$ and $\overline{GO}$ must be perpendicular.

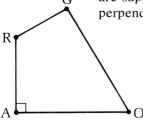

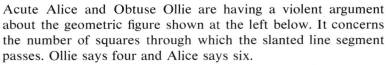

2. Given: Concave quadrilateral LYRA with LY = YR, LA = AR.
   Prove: ∠L = ∠R. (Hint: Draw diagonal $\overline{YA}$.)
   Par 5.

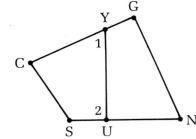

3. Given: Quadrilateral CGNS; ∠1 = ∠N.
   Prove: ∠2 = ∠G.
   Par 7.

Acute Alice and Obtuse Ollie are having a violent argument about the geometric figure shown at the left below. It concerns the number of squares through which the slanted line segment passes. Ollie says four and Alice says six.

If Ollie and Alice were shown the figure at the right, through how many squares do you think each of them would say the slanted line segment passes?

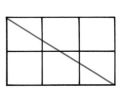
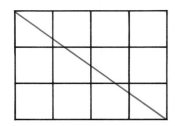

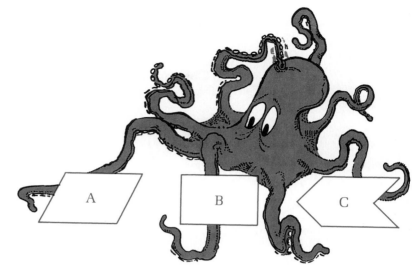

Lesson 2
Parallelograms

Experiments with octopuses show that they can distinguish between certain types of geometric shapes but not between others.* For example, of the three figures shown above, two look alike to an octopus and one looks different. Which do you suppose looks different?

If the octopus recognized a geometric figure on the basis of the number of sides it has, the answer would be C. However, from the way in which the eyes of an octopus work, it is shapes A and C that look alike and shape B that looks different!

This is rather surprising since, from a geometric point of view, figures A and B have the most properties in common. Not only are they both quadrilaterals, but their opposite sides seem to be both parallel and equal.

It is possible to prove that for any quadrilateral, one of these properties cannot exist without the other. That is, if both pairs of opposite sides of a quadrilateral are parallel, then they are also equal, and conversely. It can also be shown that any quadrilateral that has both pairs of opposite sides parallel has several other basic properties. Such a quadrilateral is called a *parallelogram*. Parallelograms are of much importance in Euclidean geometry.

---

*Niko Tinbergen, *Animal Behavior*, a book in the Life Nature Library (Time, Inc., 1965), pp. 40–41.

▶ Definition

A *parallelogram* is a quadrilateral in which both pairs of opposite sides are parallel.

If quadrilateral ABCD is a parallelogram (we will represent a parallelogram by the symbol ▱), then $\overline{AB} \parallel \overline{DC}$ and $\overline{AD} \parallel \overline{BC}$. We will now prove that it follows that AB = DC and AD = BC.

▶ Theorem 34

The opposite sides of a parallelogram are equal.

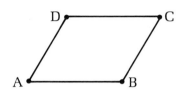

Hypothesis: ▱ ABCD.
Conclusion: AB = DC and
             AD = BC.

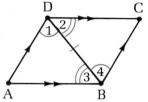

To prove this, we can draw a diagonal of the parallelogram as illustrated in the figure at the right and show that the two triangles formed must be congruent. The arrowheads on the line segments indicate that they are parallel.

Proof.

Statements	Reasons
1. ▱ABCD.	Hypothesis.
2. Draw $\overline{DB}$.	Two points determine a line.
3. $\overline{DC} \parallel \overline{AB}$.	The opposite sides of a parallelogram are parallel.
4. ∠2 = ∠3.	If two parallel lines are cut by a transversal, the alternate interior angles are equal.
5. $\overline{AD} \parallel \overline{BC}$.	Same as step 3.
6. ∠1 = ∠4.	Same as step 4.
7. DB = DB.	Reflexive.
8. △ADB ≅ △CBD.	A.S.A. postulate.
9. AB = DC and AD = BC.	Corresponding parts of congruent triangles are equal.

Parallel lines are often referred to as lines that are everywhere equidistant. We have defined the distance between two points and the distance from a point to a line. What do we mean by the distance between two lines? In the figures at the top of the next page, it seems reasonable to call it the distance from any point in one line to the other line. If the two lines are not parallel, however, it is apparent that this distance depends upon the point chosen so that the definition does not make much sense. On the other hand, it follows directly from the theorem we have just proved that for parallel lines, the distance is always the same.

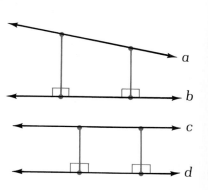

► **Corollary**
If two lines are parallel, every perpendicular segment joining one line to the other has the same length.

Because of this corollary, we are justified in making the following definition.

► Definition
The *distance between two parallel lines* is the distance from any point on one line to the other line.

Some more useful properties of parallelograms are described in the following theorems.

► **Theorem 35**
The consecutive angles of a parallelogram are supplementary.

► **Corollary**
The opposite angles of a parallelogram are equal.

► **Theorem 36**
The diagonals of a parallelogram bisect each other.

SET I

Exercises

Complete the proofs of the following theorems by supplying the reasons.

1. *Corollary to Theorem 34.* If two lines are parallel, every perpendicular segment joining one line to the other has the same length.

Hypothesis: $\ell_1 \parallel \ell_2$, $\overline{DU} \perp \ell_2$, $\overline{FY} \perp \ell_2$.
Conclusion: $DU = FY$.

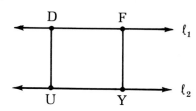

Proof.

Statements

a) $\ell_1 \parallel \ell_2$, $\overline{DU} \perp \ell_2$, $\overline{FY} \perp \ell_2$.
b) $\overline{DU} \parallel \overline{FY}$.
c) DUYF is a parallelogram.
d) $DU = FY$.

2. *Theorem 35.* The consecutive angles of a parallelogram are supplementary.

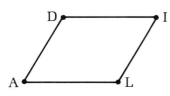

Hypothesis: DALI is a parallelogram.
Conclusion: ∡D and ∡A are supplementary.*

Proof.

Statements

a) DALI is a parallelogram.
b) $\overline{DI} \parallel \overline{AL}$.
c) ∡D and ∡A are supplementary.

3. *Corollary to Theorem 35.* The opposite angles of a parallelogram are equal.

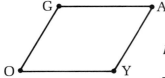

Hypothesis: GOYA is a parallelogram.
Conclusion: ∠G = ∠Y.

Proof.

Statements

a) GOYA is a parallelogram.
b) ∡G and ∡O are supplementary; ∡Y and ∡O are supplementary.
c) ∠G = ∠Y.

4. *Theorem 36.* The diagonals of a parallelogram bisect each other.

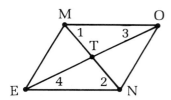

Hypothesis: MONE is a parallelogram with diagonals $\overline{MN}$ and $\overline{OE}$.
Conclusion: MN and OE bisect each other.

Proof.

Statements

a) MONE is a parallelogram with diagonals $\overline{MN}$ and $\overline{OE}$.
b) $\overline{MO} \parallel \overline{EN}$.
c) ∠1 = ∠2, ∠3 = ∠4.
d) MO = EN.
e) △MTO ≅ △NTE.
f) $\overline{MT} = TN$, OT = TE.
g) $\overline{MN}$ and $\overline{OE}$ bisect each other.

---

*We mention just one of the four pairs of consecutive angles since the other three pairs can be proved supplementary in the same way. This is a useful procedure in writing proofs that you will often see because it avoids a lot of trivial repetition.

5. Use the theorems in this lesson to complete the following statements about ☐DEGA.
a) AD = ▓▓▓▓.
b) ∠ADE = ▓▓▓▓.
c) To which angles is ∡DEG supplementary?
d) SE = ▓▓▓▓.

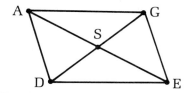

SET II

1. Every parallelogram has point symmetry.
a) Where do you think the point of symmetry of a parallelogram is?
b) Which theorem of this lesson supports your answer?
c) Explain.

Write complete proofs for each of the following.

2. Given: BRAE and EAQU are parallelograms.
   Prove: BR = UQ.
   Par 3.

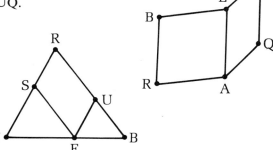

3. Given: RB = NB; $\overline{SE} \parallel \overline{RU}$ and $\overline{UE} \parallel \overline{RS}$.
   Prove: ∠N = ∠SEU.
   Par 6.

4. Given: SURT is a parallelogram;
   $\overline{TE} \perp \overline{SU}$, $\overline{UA} \perp \overline{TR}$.
   Prove: △SET ≅ △RAU.
   Par 7.

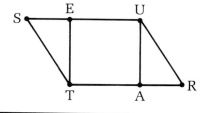

Escher's animal mosaics are among his most remarkable pictures.
You have already seen some of them on pages 148 and 227 of
this book.

In several of his mosaics, Escher began by dividing the plane
into parallelograms. One of them is based upon the two figures
shown here. Trace the grid below and see if you can figure out
how Escher filled it with just these two shapes. The position of
one of them is shown to help you in getting started.

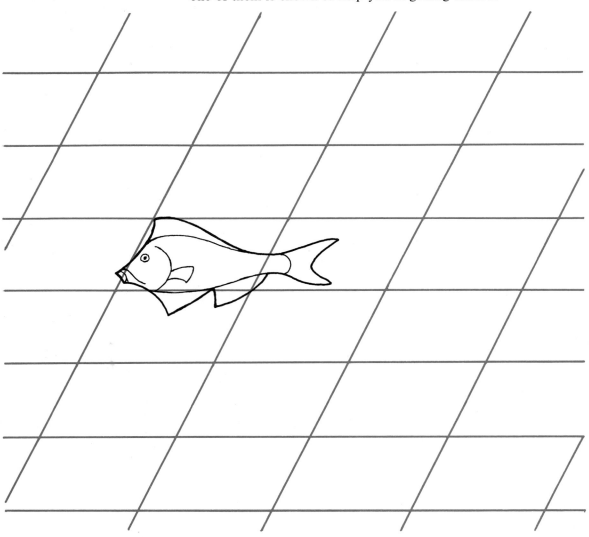

Lesson 3

Quadrilaterals That Are Parallelograms

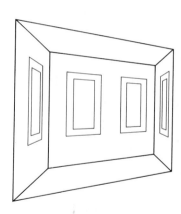

Room as "seen"

Although the dog in this room seems to be about the right size, the boy seems extraordinarily large. In reality, however, one is about as big as the other. How can this be?

We are fooled by the design of the room. Although the walls and windows seem to be rectangular, they are not. The windows in the walls at the left and right seem to be directly opposite each other, but they actually are not. Even the floorboards that seem to be parallel are not!

The first diagram at the right reveals the room's actual shape. The back wall is not rectangular because, although its left and right edges are parallel, the edges along the ceiling and floor are not. They slope toward each other at the right side of the room, so that the right wall is smaller than the left. Furthermore, the right edge of the back wall is much closer to us than the left edge. The actual floor plan of the room, in contrast with the one we think we are seeing, is shown in the second diagram. Because we assume the room is a normal one, we interpret the larger size of someone at the right to mean that he *is* larger rather than merely closer to us.

Camera

301

An important part of the illusion is our presumption that all of the quadrilaterals in the room are *parallelograms*, which they really are *not*. What do we need to know about a quadrilateral in order to be *sure* that it is a parallelogram?

Look at each of the quadrilaterals shown here in which certain parts have been marked equal or parallel. To emphasize the fact that it is the relationships among these parts that are important, and not the appearance of the quadrilaterals, the drawings have been distorted.

Is figure A a parallelogram? Yes, because both pairs of opposite sides are parallel. How about figure B? We know that the opposite sides of a parallelogram are equal, but is the converse true? If the opposite sides of a quadrilateral are equal, must it be a parallelogram? Figures C and D raise the same questions about quadrilaterals whose opposite angles are equal and whose diagonals bisect each other. Figure E illustrates a quadrilateral for which we know only that one pair of opposite sides are both equal and parallel. Is it a parallelogram?

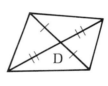

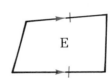

We can prove that every one of these quadrilaterals must be a parallelogram. Theorems to this effect are stated below and their proofs are left as exercises.

▶ **Theorem 37**
If both pairs of opposite sides of a quadrilateral are equal, then the quadrilateral is a parallelogram.

▶ **Theorem 38**
If two sides of a quadrilateral are both parallel and equal, then the quadrilateral is a parallelogram.

▶ **Theorem 39**
If both pairs of opposite angles of a quadrilateral are equal, then the quadrilateral is a parallelogram.

▶ **Theorem 40**
If the diagonals of a quadrilateral bisect each other, then the quadrilateral is a parallelogram.

Answer the questions about the proofs of each of the following theorems.

1. *Theorem 37.* If both pairs of opposite sides of a quadrilateral are equal, then the quadrilateral is a parallelogram.

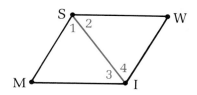

   Hypothesis: Quadrilateral SWIM
   with SW = MI
   and SM = WI.
   Conclusion: SWIM is a parallelogram.

   Proof.
   Draw $\overline{SI}$.

   a) How can we prove △SMI ≅ △IWS?

   We know that ∠1 = ∠4 and ∠2 = ∠3 because corresponding parts of congruent triangles are equal.

   b) Why is $\overline{SM} \parallel \overline{WI}$ and $\overline{SW} \parallel \overline{MI}$?
   c) Why is SWIM a parallelogram?

2. *Theorem 38.* If two sides of a quadrilateral are both parallel and equal, then the quadrilateral is a parallelogram.

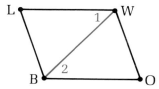

   Hypothesis: Quadrilateral BOWL
   with $\overline{LW} \parallel \overline{BO}$
   and LW = BO.
   Conclusion: BOWL is a parallelogram.

   Proof.
   Draw $\overline{BW}$.

   a) Why is ∠1 = ∠2?
   b) How can we prove △LBW ≅ △OWB?

   Therefore, LB = WO.

   c) Why is BOWL a parallelogram?

3. *Theorem 39.* If both pairs of opposite angles of a quadrilateral are equal, then the quadrilateral is a parallelogram.

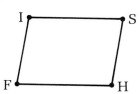

   Hypothesis: Quadrilateral FISH
   with ∠I = ∠H
   and ∠F = ∠S.
   Conclusion: FISH is a parallelogram.

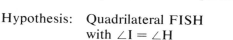

Lesson 3: Quadrilaterals That Are Parallelograms   303

Proof.
By hypothesis, $\angle I = \angle H$ and $\angle F = \angle S$.

a) Why is $\angle F + \angle I + \angle S + \angle H = 360°$?
b) Why is $\angle F + \angle I + \angle F + \angle I = 360°$?

So $2(\angle F + \angle I) = 360°$.

c) Why is $\angle F + \angle I = 180°$?

Therefore $\angle F$ and $\angle I$ are supplementary.

d) Why is $\overline{IS} \parallel \overline{FH}$?
e) Why is $\angle F + \angle H = 180°$?

Therefore $\angle F$ and $\angle H$ are supplementary, and $\overline{IF} \parallel \overline{SH}$.

f) Why is FISH a parallelogram?

4. *Theorem 40.* If the diagonals of a quadrilateral bisect each other, then the quadrilateral is a parallelogram.

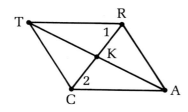

Hypothesis: Quadrilateral TRAC with diagonals $\overline{TA}$ and $\overline{RC}$ bisecting each other at K.

Conclusion: TRAC is a parallelogram.

Proof.
Since $\overline{TA}$ and $\overline{RC}$ bisect each other, $TK = KA$ and $RK = KC$.

a) How can we prove $\triangle TKR \cong \triangle AKC$?

So $\angle 1 = \angle 2$ and $TR = CA$.

b) Why is $\overline{TR} \parallel \overline{CA}$?
c) Why is TRAC a parallelogram?

5. We now have five different ways to prove that a quadrilateral is a parallelogram. One of them is to show that both pairs of opposite sides are parallel. What are the other four?

SET II

1. Given: $SG = GT$ and $KG = GN$.
   Prove: $\angle A = \angle NTI$.
   Par 5.

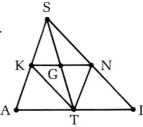

2. Given: △HOY ≅ △KEY; K is
the midpoint of $\overline{EC}$.
   Prove: HOCK is a parallelogram.
   Par 8.

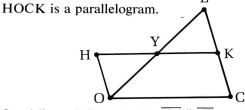

3. Given: Quadrilateral GOLF with $\overline{GO} \parallel \overline{FL}$; ∠G = ∠L.
   Prove: GOLF is a parallelogram.
   Par 5.

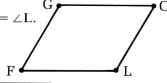

SET III

If five of the faces of a box are parallelograms, must the sixth
face also be one?

If you think the answer is yes, explain why.

If you think the answer is no, draw a picture of such a box in
which the sixth face is not.

*"Because I didn't finish reading it at the
breakfast table, that's why."*

Lesson 3: Quadrilaterals That Are Parallelograms   305

Lesson **4**

Kites and Rhombuses

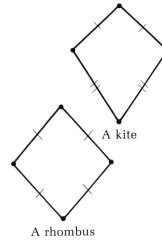

A kite

A rhombus

Charlie Brown's kite is an example of a special type of quadrilateral, or at least it was until Lucy refused to let go. The adjoining diagram illustrates the geometric figure called a kite. As you can see, it has two pairs of consecutive sides that are equal.

▶ Definition
A *kite* is a quadrilateral that has two pairs of equal consecutive sides with no side common to both pairs.

If all four sides of a quadrilateral are the same length, it is called a rhombus.

▶ Definition
A *rhombus* is an equilateral quadrilateral.

306

Two more illustrations of a kite and a rhombus are shown below. Since both figures have been drawn to scale, it is evident that if a quadrilateral is a kite, it is not necessarily a parallelogram. It is easy to prove, on the other hand, that every rhombus is a parallelogram.

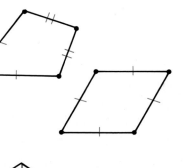

► **Theorem 41**
All rhombuses are parallelograms.

Since all rhombuses are parallelograms, they possess all of the properties of parallelograms. One of these properties that we have proved is the fact that the diagonals of a parallelogram (and hence a rhombus) bisect each other. The diagonals of a rhombus are also perpendicular.

► **Theorem 42**
The diagonals of a rhombus are perpendicular to each other.

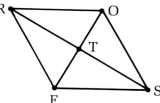

Exercises

Answer the questions about the proofs of the following theorems.

1. *Theorem 41.* All rhombuses are parallelograms.

    Hypothesis: RAIN is a rhombus.
    Conclusion: RAIN is a parallelogram.

    Proof.
    We know that RAIN is a rhombus.

    a) Why is RA = NI and RN = AI?
    b) Why is RAIN a parallelogram?

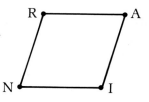

2. *Theorem 42.* The diagonals of a rhombus are perpendicular to each other.

    Hypothesis: FROS is a rhombus
    with diagonals $\overline{RS}$ and $\overline{FO}$.
    Conclusion: $\overline{RS} \perp \overline{FO}$.

    Proof.
    Since FROS is a rhombus, RF = RO and SF = SO.

    a) What relationship do points R and S have to F and O?
    b) Because of this, what relationship does $\overline{RS}$ have to $\overline{FO}$?
    c) Why?

3. The three figures shown here are rhombuses in which one or both diagonals have been drawn.
   a) Either diagonal of a rhombus divides it into two congruent triangles. What kind of triangles are they? Why?
   b) Both diagonals of a rhombus together divide it into four nonoverlapping congruent triangles. What kind of triangles are they? Why?
   c) How many lines of symmetry do you think a rhombus has?
   d) Does a rhombus have point symmetry?

4. The three adjoining figures are kites in which one or both diagonals have been drawn. Use these figures to decide whether each of the following statements is true or false.

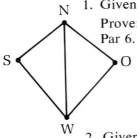

   a) Either diagonal of a kite divides it into two congruent triangles.
   b) Each diagonal of a kite divides it into two isosceles triangles.
   c) The diagonals of a kite are perpendicular to each other.
   d) A kite has line symmetry.
   e) The diagonals of a kite bisect each other.
   f) A kite has point symmetry.

SET II

1. Given: SNOW is a kite with diagonal $\overline{\text{NW}}$.
   Prove: $\overrightarrow{\text{NW}}$ bisects ∢SNO.
   Par 6.

2. Given: ∠A = ∠L, ∠H = ∠I; AI = AH.
   Prove: HAIL is a rhombus.
   Par 6.

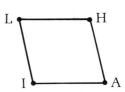

3. Given: SHWR is a rhombus;
   $\overline{SE} \perp \overline{RW}$ and $\overline{SO} \perp \overline{WH}$.
   Prove: SE = SO.
   Par 9.

SET III

The following statements appear in a German geometry book. Can you figure out what they say?

1. Ein Viereck mit vier gleichen Seiten heißt *Raute*.
2. Jede Raute ist ein Parallelogramm.
3. Die Diagonalen einer Raute stehen senkrecht aufeinander.
4. Ein Viereck, bei dem zwei Paare benachbarter Seiten gleich lang sind, heißt *Drachen*.
5. In jedem Drachen stehen die Diagonalen aufeinander senkrecht.

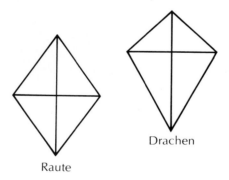

Raute

Drachen

Lesson 5

Rectangles and Squares

This pattern of dark and light squares is actually a picture of a famous man. If you look at it from several feet away, the portrait looks quite different. It was created at Bell Laboratories in an experiment to determine how much detail a picture must contain in order to be recognizable.

The basic shape used within the picture is the *square* and the picture itself is in the shape of a *rectangle*. The rectangle is undoubtedly the most commonly used of all quadrilateral shapes. We will define it in a way similar to the way in which we defined the word "rhombus."

▶ Definition
A *rectangle* is an equiangular quadrilateral.

This says that the four angles of a rectangle are equal in measure. Since a rectangle is a quadrilateral, we know that the sum of the measures of its angles is 360°. From these two facts we can conclude that the angles of a rectangle are right angles.

▶ **Theorem 43**

All four angles of a rectangle are right angles.

That rectangles, like rhombuses, are also parallelograms can be easily proved.

▶ **Theorem 44**

All rectangles are parallelograms.

What relationship do the diagonals of a rectangle have other than that they bisect each other? They need not be perpendicular, but they certainly seem to be equal.

▶ **Theorem 45**

The diagonals of a rectangle are equal.

If a quadrilateral is equilateral, we call it a rhombus; if it is equiangular, we call it a rectangle. The quadrilateral that possesses both of these properties is the square.

▶ Definition

A *square* is a quadrilateral that is both equilateral and equiangular.

From this definition, it is apparent that every square is a rhombus and that every square is a rectangle. And it also follows that every square is a parallelogram. Therefore, squares possess every property that we have proved for these other figures.

SET I

1. Use the adjoining figure to prove the theorem that all rectangles are parallelograms.

      Hypothesis: RUTH is a rectangle.
      Conclusion: RUTH is a parallelogram.
      Par 3.

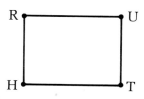

2. Answer the questions about this proof of the theorem that the diagonals of a rectangle are equal.

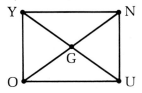

Hypothesis: YOUN is a rectangle.
Conclusion: YU = NO.

Proof.
We know that YOUN is a rectangle.
a) Why is YOUN a parallelogram?
b) Why is YO = NU?
c) Why are ∡YOU and ∡NUO right angles?

So ∠YOU = ∠NUO.

d) Why is △YOU ≅ △NUO?

So YU = NO.

3. At least two different words or phrases can be used to complete each of the following sentences. For example, the missing word in the sentence

The opposite sides of a rectangle are ▨▨▨▨.

could be either "parallel" or "equal."
   Give as many words or phrases as you can think of to complete each of the following sentences.
a) Two consecutive sides of a square are ▨▨▨▨.
b) Two consecutive angles of a rectangle are ▨▨▨▨.
c) All rectangles are ▨▨▨▨.
d) All squares are ▨▨▨▨.
e) The diagonals of a rectangle are ▨▨▨▨.
f) The diagonals of a square are ▨▨▨▨.

4. Copy the rectangle and square shown here.
a) Draw any apparent lines of symmetry in each figure.
b) Does either figure have point symmetry?

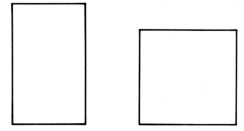

The figures for the following exercises are not necessarily drawn accurately.

1. If one angle of a parallelogram is a right angle, must the parallelogram be a rectangle? Base your explanation on the adjoining figure.

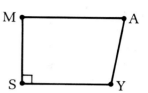

2. Given: MARI is a parallelogram; MR = AI.
   Prove: ∢MIR is a right angle.
   Par 10.

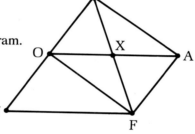

3. The results of Exercises 2 and 1 can be combined in that order to prove a theorem. What is it?

4. Given: KOFA is a rectangle;
   OUFA is a parallelogram.
   Prove: △KUF is isosceles.
   Par 6.

Dilcue had a nightmare due to studying too hard for his geometry midterm. He dreamed that four congruent right triangles came marching toward him and surrounded him.

The triangles lined their sides up so that LOUG was a square. Dilcue noticed that EHRI also seemed to be a square, but was too terrified to be able to think clearly enough to figure out why. Can you?

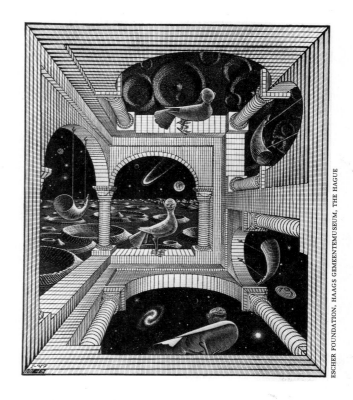

Lesson 6

Trapezoids

This woodcut by Escher titled *Another World* shows the interior of a strange building in which the walls, floor, and ceiling are interchangeable. The view through the windows at the top and upper right indicates that we are looking down from the ceiling. The view through the windows at the bottom and lower right, on the other hand, is from the floor up toward the sky. And through the center and left-hand windows we are looking at the horizon. Is the rectangular region in the center a *wall, floor,* or *ceiling*? It depends upon your point of view!

The design of the drawing reveals that rectangles seen in perspective may look like another type of quadrilateral instead: the *trapezoid,* in which just one pair of sides are parallel.

▶ Definition
A ***trapezoid*** is a quadrilateral that has exactly one pair of parallel sides.

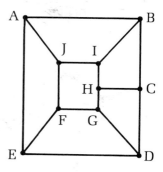

315

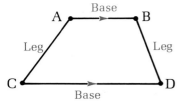

The parallel sides of a trapezoid are called its *bases* and the nonparallel sides are called its *legs*. The pairs of angles that include each base are called *base angles*: one pair of base angles in trapezoid ABCD is ∡A and ∡B and the other pair is ∡C and ∡D.

The base angles of a trapezoid do not ordinarily have any special relationship. If the legs of a trapezoid are equal, however, then we can prove that each pair of base angles are equal.

▶ Definition

An *isosceles trapezoid* is a trapezoid whose legs are equal.

▶ Theorem 46

The base angles of an isosceles trapezoid are equal.

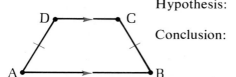

Hypothesis: Isosceles trapezoid ABCD
with bases $\overline{AB}$ and $\overline{DC}$.
Conclusion: ∠A = ∠B and ∠C = ∠D.

To prove this, we will add two line segments to the figure to form a pair of congruent right triangles.

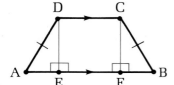

Proof.
Draw $\overline{DE} \perp \overline{AB}$ and $\overline{CF} \perp \overline{AB}$. (Through a point not on a line, there is exactly one perpendicular to the line.) Since ∡DEA and ∡CFB are right angles, △DEA and △CFB are right triangles. We know that AD = BC because the legs of an isosceles trapezoid are equal. Since $\overline{DC} \parallel \overline{AB}$ (the bases of a trapezoid are parallel), DE = CF because if two lines are parallel, every perpendicular segment joining one line to the other has the same length. Therefore, △AED ≅ △BFC (H. L.), so ∠A = ∠B.

Now in the original figure, ∡D and ∡A are supplementary, as are ∡C and ∡B. (If two parallel lines are cut by a transversal, the interior angles on the same side of the transversal are supplementary.) Therefore, ∠D = ∠C because supplements of equal angles are equal.

By using this theorem, it is easy to prove that the diagonals of an isosceles trapezoid, like those of a rectangle, are equal.

▶ Corollary

The diagonals of an isosceles trapezoid are equal.

1. Quadrilateral FORD is an isosceles trapezoid with bases $\overline{FD}$ and $\overline{OR}$. What relationships do each of the following have?
   a) $\overline{FD}$ and $\overline{OR}$.
   b) $\overline{FO}$ and $\overline{DR}$.
   c) ∡O and ∡R.
   d) ∡F and ∡O.

Exercises

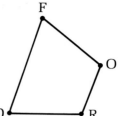

2. The theorem that the base angles of an isosceles trapezoid are equal might be restated in the following way:

   If the legs of a trapezoid are equal, then its base angles are equal.

   a) What is the converse of this statement?
   b) What are the hypothesis and conclusion of the converse in terms of the adjoining figure?
   c) One way to prove the converse begins by drawing a line through U parallel to $\overline{BK}$. What permits us to do this?
   d) Why is ∠UCI = ∠K?
   e) Since ∠K = ∠I, why is ∠UCI = ∠I?
   f) Why is UC = UI?
   g) Why is KCUB a parallelogram?
   h) Why is BK = UC?
   i) Why is BK = UI?

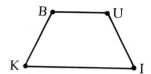

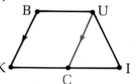

1. Can the diagonals of a trapezoid bisect each other? Explain.

2. Use the figure below to prove the corollary to Theorem 46.

   The diagonals of an isosceles trapezoid are equal.

   Hypothesis: OLDS is an isosceles trapezoid.
   Conclusion: OD = SL.
   Par 6.

3. Given: △CYE with ∠C = ∠VHE; CH = YV.
   Prove: △CYE is isosceles.
   Par 8.

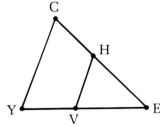

SET III

On a radio quiz program* several years ago, one of the questions was to define the word "trapezoid." A contestant answered that it was "a triangle with its top cut off" and, although it was apparent from the announcer's response that he hadn't expected this answer, he accepted it as correct.

Can you reword the contestant's answer so that it is mathematically correct yet still expresses what he had in mind?

*"Testing One Two Three," KNX, Los Angeles.

© 1966 UNITED FEATURE SYNDICATE, INC.

Lesson 7

The Midsegment Theorem

1200 miles
Vancouver   Winnipeg

A   ? miles   B

Mexico City

Each autumn, millions of birds go south for the winter. Some fly across entire continents, traveling as far as five thousand miles in one migration. Even more incredible is the fact that small birds weighing less than an ounce can fly for more than sixty hours without stopping to rest!*

The adjoining map shows the flight paths of two flocks of birds migrating from the Canadian cities of Vancouver and Winnipeg to Mexico City. Suppose that the birds leave Vancouver and Winnipeg at the same time, that they fly at steady rates, and that they arrive in Mexico City at the same time. From the fact that Vancouver is approximately 1200 miles due west of Winnipeg, can we draw any conclusion about the relative positions of the two flocks of birds at the halfway points of their flights? (These points are labeled A and B on the map.) The flock at A appears to be due west of the one at B and the distance between them about one half of what it was at the beginning.

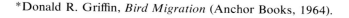

*Donald R. Griffin, *Bird Migration* (Anchor Books, 1964).

319

In terms of the triangle determined by the three cities, the line segment that joins the midpoints of the two sides of the triangle appears to have two relationships to the third side: it seems to be parallel to it and about half as long. We will call such a segment a *midsegment* of the triangle.

► Definition

A *midsegment* of a triangle is a line segment that joins the midpoints of two of its sides.

We will now prove our conjecture about the relationship of a triangle and one of its midsegments.

► **Theorem 47**   (The Midsegment Theorem)

A midsegment of a triangle is parallel to the third side and half as long.

Hypothesis: △ABC with midsegment DE.

Conclusion: $\overline{DE} \parallel \overline{BC}$ and $DE = \frac{1}{2}BC$.

It is impossible to prove either part of the conclusion without adding something to the figure. Our method will be to reflect △ADE through point E as shown in the figure below. In doing this, a parallelogram is formed that has the same base, $\overline{BC}$, as the original triangle. Both conclusions about the midsegment of the triangle can be derived from the fact that it is part of $\overline{DF}$, the side opposite the base of this parallelogram.

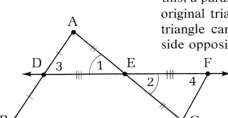

Proof.

To reflect △ADE through point E, draw $\overleftrightarrow{DE}$, choose point F on the ray opposite $\overrightarrow{ED}$ so that EF = DE, and draw $\overline{CF}$.

In △ADE and △CFE,

$$DE = EF,$$
$$\angle 1 = \angle 2 \text{ (since they are vertical angles),}$$

and

$$AE = EC \text{ (since E is the midpoint of } \overline{AC}).$$

So △ADE ≅ △CFE by the S.A.S. Congruence Postulate.

Now

$$BD = DA \text{ (since D is the midpoint of } \overline{BA}),$$

and

$$DA = CF \text{ (they are corresponding sides}$$
$$\text{of the congruent triangles),}$$

so BD = CF (transitive). Furthermore, $\angle 3 = \angle 4$ (they are corresponding angles of the triangles), so $\overleftrightarrow{BA} \parallel \overleftrightarrow{CF}$ (two lines that form equal alternate interior angles with a transversal are parallel). Hence $\overline{BD} \parallel \overline{CF}$.

Since $BD = CF$ and $\overline{BD} \parallel \overline{CF}$, BCFD is a parallelogram (if two sides of a quadrilateral are both parallel and equal, it is a parallelogram). Therefore, $\overline{DF} \parallel \overline{BC}$ and $DF = BC$ (the opposite sides of a parallelogram are both parallel and equal). Hence $\overline{DE} \parallel \overline{BC}$.

We also know that $DE = \frac{1}{2}DF$ because E is the midpoint of $\overline{DF}$ and the midpoint of a line segment divides it into segments half as long. Since $DF = BC$, we have $DE = \frac{1}{2}BC$ (substitution).

SET I

1. Segments $\overline{ME}$ and $\overline{MI}$ are midsegments of $\triangle HUL$. What conclusions can you draw about each of the following?
   a) $\overline{MI}$ and $\overline{HL}$.
   b) $\angle IMU$ and $\angle H$.
   c) HE and EL.
   d) ME and UL.
   e) Quadrilateral MULE.
   f) Quadrilateral MILE.

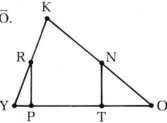

2. Complete the following proof by supplying the reasons.

   Given: $\triangle YKO$ with R and N the midpoints of sides $\overline{YK}$ and $\overline{KO}$; $\overline{RP} \perp \overline{YO}$ and $\overline{NT} \perp \overline{YO}$.
   Prove: $RP = NT$.

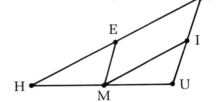

Proof.

<u>   Statements   </u>

a) $\triangle YKO$ with R and N the midpoints of sides $\overline{YK}$ and $\overline{KO}$.
b) Draw $\overline{RN}$.
c) $\overline{RN}$ is a midsegment of $\triangle YKO$.
d) $\overline{RN} \parallel \overline{YO}$.
e) $\overline{RP} \perp \overline{YO}$ and $\overline{NT} \perp \overline{YO}$.
f) $RP = NT$.

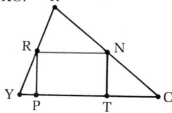

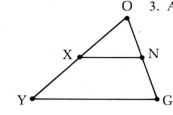

O 3. Answer the questions about the following indirect proof.

Given: In △OYG, X is the midpoint of $\overline{OY}$; $\overline{XN} \parallel \overline{YG}$.

Prove: N is the midpoint of $\overline{OG}$.

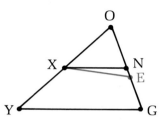

Proof.

Suppose N is *not* the midpoint of $\overline{OG}$. Then let E be the midpoint of $\overline{OG}$.

a) What permits us to assume this?

Draw $\overline{XE}$.

b) Why is $\overline{XE} \parallel \overline{YG}$?

This contradicts the given information that $\overline{XN} \parallel \overline{YG}$.

c) Why?

Therefore, our original assumption that N is not the midpoint of $\overline{OG}$ is wrong and the proof is complete.

___

SET II

1. Copy this quadrilateral and use your straightedge and compass to bisect each of its sides. Starting with $\overline{CL}$, label the midpoints of the sides clockwise around the figure H, O, I, and E, respectively. Draw quadrilateral HOIE.

a) What special type of quadrilateral does HOIE seem to be?

Draw $\overline{CR}$.

b) Why is $\overline{HO} \parallel \overline{CR}$ and $\overline{EI} \parallel \overline{CR}$?
c) Why is $\overline{HO} \parallel \overline{EI}$?
d) Why is $HO = \frac{1}{2}CR$ and $EI = \frac{1}{2}CR$?
e) Why is $HO = EI$?
f) How do the conclusions in parts c and e prove that your answer to part a is correct?

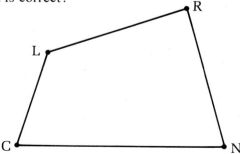

2. Copy the *concave* quadrilateral shown here and carry out the same procedure as described in Exercise 1.
   a) What seems to be the result this time?
   b) Does the argument presented in parts b–f of Exercise 1 apply to the result?

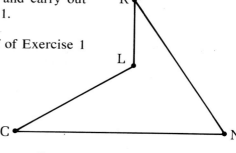

Write complete proofs for each of the following.

3. Given: $\overline{UE}$ and $\overline{EA}$ are midsegments of $\triangle BTN$.
   Prove: $\angle UEA = \angle T$.
   Par 4.

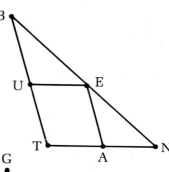

4. Given: Isosceles $\triangle AGO$ with $AG = GO$; midsegment $\overline{RN}$.
   Prove: $\triangle ARN$ is also isosceles.
   Par 8.

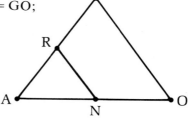

SET III

Obtuse Ollie says that if you draw a quadrilateral and connect the midpoints of its opposite sides with two line segments, they will always bisect each other. He drew the figure shown here to support his claim.

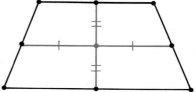

1. Do you think Ollie is correct?

2. If not, can you draw a figure that contradicts his conclusion?

Chapter 8/Summary and Review

Basic Ideas

Theorems

33. The sum of the measures of the angles of a quadrilateral is 360°. 292
34. The opposite sides of a parallelogram are equal.   296
    Corollary.   If two lines are parallel, every perpendicular segment joining one line to the other has the same length.   297
35. The consecutive angles of a parallelogram are supplementary. 297
    Corollary.   The opposite angles of a parallelogram are equal. 297
36. The diagonals of a parallelogram bisect each other.   297
37. If both pairs of opposite sides of a quadrilateral are equal, then the quadrilateral is a parallelogram.   302
38. If two sides of a quadrilateral are both parallel and equal, then the quadrilateral is a parallelogram.   302
39. If both pairs of opposite angles of a quadrilateral are equal, then the quadrilateral is a parallelogram.   302
40. If the diagonals of a quadrilateral bisect each other, then the quadrilateral is a parallelogram.   302
41. All rhombuses are parallelograms.   307
42. The diagonals of a rhombus are perpendicular to each other. 307
43. All four angles of a rectangle are right angles.   311
44. All rectangles are parallelograms.   311
45. The diagonals of a rectangle are equal.   311

<small>SET</small> I

Exercises

1. This painting by a contemporary Venezuelan artist, Jésus-
   Raphaël Soto, contains several types of quadrilaterals.
   a) Name the types you see.
      Which of the types you have named possess the following
   properties?
   b) The sum of the measures of its angles is 360°.
   c) Its consecutive angles are supplementary.
   d) Its diagonals are perpendicular to each other.
   e) Its diagonals are equal.

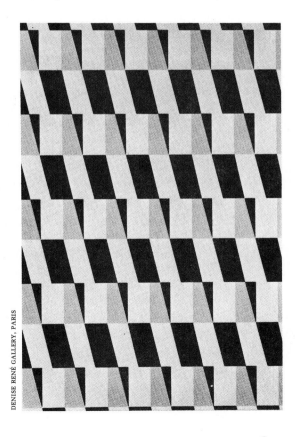

DENISE RENÉ GALLERY, PARIS

2. Answer the following questions about this proof that the bisectors of two consecutive angles of a parallelogram are perpendicular to each other.

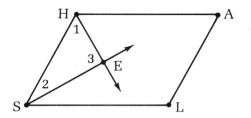

Given: SHAL is a parallelogram;
$\overrightarrow{HE}$ bisects ∡SHA
and $\overrightarrow{SE}$ bisects ∡HSL.
Prove: $\overrightarrow{HE} \perp \overrightarrow{SE}$.

Proof.

a) Since SHAL is a parallelogram, ∡SHA and ∡HSL are supplementary. Why?

Therefore, $\angle SHA + \angle HSL = 180°$.

b) Since $\overrightarrow{HE}$ bisects ∡SHA and $\overrightarrow{SE}$ bisects ∡HSL, $\angle 1 = \frac{1}{2}\angle SHA$ and $\angle 2 = \frac{1}{2}\angle HSL$. Why?

c) We also know that $\frac{1}{2}\angle SHA + \frac{1}{2}\angle HSL = 90°$. Why?

d) Therefore, $\angle 1 + \angle 2 = 90°$. Why?

e) What can we conclude from this and the fact that $\angle 1 + \angle 2 + \angle 3 = 180°$?

f) How does this prove that $\overrightarrow{HE} \perp \overrightarrow{SE}$?

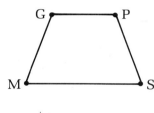

3. Every isosceles trapezoid has line symmetry. As to why, answer the following questions about this explanation.

Suppose GPSM is an isosceles trapezoid with bases $\overline{GP}$ and $\overline{MS}$. Let Y and U be the midpoints of the bases.

a) What permits us to choose these points?

Draw $\overleftrightarrow{YU}$, $\overline{GU}$, and $\overline{PU}$.

b) How can it be shown that $\triangle GMU \cong \triangle PSU$?

Since $\triangle GMU \cong \triangle PSU$, GU = PU. And GY = YP since Y is the midpoint of $\overline{GP}$. So U and Y are equidistant from G and P.

c) Why is $\overleftrightarrow{YU}$ the perpendicular bisector of $\overline{GP}$?

d) How can it be shown that $\overleftrightarrow{YU} \perp \overline{MS}$?

Since U is the midpoint of $\overline{MS}$, it follows that $\overleftrightarrow{YU}$ is also the perpendicular bisector of $\overline{MS}$.

e) Why are P and S the reflections of G and M through $\overleftrightarrow{YU}$?

It follows from this that $\overline{GP}, \overline{PS}, \overline{SM}$, and $\overline{MG}$ are the reflections through $\overleftrightarrow{YU}$ of $\overline{PG}, \overline{GM}, \overline{MS}$, and $\overline{SP}$, respectively.

f) Why, then, does GPSM have line symmetry?

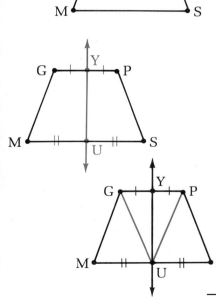

1. Use this diagram to explain why the diagonals of every kite are perpendicular to each other.

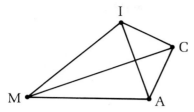

2. Given: $\overline{UR}$ and $\overline{RZ}$ are midsegments of $\triangle AQT$.
   Prove: $\triangle AUR \cong \triangle RZT$.
   Par 6.

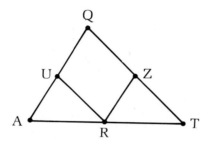

3. Given: TALC is a parallelogram; $\angle 1 = \angle 2$.
   Prove: TALC is a rhombus.
   Par 9.

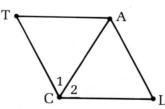

4. Explain by means of indirect reasoning why the line segments $\overline{RE}$ and $\overline{AL}$ drawn in $\triangle MRL$ cannot bisect each other.

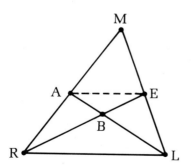

If you have ever driven by an orchard in which the trees are evenly spaced in parallel rows, you may have noticed that the trees seem to line up in other parallel rows as your perspective of them changes.

The figures below represent overhead views of the same rows of trees as seen from two perspectives. On the assumptions that $a \parallel b \parallel c \parallel d$, $\ell_1 \parallel \ell_2$, and the trees on ℓ_1 are evenly spaced, can you explain why $f \parallel g \parallel h$?

COURTESY OF SUNKIST GROWERS, INC.

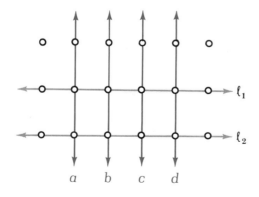

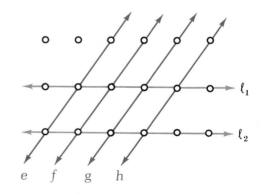

Chapter **9**

AREA

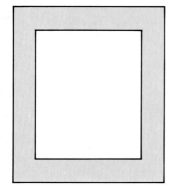

Lesson **1**

Polygonal Regions and Area

Which do you think is larger: the photograph or its frame? Another way that this question could be asked is, Which region has the greater *area*? If you picked the photograph, you are correct because it covers an area of 10 square centimeters on this page and the frame covers an area of 9 square centimeters.

It is important to realize that when we use the word "area" in geometry, we mean a *number* and that this number measures a *region*. The photograph is an example of a rectangular region and its frame is a region bounded by two rectangles. It would be convenient to have a general name for these regions and others whose areas we will determine.

First, we will consider the difference between a triangle and a triangular region. Remember that we defined a triangle as a set of three *line segments*; accordingly, the points inside a triangle are not part of it. A *triangular region*, on the other hand, includes both the triangle and all the points in its interior.

The cat-and-mouse photograph is by Charles Barr in *Laughing Camera 3* (Hanns Reich Verlag).

► Definition

A *triangular region* is the union of a triangle and its interior.

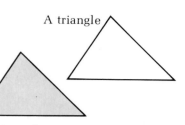

A triangle

The photograph and the frame can be divided into triangular regions and, as a result, are called *polygonal regions*.

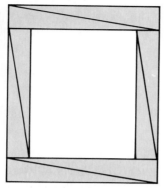

A triangular region

► Definition

A *polygonal region* is the union of a finite number of triangular regions in a plane that have no interior points in common.

A polygonal region can be divided into triangular regions in many different ways. For example, the photograph could also be divided like the one at the right.

We said that the photograph and frame have areas of 10 and 9 square centimeters, respectively. It seems reasonable to assume that *every* polygonal region has an area and that, for a given unit of measure, it is uniquely determined.

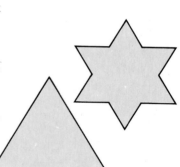

► Postulate 16 (The Area Postulate)

To every polygonal region there corresponds a unique positive number.

This number is called the *area* of the region. We will represent the word "area" by α, the first letter of the Greek alphabet.

How do you think the areas of these two polygonal regions compare?

One way of cutting them up into triangular regions is shown below.

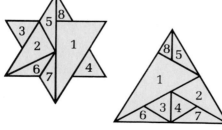

Each pair of identically numbered regions seems to be bounded by a pair of congruent triangles. If this is the case, it seems reasonable that we could cut out the eight regions of one figure and rearrange them to form the other. Our thinking is based upon the following assumptions.

▶ **Postulate 17** (The Area Addition Postulate)
If a polygonal region is the union of two or more polygonal regions with no interior points in common, then its area is the sum of their areas.

▶ **Postulate 18**
Triangular regions bounded by congruent triangles have equal areas.

Now that you know that areas are measures of *polygonal regions*, we will take the liberty of using such expressions as "the area of a square" when we mean "the area of a polygonal region bounded by a square." Postulate 18 could be more simply stated: Congruent triangles have equal areas. Also, we will not bother to shade every polygonal region with whose area we are concerned.

Exercises

SET I

1. Copy the figures below and then show that each is a polygonal region by dividing it into triangular regions. In each case, try to form as few triangular regions as possible.

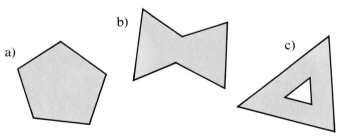

a)   b)   c)

2. In this figure, $\triangle PRE \cong \triangle NIC$. Copy and complete the following equations, explaining your answer in each case.
   a) $\alpha\triangle PRE = \alpha\text{\rule{0.6cm}{0.3cm}}$.
   b) $\alpha PNCE = \alpha\text{\rule{0.6cm}{0.3cm}} + \alpha\text{\rule{0.6cm}{0.3cm}} + \alpha\text{\rule{0.6cm}{0.3cm}}$.

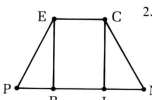

3. In this figure, $\alpha\triangle SUL = \alpha\triangle TAN$.
   a) Can you conclude from this that $\triangle SUL \cong \triangle TAN$?
   b) Is the converse of Postulate 18 true?

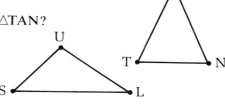

4. Complete the following proof by supplying the reasons.

> Given: $\square CZAR$ with diagonal $\overline{RZ}$.
> Prove: $\alpha\triangle CRZ = \frac{1}{2}\alpha\square CZAR$.

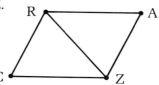

Proof.

<u>Statements</u>

a) $\square CZAR$ with diagonal $\overline{RZ}$.
b) $CR = ZA$, $RA = CZ$.
c) $\angle C = \angle A$.
d) $\triangle CRZ \cong \triangle AZR$.
e) $\alpha\triangle CRZ = \alpha\triangle AZR$.
f) $\alpha\triangle CRZ + \alpha\triangle AZR = \alpha\square CZAR$.
g) $\alpha\triangle CRZ + \alpha\triangle CRZ =$
   $\quad 2\alpha\triangle CRZ = \alpha\square CZAR$.
h) $\alpha\triangle CRZ = \frac{1}{2}\alpha\square CZAR$.

SET II

1. In this lesson, we referred to the *interior* of a triangle without formally defining the word. This is a case in which the meaning is so evident from a diagram that a definition confuses more than it helps us understand. The following exercises suggest one way of developing a formal definition.

   Copy this figure in which point P represents any point in the interior of the triangle.

   a) If a line is drawn through P, in how many points will it intersect the triangle?
   b) Would your answer to part a still be true if P were some other place inside the triangle?
   c) Let X and Y represent the points in which a given line through P intersects the triangle. What betweenness relationship must exist among X, Y, and P?

   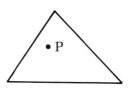

   The interior of a triangle, then, could be defined as the set of all points P such that every line through P intersects the triangle in two points and the betweenness relationship that you wrote for part c holds.

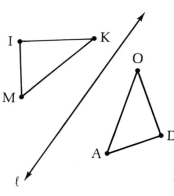

2. Given: △MIK is the reflection
            of △ADO through line ℓ.
   Prove: α△MIK = α△ADO.
   Par 3.

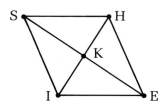

3. Quadrilateral SHEI is a rhombus with diagonals $\overline{SE}$ and $\overline{HI}$.
   Explain why $\alpha\triangle IKE = \frac{1}{4}\alpha SHEI$.

4. Given: HCLP and HAIP are parallelograms.
   Prove: αHCLP = αHAIP.
   Par 9.

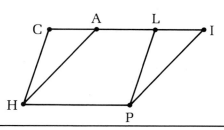

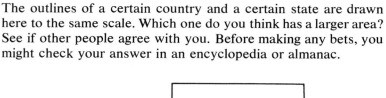

SET III

The outlines of a certain country and a certain state are drawn
here to the same scale. Which one do you think has a larger area?
See if other people agree with you. Before making any bets, you
might check your answer in an encyclopedia or almanac.

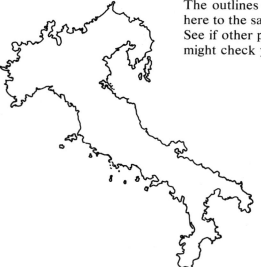

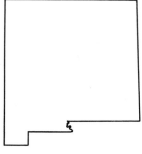

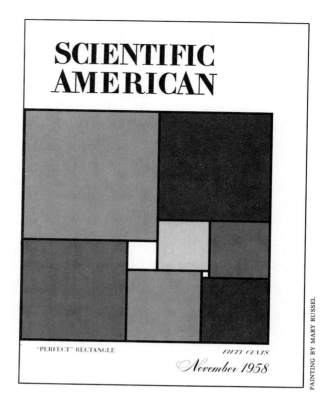

SCIENTIFIC AMERICAN

"PERFECT" RECTANGLE   *FIFTY CENTS*

November 1958

PAINTING BY MARY RUSSEL

Lesson 2

Squares and Rectangles

The picture on this magazine cover illustrates a geometric puzzle about squares. The puzzle is this: Is it possible to divide a square into smaller squares so that no two have the same area? The adjoining figures illustrate three squares that have been divided into smaller squares, but, in each case, some of them have equal areas.

The picture on the magazine cover seems to show a square that has been divided into nine smaller squares, all having different areas. The figure itself, however, is not a square but a rectangle, because one pair of opposite sides is slightly longer than the other pair.

Many other rectangles that can be divided into unequal squares like this one were discovered before a way to divide a square was found. Since it involves 39 smaller squares, ranging in side lengths from 7 units to 2,378 units, it would be difficult to illustrate here!

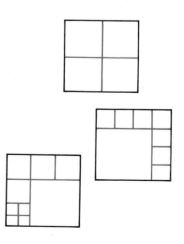

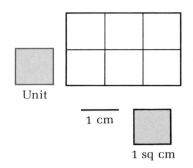

Unit

1 cm

1 sq cm

If we divide a rectangle or a square into a set of smaller squares that are all *equal* in area, we might find its area by using one of them as our unit and counting their number. For example, the area of the rectangle shown here is 6 units.

Our unit of area can be any shape and size we wish. However, area is conventionally measured in *square* units that are related to units of *distance* in a simple way. Each side of the area unit that we used to measure the rectangle above is 1 centimeter long. The area unit is called 1 *square centimeter* and the area of the rectangle is 6 square centimeters.

What is the area of the rectangle shown here? We can find it by either counting the squares or by multiplying its length and width, since they indicate the number of squares in each row and the number of rows.

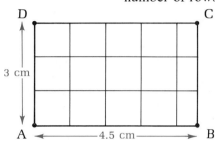

α Rectangle ABCD $= 3$ cm $\cdot$ 4.5 cm

$= 13.5$ sq cm

If the dimensions of the rectangle are more complicated numbers, then the square-counting method can become very difficult and the multiplication shortcut not so obvious. It seems reasonable, however, that the shortcut is not limited to certain kinds of numbers. Hence

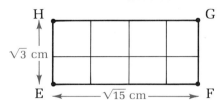

α Rectangle EFGH $= \sqrt{15}$ cm $\cdot \sqrt{3}$ cm

$= \sqrt{45}$ sq cm

As shown in the adjoining figure, two consecutive sides of a rectangle may be called its base and altitude. The letters b and h are usually used to represent their respective lengths (h for "height," another word for "altitude").

We now state our assumption that the multiplication shortcut works for every rectangle as a postulate.

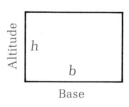

▶ **Postulate 19**

The area of a rectangle is the product of the lengths of its base and its altitude.

Since a square is a rectangle whose base and altitude are equal, the following corollary is obvious.

► **Corollary**

The area of a square is the square of the length of its side.

The areas of these squares are 9 square units and 25 square units, respectively. Their *perimeters*, on the other hand, are 12 units and 20 units, respectively.

← 3 units →

► **Definition**

The ***perimeter*** of a triangle or quadrilateral is the sum of the lengths of its sides.

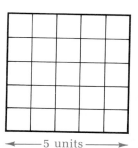

We will represent the word "perimeter" by ρ, another letter of the Greek alphabet.

The following table shows the relationships of the perimeters and areas of squares and rectangles to their dimensions.

← 5 units →

Quadrilateral	Perimeter	Area
Square, side s	$4s$	s^2
Rectangle, base b, altitude h	$2b + 2h$	bh

SET I

Exercises

1. Obtuse Ollie confuses relationships between units of *length* with those between units of *area*.
   a) He knows that 1 yard is equal to 3 feet and thinks that 1 square yard is equal to 3 square feet. Draw a diagram that could be used to explain to him why 1 square yard is not equal to 3 square feet.
   b) Ollie also thinks that 1 square foot is equal to 12 square inches. Explain why this is not true.

2. An ice rink for professional hockey is rectangular in shape with rounded corners. On the assumption that the rink is a simple rectangle 200 feet long and 85 feet wide, find its area in:
   a) square feet.
   b) square inches.
   c) square yards.

Lesson 2: Squares and Rectangles 337

3. If the area of a square is 100 square inches, then its sides must be 10 inches long because $\sqrt{100} = 10$. If the area of a square is 50 square inches, how long are its sides? They are $\sqrt{50}$ inches long, but how long is that? You learned in algebra that if $a \geq 0$ and $b \geq 0$, $\sqrt{ab} = \sqrt{a}\sqrt{b}$. Therefore, $\sqrt{50} = \sqrt{25 \cdot 2} = \sqrt{25}\sqrt{2} = 5\sqrt{2}$. Now $\sqrt{2} \approx 1.41$ ($\approx$ means "is approximately equal to"), so $\sqrt{50} \approx 5(1.41)$, or 7.05. The sides of a square whose area is 50 square inches are approximately 7.05 inches long.
   a) Why does an answer slightly larger than 7 seem correct?
   b) What is the area of a square whose sides are *exactly* 7.05 inches long?

4. Use the method illustrated in the previous exercise and the table at the left to solve the following problems. Find the approximate lengths of the sides of squares having the following areas.
   a) 2 square centimeters.
   b) 20 square centimeters.
   c) 200 square centimeters.
   d) 2000 square centimeters.

5. Sketch two squares, one with sides twice the length of the other.
   a) How do their perimeters compare?
   b) How do their areas compare?

SET II

1. Draw figures to illustrate the following exercises and find their areas.
   a) A square whose perimeter is 20 feet.
   b) A square whose perimeter (in feet) and area (in square feet) are the same number. (Let $x =$ the length of a side of the square.)
   c) A rectangle whose perimeter is 20 feet and whose altitude is 3 feet.
   d) A rectangle whose perimeter is 20 feet and whose altitude is 1 foot.

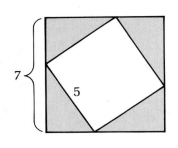

7 {
5

Find the area of the shaded region in each of these figures.

2. Assume that both quadrilaterals are squares.

3. Assume that the "frame" and "holes" are squares and that all the holes are alike.

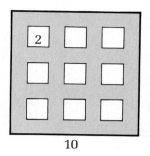

2

10

4. Assume that this figure contains two overlapping squares.

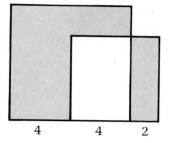

4   4   2

5. According to the *Guinness Book of World Records*, the largest painting ever made was *Panorama of the Mississippi*, completed by John Banvard in 1846. Before being destroyed by fire, it was 5000 feet long and 12 feet high! If a flat frame 1 foot wide were put around a painting this large, what would be the area of the front of the frame?

SET III

This diagram shows the arrangement of the squares on the *Scientific American* cover pictured at the beginning of this lesson. If the area of square C is 64 and the area of square D is 81, can you figure out the areas of the other seven squares? Trace the diagram on your paper so that you can mark it as you find each area.

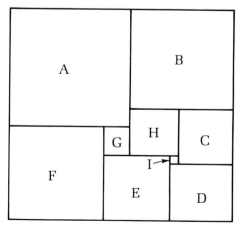

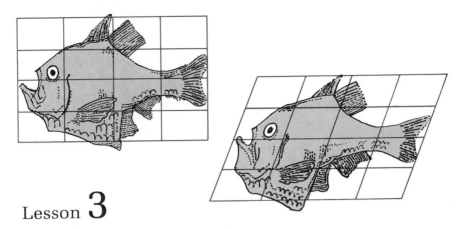

Lesson 3

Parallelograms and Triangles

One of the ancient Greeks, perhaps Plato or Pythagoras, once said that the book of nature is written in the characters of geometry. A remarkable application of geometry to the study of biological shapes was made by the great British scientist D'Arcy Thompson. In his book *On Growth and Form,* he showed many examples of how one species of animal can be considered to be a geometric transformation of another.* For instance, the fish shown on the rectangular grid at the left above is of the species *Argyropelecus olfersi.* If the rectangles are transformed into the parallelograms shown at the right, the shape of the fish becomes that of a species in an entirely different genus, *Sternoptyx diaphana*!

How do you think the pictures of the two fish compare in size? The answer to this question depends upon how the rectangles in the first grid compare in area with the parallelograms in the second. In this lesson we will prove that the formulas for the areas of these two figures are the same.

Any side of a parallelogram may be chosen as its *base.* A line segment joining the lines containing this side and the opposite side that is perpendicular to both of them is called an *altitude* of the parallelogram. (Since the opposite sides of a parallelogram are parallel and we have proved that every perpendicular segment joining one of two parallel lines to the other has the same length, every altitude of a parallelogram corresponding to the same base has the same length.)

Altitude

Base

---

*D'Arcy Thompson, "On the theory of transformations, or the comparison of related forms," *On Growth and Form,* edited by J. T. Bonner. Cambridge University Press © 1961.

340

We have assumed that the area of a rectangle is the product of the lengths of its base and altitude. The figures below suggest why this is also true for a parallelogram.

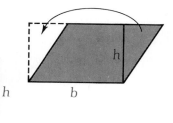

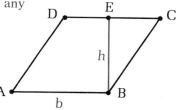

► **Theorem 48**

The area of a parallelogram is the product of the lengths of any base and corresponding altitude.

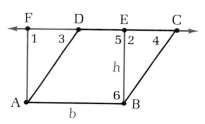

      Hypothesis: $\square ABCD$ with base $\overline{AB}$ of length b and altitude $\overline{EB}$ of length h.
      Conclusion: $\alpha ABCD = bh$.

Proof.

Draw $\overleftrightarrow{DC}$ and, through A, draw $\overline{AF} \perp \overleftrightarrow{DC}$. In $\triangle AFD$ and $\triangle BEC$,

    $\angle 1 = \angle 2$ (they are right angles),
    $\angle 3 = \angle 4$ (they are equal corresponding angles formed by $\overline{AD} \parallel \overline{BC}$ and transversal $\overleftrightarrow{DC}$),

and

    $AD = BC$ (they are equal opposite sides of $\square ABCD$).

Therefore, $\triangle AFD \cong \triangle BEC$ (A.A.S.) and so $\alpha \triangle AFD = \alpha \triangle BEC$.
    By the addition postulate of equality, we know that

$$\alpha \triangle AFD + \alpha ADEB = \alpha \triangle BEC + \alpha ADEB.$$

We also know from the Area Addition Postulate that

$$\alpha ABCD = \alpha \triangle BEC + \alpha ADEB$$

and

$$\alpha ABEF = \alpha \triangle AFD + \alpha ADEB.$$

So $\alpha ABCD = \alpha ABEF$ by substitution.

Quadrilateral ABEF is a rectangle because it is equiangular ($\angle 1$ is a right angle because $\overline{AF} \perp \overleftrightarrow{DC}$, $\angle FAB$ is a right angle because in a plane, a line perpendicular to one of two parallel lines is perpendicular to the other, and $\angle 5$ and $\angle 6$ are right angles because $\overline{EB}$ is an altitude). So $\alpha ABEF = bh$, and hence, $\alpha ABCD = bh$.

This proof is based upon the assumption that point E, the other endpoint of the altitude drawn from point B of $\square ABCD$ to side $\overline{DC}$, lies between D and C. In other parallelograms, this is not necessarily true and we will consider how our proof would need to be modified to cover the other possibilities in one of the exercises.

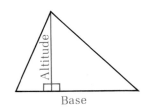

The area of a triangle can also be expressed in terms of the lengths of a base and corresponding altitude. Any side can be chosen as the base. The corresponding *altitude* is the line segment from the opposite vertex that is perpendicular to the line of the base.

To derive a formula for the area of a triangle, we can construct a parallelogram around it and show that the triangle is half its area.

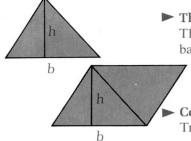

► **Theorem 49**
The area of a triangle is half the product of the lengths of any base and corresponding altitude.

This theorem has two useful corollaries.

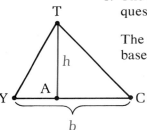

► **Corollary 1**
Triangles with equal bases and equal altitudes have equal areas.

► **Corollary 2**
If two triangles have equal altitudes, then the ratio of their areas is equal to the ratio of the lengths of their bases.

Exercises

SET I

1. The proof of Theorem 49 is very easy. Answer the following questions about it.

   The area of a triangle is half the product of the lengths of any base and corresponding altitude.

   Hypothesis: $\triangle YTC$ with base $\overline{YC}$ of length b and altitude $\overline{TA}$ of length h.

   Conclusion: $\alpha \triangle YTC = \frac{1}{2}bh$.

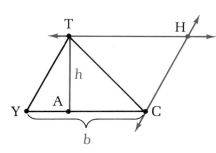

   Proof.
   Through T, draw $\overleftrightarrow{TH} \parallel \overline{YC}$ and through C, draw $\overleftrightarrow{CH} \parallel \overline{YT}$.
   a) What permits us to do this?
   b) What kind of quadrilateral is formed?
   c) What is its area?
   d) What relationship do $\triangle YTC$ and $\triangle HCT$ have?
   e) $\alpha \triangle YTC + \alpha \triangle HCT = \alpha \square YCHT$. Why?

   Since $\alpha \triangle YTC = \alpha \triangle HCT$ and $\alpha \square YCHT = bh$, it follows that $2\alpha \triangle YTC = bh$.
   f) Why is $\alpha \triangle YTC = \frac{1}{2}bh$?

2. The reason why Corollary 1 to Theorem 49 is true is obvious.

   Triangles with equal bases and equal altitudes have equal areas.

   Suppose $\triangle$WHA and $\triangle$LER have bases of length b and altitudes of length h.
   a) $\alpha\triangle$WHA $= \frac{1}{2}bh$ and $\alpha\triangle$LER $= \frac{1}{2}bh$. Why?
   b) Therefore $\alpha\triangle$WHA $= \alpha\triangle$LER. Why?

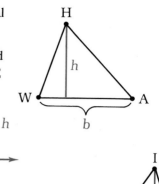

3. Corollary 2 to Theorem 49 is not quite so self-evident.

   If two triangles have equal altitudes, then the ratio of their areas is equal to the ratio of the lengths of their bases.

   Suppose $\triangle$DIN and $\triangle$GHY have bases of lengths a and b and altitudes of length h.
   a) What is the area of each triangle?

   b) Why is $\dfrac{\alpha\triangle\text{DIN}}{\alpha\triangle\text{GHY}} = \dfrac{\frac{1}{2}ah}{\frac{1}{2}bh}$?

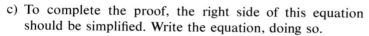

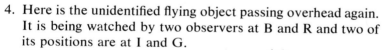

   c) To complete the proof, the right side of this equation should be simplified. Write the equation, doing so.

4. Here is the unidentified flying object passing overhead again. It is being watched by two observers at B and R and two of its positions are at I and G.
   a) If the object flies in a line parallel to $\overleftrightarrow{BR}$, the line of the observers, then its distances to that line are always equal. Why?
   b) Why does the triangle determined by the object and the two observers always have the same area? In other words, why is $\alpha\triangle$BGR $= \alpha\triangle$BIR?

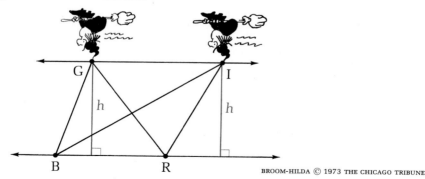

Lesson 3: Parallelograms and Triangles   343

5. Now suppose the flying object is being watched by four observers along $\overleftrightarrow{KC}$.
   a) If KE = 100 ft and TC = 50 ft, how does $\alpha\triangle KHE$ compare with $\alpha\triangle THC$?
   b) Why?

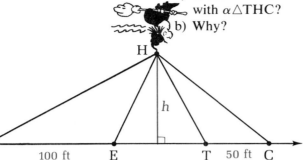

SET II

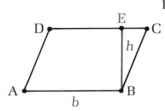

1. In our proof that the area of a parallelogram is the product of the lengths of any base and corresponding altitude, we assumed something from our figure that is not true of all parallelograms. We assumed that point E, the other endpoint of the altitude drawn from point B of $\square$ABCD to side $\overline{DC}$, lies between D and C. As a result, $\overline{BE}$ cuts the parallelogram up into a trapezoid and a triangle.

   The figure below shows that point E might have four other positions with respect to D and C.

   The simplest of these possibilities is the fourth one, for which the complete figure is shown here. In this case, the altitude from B to $\overleftrightarrow{DC}$ is simply the side $\overline{BC}$.
   a) What kind of figure must $\square$ABCD then be?
   b) How does this show that its area is bh?

   The fifth possibility is similar to the one we considered in our proof. The two triangles numbered 1 can be proved congruent so their areas are equal. Also, ABEF is a rectangle, so αABEF = bh.
   c) How can we show that αABCD = αABEF?

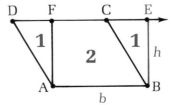

344   Chapter 9: AREA

The third possibility is shown here.

d) The parts in this figure are very much like those of the one we have just considered. What is the one minor difference?

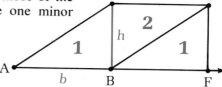

We have saved the most complicated possibility, the second one, until last. In three of the other possibilities, we could, in effect, "cut" the parallelogram into two pieces and rearrange them to form a rectangle. And the simplest possibility didn't even require doing this.

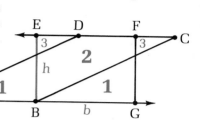

In the possibility under consideration, we have, in effect, to cut the parallelogram into *three* pieces before it can be rearranged into a rectangle. By means of methods you already know, we can prove in this figure that the pair of triangles numbered 1 are congruent, and also the pair of triangles numbered 3. Since BGFE is a rectangle, αBGFE $= bh$.

e) How does it follow that αABCD $= bh$?

2. Explain why the area of a right triangle is equal to half the product of the lengths of its legs.

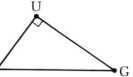

3. In this figure, FRIE is a rectangle and FGTE is a parallelogram. If FR = FG, how do you think FRIE and FGTE compare in area? Explain your answer.

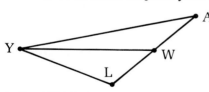

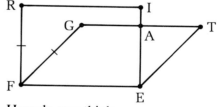

4. In △YAL, W is the midpoint of side $\overline{AL}$. How do you think △YAW and △YWL compare in area? Explain your answer.

SET III

Dilcue sat up all night trying to figure out the following problem. Even his tutor couldn't do it!

In △LUN, A and C are the midpoints of $\overline{LU}$ and $\overline{LN}$, respectively. The segments $\overline{UC}$ and $\overline{AN}$ intersect at H. Can you explain why the shaded regions have equal areas?

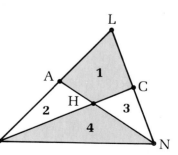

Lesson **4**

Trapezoids

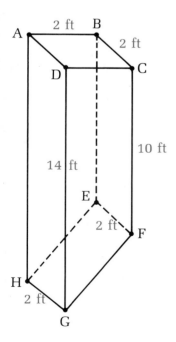

Most swimming pools vary in depth from one end to the other. Suppose Irwin's pool is actually 10 feet deep at one end and 14 feet deep at the other, so that the number he has told Broom-Hilda is its average depth. The adjoining figure represents the pool's shape. The top of the pool is the square ABCD; since its sides are only 2 feet long, the surface area of the water is just 4 square feet.

Irwin probably would have preferred that this area be greater if he'd had more money to spend on the pool. The cost of the materials used is actually determined by the areas of the four vertical walls and the floor of the pool. Two of the walls are rectangular in shape and so their areas are easy to find:

$$\alpha BCFE = 2 \text{ ft} \cdot 10 \text{ ft} = 20 \text{ sq ft.}$$
$$\alpha ADGH = 2 \text{ ft} \cdot 14 \text{ ft} = 28 \text{ sq ft.}$$

The other two walls, ABEH and DCFG, are not rectangles but trapezoids. Since they have the same dimensions, only one of them is shown here; it has been turned so that the parallel sides are horizontal.

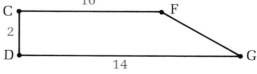

346

What is the area of this wall? Since we have not yet developed a formula for the area of a trapezoid, we might take advantage of the fact that $\overline{CD}$ is perpendicular to both bases ($\overline{CF}$ and $\overline{DG}$) and draw a line through F perpendicular to $\overline{DG}$. This divides the trapezoid into a rectangle and a triangle, whose areas we know how to find.

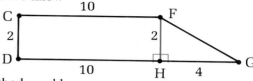

$$\alpha CFGD = \alpha CFHD + \alpha \triangle FGH$$
$$= 10 \cdot 2 + \tfrac{1}{2}(4 \cdot 2) = 20 + 4 = 24 \text{ sq ft.}$$

If $\overline{CD}$ were not perpendicular to the bases, this method would not work. Instead, we might draw a diagonal of the trapezoid to form two triangles. An *altitude* of a trapezoid is a line segment joining the lines containing the bases that is perpendicular to both of them. The altitudes of the two triangles, $\overline{CD}$ and $\overline{FH}$, are also altitudes of the trapezoid.

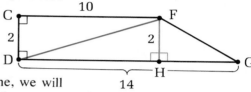

$$\alpha CFGD = \alpha \triangle CFD + \alpha \triangle DFG$$
$$= \tfrac{1}{2}(10 \cdot 2) + \tfrac{1}{2}(14 \cdot 2)$$
$$= 10 + 14 = 24 \text{ sq ft.}$$

Since this method is more general than the first one, we will use it to derive a formula for the area of any trapezoid.

▶ **Theorem 50**

The area of a trapezoid is half the product of the length of its altitude and the sum of the lengths of its bases.

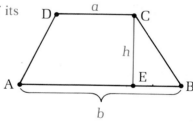

Hypothesis: Trapezoid ABCD with bases $\overline{AB}$ and $\overline{DC}$ of lengths a and b and altitude $\overline{CE}$ of length h.

Conclusion: $\alpha ABCD = \tfrac{1}{2}h(a + b)$.

Proof.

Draw $\overline{AC}$ to divide trapezoid ABCD into two triangles: $\triangle ABC$ and $\triangle ADC$. Altitude $\overline{CE}$ of the trapezoid is also an altitude of $\triangle ABC$, so

$$\alpha \triangle ABC = \tfrac{1}{2}bh.$$

Draw $\overleftrightarrow{DC}$ and, through A, draw $\overline{AF} \perp \overleftrightarrow{DC}$; $\overline{AF}$ is the altitude corresponding to base $\overline{DC}$ of $\triangle ADC$. Since $\overleftrightarrow{FC} \parallel \overleftrightarrow{AB}$, AF = CE = h (if two lines are parallel, every perpendicular segment joining one line to the other has the same length).

$$\alpha \triangle ADC = \tfrac{1}{2}ah.$$

By the Area Addition Postulate,

$$\alpha ABCD = \alpha \triangle ADC + \alpha \triangle ABC$$
$$= \tfrac{1}{2}ah + \tfrac{1}{2}bh = \tfrac{1}{2}h(a + b).$$

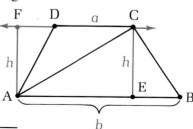

Exercises

· SET I

1. Find the area of each of these figures.

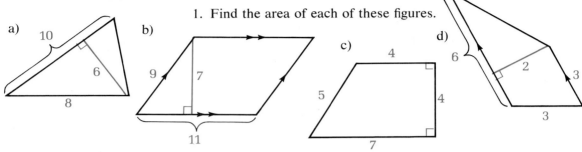

a) 10, 6, 8

b) 9, 7, 11

c) 4, 5, 4, 7

d) 6, 2, 3, 3

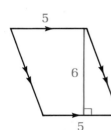

5, 6, 5

2. Obtuse Ollie didn't look very carefully at this parallelogram and carelessly used the trapezoid formula to figure out its area.
   a) What answer did he get?
   b) What is the actual area of the figure?
   c) Can you explain?

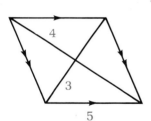

4, 3, 5

3. At first Acute Alice didn't think there were enough numbers on this rhombus to be able to find its area.
   a) What else do you suppose she thought she needed to know?

   After thinking it over, however, she figured out another way to find the area.
   b) Can you explain?

SET II

Find the area of the shaded region in each of these figures.

1. Assume that KANT is a rhombus.

2. Assume that HUME is a rectangle.

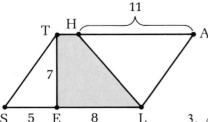

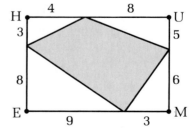

3. Assume that TALS is a parallelogram and that $\overline{TE} \perp \overline{SL}$.

4. In this figure, $\overline{LO}$ is a midsegment of right $\triangle PAT$. The lengths of $\overline{PL}$ and $\overline{LO}$ are x and y, respectively.
   a) Copy the figure and mark on it the lengths of $\overline{LA}$ and $\overline{AT}$.
   b) Write formulas for the areas of $\triangle PLO$ and ATOL.
   c) What is the relationship between $\alpha\triangle PLO$ and $\alpha ATOL$?

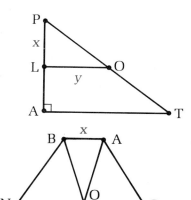

5. In this figure, BACN is a trapezoid in which O is any point on $\overline{NC}$. The lengths of $\overline{BA}$ and $\overline{NC}$ are x and $3x$, respectively.
   a) Copy the figure, draw an altitude of the trapezoid, and mark its length y.
   b) Write formulas for the areas of $\triangle BAO$ and BACN.
   c) What is the relationship between $\alpha\triangle BAO$ and $\alpha BACN$?

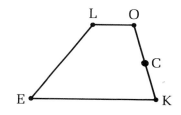

6. Point C is the midpoint of one leg of trapezoid LOKE.
   a) Trace the figure and construct the reflection image of LOKE through point C. Label the images of points L and E points L′ and E′, respectively.
   b) What kind of figure seems to be formed?
   c) Draw an altitude of trapezoid LOKE and label its length h. Also label the lengths of the bases of trapezoid LOKE a and b.
   d) Using the fact that distance is preserved in a point reflection, also label the lengths of $\overline{OE'}$ and $\overline{KL'}$.
   e) Write a formula for the area of LE′L′E in terms of a, b, and h.
   f) On the assumption that LOKE and L′KOE′ have equal areas, what relationship does αLOKE have to αLE′L′E?
   g) Use your answers to parts e and f to write a formula for αLOKE.

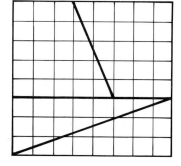

SET III

*A Checkerboard Puzzle.** This square "checkerboard" has been divided into two trapezoids and two right triangles. Copy it on graph paper, cut out the four pieces and try to rearrange them to form a rectangle having a different shape. If you succeed, tape the pieces to your paper.

How does the area of the rectangle you have formed compare with the area of the original square? What basic property of area does this seem to contradict?

*An entire chapter of the book *Mathematics, Magic and Mystery* by Martin Gardner (Dover, 1956) is devoted to puzzles of this type.

Lesson 5

The Pythagorean Theorem

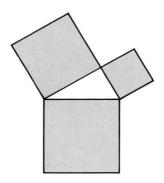

At the beginning of this century, many scientists believed that intelligent creatures might live on Mars. Among them was the American astronomer Percival Lowell, whose work led to the discovery of Pluto. Mr. Lowell thought that he could see canals on the surface of Mars through his telescope and speculated that they might have been dug by a Martian civilization to irrigate their dry land with water melted from polar ice caps.

To let the Martians know that there was also intelligent life on the earth (in a time long before any kind of space travel was possible), it was proposed that gigantic geometric figures be used to convey a message. For instance, broad lanes of trees might be planted in Siberia to form a huge right triangle. Or canals might be dug in the Sahara desert to do the same thing; kerosene could be poured on the water in them and set on fire at night for the Martians to see through their telescopes! A geometric figure felt to be especially appropriate for this is shown here. It illustrates what is perhaps the most famous theorem in all of geometry—the Pythagorean Theorem.

The theorem says that if squares are constructed on the three sides of a right triangle, the area of the square on the hypotenuse is equal to the sum of the areas of the squares on the two legs. This is true for *every* right triangle, regardless of its shape. If a and b are the lengths of the legs and c is the length of the hypotenuse, then

$$c^2 = a^2 + b^2.$$

The following statement of this theorem emphasizes the relationship of the lengths of the triangle's sides rather than the areas of the squares on them.

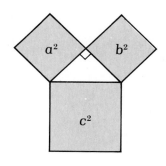

▶ **Theorem 51** (The Pythagorean Theorem)
In a right triangle, the square of the hypotenuse is equal to the sum of the squares of the legs.

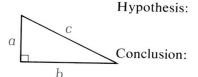

Hypothesis: A right triangle with hypotenuse c and legs a and b.
Conclusion: $c^2 = a^2 + b^2$.

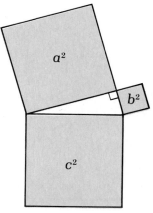

Many different proofs have been developed for the Pythagorean Theorem—more, in fact, than for any other theorem of geometry. One proof is based upon the two figures shown below. Each is a square with sides of length $a + b$. The first has been subdivided into four right triangles congruent to the original triangle and *a square whose sides are equal to its hypotenuse.* The second figure also contains four right triangles congruent to the original triangle. The theorem follows from the fact that the rest of the figure consists of *two squares whose sides are equal to the legs of the triangle.*

Expressing this algebraically in terms of the total areas of the two figures, we have:

$$(a + b)^2 = 4(\tfrac{1}{2}ab) + c^2$$

and

$$(a + b)^2 = 4(\tfrac{1}{2}ab) + a^2 + b^2.$$

Hence,

$$4(\tfrac{1}{2}ab) + c^2 = 4(\tfrac{1}{2}ab) + a^2 + b^2$$

and

$$c^2 = a^2 + b^2.$$

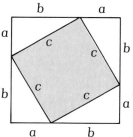

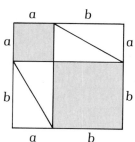

Lesson 5: The Pythagorean Theorem   351

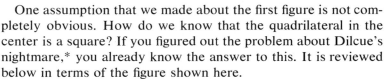

One assumption that we made about the first figure is not completely obvious. How do we know that the quadrilateral in the center is a square? If you figured out the problem about Dilcue's nightmare,* you already know the answer to this. It is reviewed below in terms of the figure shown here.

The four right triangles are congruent by S.A.S., so EF = FG = GH = HE. Hence, EFGH is equilateral.

Angles 3 and 4 are complementary because they are the acute angles of right $\triangle$EBF, so $\angle 3 + \angle 4 = 90°$. Since they are corresponding angles of congruent triangles, $\angle 1 = \angle 4$. Therefore, $\angle 3 + \angle 1 = 90°$ by substitution. Since $\angle 1 + \angle 2 + \angle 3 = 180°$, $\angle 2 = 90°$ by subtraction. In the same way, it can be shown that the measures of the other three angles of EFGH are 90°. Since EFGH is equiangular as well as equilateral, it is a square.

In the figure shown here, squares have been drawn on the sides of $\triangle$ABC and the squares into which each has been subdivided are all the same size. Is $\triangle$ABC a right triangle?

It is tempting to say yes, since

$$3^2 + 4^2 = 5^2.$$

However, to conclude that $\triangle$ABC must be a right triangle because the square of one of its sides is equal to the sum of the squares of the other two sides is to assume that the *converse* of the Pythagorean Theorem is true. Conveniently, it is.

▶ Theorem 52
If the square of one side of a triangle is equal to the sum of the squares of the other two sides, then the triangle is a right triangle.

Exercises

SET I

1. Use the Pythagorean Theorem and the adjoining table to find the length of the hypotenuses of right triangles having legs of the following lengths.
   a) 5 and 12.
   b) 6 and 8.
   c) 8 and 15.

Number	Square
13	169
14	196
15	225
16	256
17	289
18	324
19	361
20	400

2. Make scale drawings on graph paper of the three right triangles of Exercise 1 and measure their hypotenuses to check your answers.

*See pages 313–314.

Before doing the following exercises, study this example.

Find x.

Solution.
$$x^2 + 11^2 = 14^2$$
$$x^2 + 121 = 196$$
$$x^2 = 75$$
$$x = \sqrt{75} = \sqrt{3 \cdot 25} = \sqrt{3}\sqrt{25} = 5\sqrt{3}.$$

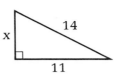

3. Find the length of the side labeled x in each of the right triangles shown here. Express all irrational answers in simple radical form.

a)

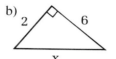

b)

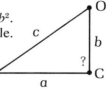

c)

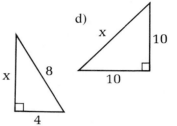

d)

4. To prove the converse of the Pythagorean Theorem, we can construct a right triangle and then prove that it is congruent to the given triangle.

If the square of one side of a triangle is equal to the sum of the squares of the other two sides, then the triangle is a right triangle.

Hypothesis: $\triangle ROC$ with $c^2 = a^2 + b^2$.
Conclusion: $\triangle ROC$ is a right triangle.

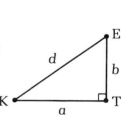

Proof.
Construct a right $\triangle KET$ with legs a and b. (First, draw $\overline{KT}$ with length a, then draw a line perpendicular to $\overline{KT}$ at T to form a right angle, then draw $\overline{TE}$ with length b, then draw $\overline{KE}$.) Let the hypotenuse of $\triangle KET$ have length d.
a) Why is $d^2 = a^2 + b^2$?
b) Why is $c^2 = a^2 + b^2$?
c) How does it follow that $d^2 = c^2$?
d) Why is $d = c$?
e) Why is $\triangle ROC \cong \triangle KET$?
f) What can we conclude about $\angle C$, and hence, $\triangle ROC$, as a result?

5. Obtuse Ollie thinks the converse of the Pythagorean Theorem would be clearer if it were worded this way:

If the square of the hypotenuse of a triangle is equal to the sum of the squares of the two legs, then the triangle is a right triangle.

a) Acute Alice thinks this way of saying it is illogical. Why?

b) Obtuse Ollie says: "Okay. Suppose you know the square of one side of a triangle is equal to the sum of the squares of the other two sides. How can you tell which side is the hypotenuse since you don't know where the right angle is?" How would you answer him?

6. Which of the following sets of numbers could be the lengths of the sides of a right triangle?
   a) 8, 9, 12.
   b) $1\frac{1}{5}, 1\frac{3}{5}, 2$.
   c) $\sqrt{2}, \sqrt{3}, \sqrt{5}$.

SET II

In the following exercises, several famous proofs of the Pythagorean Theorem are presented. The approach in each case is very informal, with many of the details omitted.

1. President Garfield invented an original proof for the Pythagorean Theorem in 1876 when he was a member of the House of Representatives. It is based upon the figure shown here.

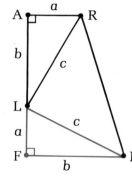

   First, right △LFE is copied as shown so that △RAL ≅ △LFE and $\overline{RE}$ is drawn to form △RLE. By an argument similar to that in the proof of the Pythagorean Theorem considered in this lesson, we can show that ∡RLE is a right angle so that △RLE is a right triangle.
   a) Find the areas of the three triangles in terms of a, b, and c.
   b) What kind of quadrilateral does AREF seem to be?
   c) Find its area in terms of a and b.
   d) Use the Area Addition Postulate to write an equation relating the areas of the three triangles to the area of AREF.
   e) To finish the proof, simplify this equation.

2. Bhaskara, a Hindu mathematician of the twelfth century, created a proof of the Pythagorean Theorem based upon this figure. Three copies of the right triangle have been added to it to form two squares.

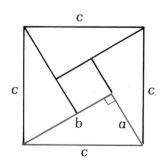

   a) The length of a side of the large square is c. What is the length of a side of the small square?
   b) What is the area of the small square?
   c) What is the total area of the four triangles?
   d) Use the Area Addition Postulate to write an equation relating the areas of the small square and four triangles to the area of the large square.
   e) Finish the proof by simplifying this equation.

354   Chapter 9: AREA

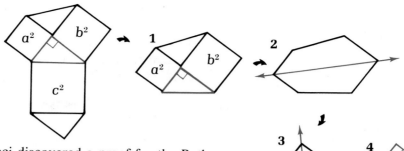

3. Leonardo da Vinci discovered a proof for the Pythagorean Theorem based upon the figure at the left above. In it, squares have been drawn on the three sides of the right triangle and some line segments added to make two copies of the triangle. To follow the proof, look at the sequence of figures numbered 1 through 4, based upon parts of da Vinci's figure.

a) What kind of symmetry does Figure 1 seem to have?

b) Part of Figure 1 has been redrawn as Figure 2. What relationship do the two quadrilaterals in Figure 2 seem to have?

c) If the two quadrilaterals in Figure 2 were cut out, they could be put together to form Figure 3. How?

d) What kind of symmetry does Figure 3, and therefore Figure 4, seem to have?

Since Figures 2 and 3 consist of the same parts, Figures 1 and 4 have the same area.

e) What parts does Figure 1 consist of?

f) What parts does Figure 4 consist of?

g) How does this prove the theorem?

SET III

Any pair of squares, no matter what their relative size, can be cut into five pieces that will fit together to form a third square. The cuts are illustrated in the figure shown in the margin.

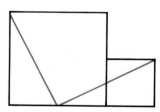

Trace the two squares below. Can you figure out where to make the cuts so that a third square can be formed from the pieces?

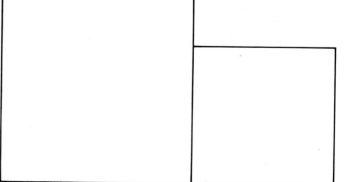

Lesson **6**

Heron's Theorem

The science of crystallography deals with the structure of crystals. Photographs of crystals taken through a microscope reveal many geometric patterns. The one shown here is of diamond crystals and includes a large number of triangles.

The triangles look as if they are equilateral and since they have different sizes they have different areas. The length of one side of an equilateral triangle determines its area just as does one side of a square. We know this because two triangles are congruent if the sides of one are equal to the sides of the other (S.S.S.) and thus must have equal areas.

It also follows from this that the area of *any* triangle is determined by the lengths of its sides. A formula based upon this fact was derived by Heron of Alexandria, a mathematician of the first century A.D. Archimedes is thought to have actually discovered the formula, but it is Heron's proof that has survived.

The formula includes a number called the *semiperimeter* of the triangle.

▶ Definition

The *semiperimeter* of a triangle is the number that is half its perimeter.

Hence, if the sides of a triangle have lengths a, b, and c, and s is the semiperimeter of the triangle,

$$s = \frac{a+b+c}{2}.$$

The proof of Heron's theorem is quite complicated and requires some rather difficult algebra.

▶ **Theorem 53**   (Heron's Theorem)
The area of a triangle with sides of lengths a, b, and c and semiperimeter s is $\sqrt{s(s-a)(s-b)(s-c)}$.

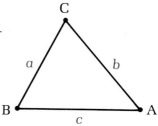

      Hypothesis: $\triangle ABC$ with sides
                   of lengths a, b, and c
      Conclusion: $\alpha\triangle ABC = \sqrt{s(s-a)(s-b)(s-c)}$.

Proof.
Draw $\overline{CD} \perp \overline{BA}$, forming right triangles CBD and CAD.* Let $CD = h$ and $BD = x$, so that $DA = c - x$.

$$\alpha\triangle ABC = \tfrac{1}{2}ch.$$

To express h in terms of a, b, and c, we apply the Pythagorean Theorem:

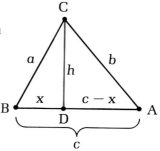

$$h^2 + x^2 = a^2$$
$$h^2 = a^2 - x^2$$

and

$$h^2 + (c-x)^2 = b^2$$
$$h^2 = b^2 - (c-x)^2$$
$$= b^2 - (c^2 - 2cx + x^2)$$
$$= b^2 - c^2 + 2cx - x^2.$$

Hence,

$$a^2 - x^2 = b^2 - c^2 + 2cx - x^2 \quad \text{(Substitution)}$$
$$a^2 = b^2 - c^2 + 2cx \quad\quad\quad \text{(Addition)}$$
$$a^2 - b^2 + c^2 = 2cx \quad\quad\quad \text{(Subtraction)}$$
$$x = \frac{a^2 - b^2 + c^2}{2c} \quad\quad\quad \text{(Division)}$$

Since $h^2 = a^2 - x^2$,

$$h^2 = (a - x)(a + x)$$
$$= \left(a - \frac{a^2 - b^2 + c^2}{2c}\right)\left(a + \frac{a^2 - b^2 + c^2}{2c}\right)$$
$$= \left(\frac{2ac - (a^2 - b^2 + c^2)}{2c}\right)\left(\frac{2ac + (a^2 - b^2 + c^2)}{2c}\right).$$

   *Our proof assumes that $\overline{CD}$ lies inside the triangle; comparable proofs can be written for the cases in which $\overline{CD}$ lies outside the triangle or coincides with one of its sides.

$$4c^2h^2 = (2ac - a^2 + b^2 - c^2)(2ac + a^2 - b^2 + c^2)$$
$$= [b^2 - (a^2 - 2ac + c^2)][(a^2 + 2ac + c^2) - b^2]$$
$$= [b^2 - (a-c)^2][(a+c)^2 - b^2]$$
$$= [b - (a-c)][b + (a-c)][(a+c) - b][(a+c) + b]$$
$$= (b + c - a)(a + b - c)(a + c - b)(a + b + c).\dagger$$

Since s is the semiperimeter, $s = \dfrac{a+b+c}{2}$.

$$a + b + c = 2s.$$
$$(a + b + c) - 2a = 2s - 2a$$
$$b + c - a = 2(s - a)$$
$$(a + b + c) - 2b = 2s - 2b$$
$$a + c - b = 2(s - b)$$
$$(a + b + c) - 2c = 2s - 2c$$
$$a + b - c = 2(s - c).$$

By substitution of these results into the equation marked †,

$$4c^2h^2 = [2(s-a)][2(s-c)][2(s-b)][2s]$$
$$= 16s(s-a)(s-b)(s-c)$$
$$c^2h^2 = 4s(s-a)(s-b)(s-c).$$

Taking square roots,
$$ch = 2\sqrt{s(s-a)(s-b)(s-c)}.$$

But $\alpha\triangle ABC = \frac{1}{2}ch$, so
$$\alpha\triangle ABC = \sqrt{s(s-a)(s-b)(s-c)}.$$

By means of Heron's Theorem, we can derive a formula for the area of an equilateral triangle.

▶ **Corollary**
The area of an equilateral triangle with sides of length a is $\dfrac{a^2}{4}\sqrt{3}$.

Proof.
The perimeter of the triangle is $3a$, so the semiperimeter is $\dfrac{3a}{2}$.

$$\alpha_{\text{eq. triangle}} = \sqrt{s(s-a)(s-b)(s-c)}$$

$$\alpha_{\text{eq. triangle}} = \sqrt{\frac{3a}{2}\left(\frac{3a}{2} - a\right)\left(\frac{3a}{2} - a\right)\left(\frac{3a}{2} - a\right)}$$

$$\alpha_{\text{eq. triangle}} = \sqrt{\frac{3a}{2} \cdot \frac{a}{2} \cdot \frac{a}{2} \cdot \frac{a}{2}} = \sqrt{\frac{3a^4}{16}} = \frac{a^2}{4}\sqrt{3}$$

1. Find the area of a 3-4-5 right triangle
   a) by using the formula $\alpha_{\text{triangle}} = \frac{1}{2}bh$.
   b) by using Heron's Theorem.

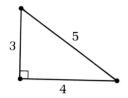

Exercises

2. Use Heron's Theorem to find the areas of the triangles whose sides are the following lengths. Express each answer in simple radical form.
   a) 8, 15, 17.
   b) 7, 20, 23.
   c) 3, 50, 51.

3. Suppose a triangle has sides of lengths 4, 6, and 10.
   a) Try to use Heron's Theorem to find its area.
   b) The result is quite peculiar. Can you explain why it turns out as it does?

4. Find the area of an equilateral triangle whose sides have each of the following lengths. Express each answer in simple radical form.
   a) 2.
   b) 10.
   c) $\sqrt{20}$.

SET II

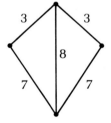

1. Find the area of this kite.

2. Find the area of this quadrilateral.

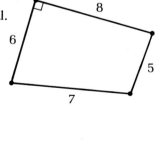

3. Find h.

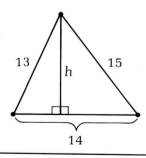

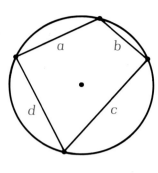

A Hindu mathematician of the seventh century, named Brahmagupta, discovered a formula for the area of a quadrilateral whose vertices lie on a circle. It is

$$\alpha_{\text{circ. quad.}} = \sqrt{(s-a)(s-b)(s-c)(s-d)},$$

where s is the semiperimeter of the quadrilateral, defined in the same way as it is for a triangle.

1. If the vertices of the quadrilateral move on the circle so that d shrinks to 0, what does the quadrilateral become?

2. If $d = 0$, what does Brahmagupta's formula become?

3. Use Brahmagupta's formula to find the areas of the quadrilaterals below.

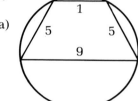

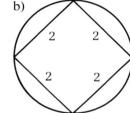

4. No one has ever discovered a formula for the area of a general quadrilateral in terms of the lengths of its sides and no one ever will. Can you explain why?

Chapter 9 / Summary and Review

Basic Ideas

Postulates

16. *The Area Postulate.*   To every polygonal region there corresponds a unique positive number.   331
17. *The Area Addition Postulate.*   If a polygonal region is the union of two or more polygonal regions with no interior points in common, then its area is the sum of their areas.   332
18. Triangular regions bounded by congruent triangles have equal areas.   332
19. The area of a rectangle is the product of the lengths of its base and altitude.   336

Theorems

Corollary to Postulate 19. The area of a square is the square of the length of its side. 337

48. The area of a parallelogram is the product of the lengths of any base and corresponding altitude. 341

49. The area of a triangle is half the product of the lengths of any base and corresponding altitude. 342

Corollary 1. Triangles with equal bases and equal altitudes have equal areas. 342

Corollary 2. If two triangles have equal altitudes, then the ratio of their areas is equal to the ratio of the lengths of their bases. 342

50. The area of a trapezoid is half the product of the length of its altitude and the sum of the lengths of its bases. 347

51. *The Pythagorean Theorem.* In a right triangle, the square of the hypotenuse is equal to the sum of the squares of the legs. 351

52. If the square of one side of a triangle is equal to the sum of the squares of the other two sides, then the triangle is a right triangle. 352

53. *Heron's Theorem.* The area of a triangle with sides of lengths a, b, and c and semiperimeter s is $\sqrt{s(s-a)(s-b)(s-c)}$. 357

Corollary. The area of an equilateral triangle with sides of length a is $\dfrac{a^2}{4}\sqrt{3}$. 358

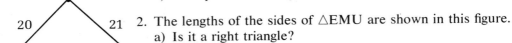

Exercises

SET I

1. Find the areas of the following figures.
   a) A square whose perimeter is 100.
   b) A rectangle whose base is 5 and whose perimeter is 16.
   c) An equilateral triangle if one of its midsegments is 4.

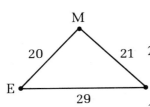

2. The lengths of the sides of △EMU are shown in this figure.
   a) Is it a right triangle?
   b) What relationships do ∡E and ∡U have?

3. Triangle AUK is a right triangle whose hypotenuse is 41 and whose shorter leg is 9.
   a) Find the length of its longer leg.
   b) Find the area of the triangle.

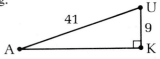

4. Acute Alice's mother wants new carpeting for her living room. The shape and dimensions of the room are shown here and the kind of carpeting she wants costs $10 per square yard.

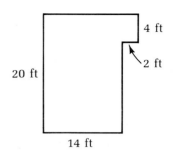

20 ft

4 ft

2 ft

14 ft

Obtuse Ollie figures that the area of the room is 288 square feet. Since 1 yard = 3 feet, he divides by 3 to get 96 square yards and since the carpet costs $10 per square yard, he divides by 10 to get 9.6. He tells Alice's mother the carpeting will cost $9.60.

a) What's wrong with Ollie's calculations?
b) What is the actual cost of the carpeting?

SET II

Find the area of the shaded region in figures 1–6.

1. STAR and LING are rectangles.

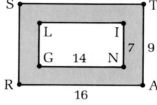

2. Squares have been drawn on the three sides of right △HLI.

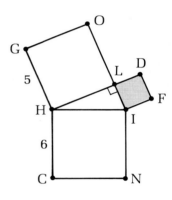

3. SWIF is a square and △FIT is equilateral.

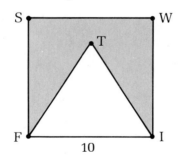

4. △LFO is a right triangle; A, C, and N are the midpoints of its sides.

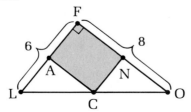

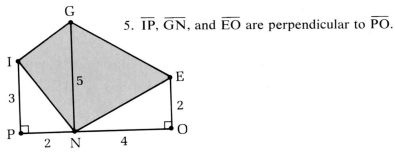

5. $\overline{IP}$, $\overline{GN}$, and $\overline{EO}$ are perpendicular to $\overline{PO}$.

6. $\overline{NV} \perp \overline{AE}$.

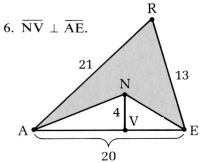

7. Point N is the midpoint of side $\overline{EA}$ of $\square$CRAE. Explain why $\alpha\triangle$CEN $= \alpha\triangle$RNA.

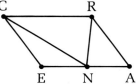

8. Suppose that HERO is a kite with HE $=$ HO and RE $=$ RO. Explain why αHERO $= \frac{1}{2}$HR $\cdot$ EO.

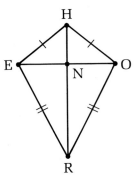

9. The most famous proof of the Pythagorean Theorem is the one that originally appeared in Euclid's *Elements*. A reproduction of a manuscript of an Arabic translation of this proof made more than one thousand years ago is shown on the next page.

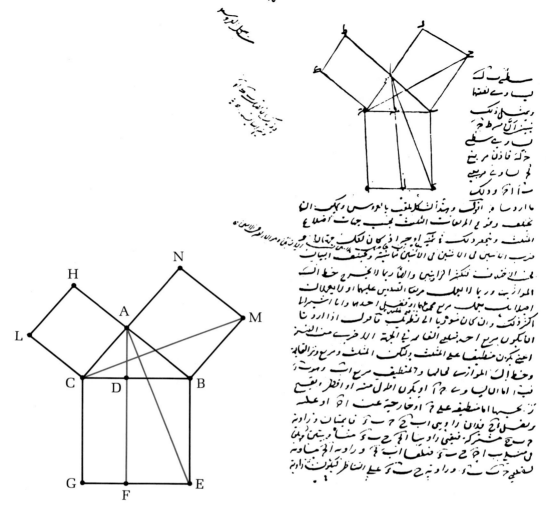

The figure upon which it is based contains a right triangle, △ABC, with squares upon its sides; to it have been added three line segments: $\overline{AF}$ (drawn parallel to $\overline{BE}$), $\overline{MC}$, and $\overline{AE}$.

Euclid's proof is quite complex, so we will consider only its main idea. As a test of your reasoning, you may want to try to justify some of the details. It can be proved that $\alpha MBAN = 2\alpha\triangle MBC$, $\alpha DBEF = 2\alpha\triangle ABE$, and $\triangle MBC \cong \triangle ABE$.

a) What can you conclude about $\alpha MBAN$ and $\alpha DBEF$? Explain.

By forming some additional triangles, it can also be proved that $\alpha HACL = \alpha DCGF$.

b) How does the Pythagorean Theorem follow from this fact and your conclusion to part a?

A challenging puzzle consisting of twenty identical triangular pieces is shown in the first photograph below. The object is to arrange them to form a square. The second photograph shows sixteen of the pieces put together in this fashion. It is not a solution to the puzzle, however, because four pieces have not been used.

A pattern for one of the triangular pieces is shown here. Cut out twenty copies of it and see if you can solve the puzzle. You can get some clues on how to put the pieces together by using the Pythagorean Theorem and a few facts about area.

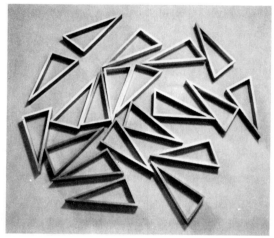

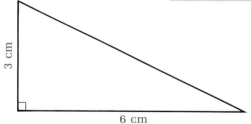

Chapter 10

SIMILARITY

Lesson **1**
Ratio and Proportion

The two cars parked on the street in this photograph look very much alike. The band of the crosswalk running underneath them, however, reveals that something is wrong. Since the band's width is so much narrower below the dark car than below the light one, the two cars must be parked quite a distance apart. Why doesn't the one closer to us, then, look much larger than the one parked across the street?

In reality, the closer car is only a small model. The camera was carefully placed with respect to the two cars so that they give the illusion of being the same size. Without the clues present in the photograph, it would be difficult to tell the actual car and the model apart.

Suppose the actual car is 6 feet 8 inches wide and 18 feet 4 inches long. To look like the actual car, how long must the model be if it is 4 inches wide? Since 6 ft 8 in. = 80 in., the actual car is 20 times wider than the model (80 in. = 20 · 4 in.). If the two have the same shape, it must also be 20 times longer. Since 18 ft 4 in. = 220 in., $20x = 220$, and $x = 11$. The model is 11 inches long.

Another way to solve this problem is to write a proportion:

$$\frac{80}{4} = \frac{220}{x}.$$

This proportion is based upon the fact that since the car and the model have the *same shape*, their dimensions have the *same ratio*. In geometry, figures that have the same shape are said to be *similar* to each other.

Before studying the properties of similar figures, we will review the meaning of *ratio* and *proportion*.

▶ Definitions

The **ratio** of the numbers a to b is the number $\frac{a}{b}$. (Note that b cannot be 0 since division by 0 is undefined.)

A **proportion** is an equality between two ratios.

We can represent a proportion symbolically as

$$\frac{a}{b} = \frac{c}{d}.$$

The numbers a, b, c, and d are called the *first, second, third,* and *fourth terms* of the proportion, respectively. The second and third terms, b and c, are also called the *means*, and the first and fourth terms, a and d, are called the *extremes* of the proportion.

Proportions have many interesting properties. Consider, for example, the proportion involving the dimensions of the cars in this lesson:

$$\frac{80}{4} = \frac{220}{11}.$$

If we multiply the extremes of this proportion, $80 \cdot 11$, and the means, $4 \cdot 220$, we get the same number: 880. If we "turn both ratios of the proportion upside-down," the resulting proportion is still correct:

$$\left(\frac{80}{4} = \frac{220}{11}\right) \quad \leadsto \quad \frac{4}{80} = \frac{11}{220}.$$

We can also interchange the two means of the proportion,

$$\frac{80}{4} = \frac{220}{11} \quad \leadsto \quad \frac{80}{220} = \frac{4}{11},$$

or add the denominators to the numerators,

$$\frac{80 +}{4} = \frac{220 +}{11} \quad \leadsto \quad \frac{80 + 4}{4} = \frac{220 + 11}{11} \quad \text{or} \quad \frac{84}{4} = \frac{231}{11}.$$

These results are correct, not just for the proportion in this example, but for *every* proportion. They are included in the following theorems, stated in both words and symbols. The letters represent real numbers and, since division by zero is undefined, it is understood that no denominators can be zero. Each theorem is given a name, which you may use in referring to it.

▶ **Proportion Theorem 1** (The Means-Extremes Theorem)

In a proportion, the product of the means is equal to the product of the extremes.

$$\text{If } \frac{a}{b} = \frac{c}{d}, \text{ then } ad = bc.$$

▶ **Proportion Theorem 2** (The Upsidedownable Theorem)

In a proportion, the ratios may be inverted.

$$\text{If } \frac{a}{b} = \frac{c}{d}, \text{ then } \frac{b}{a} = \frac{d}{c}.$$

▶ **Proportion Theorem 3** (The Interchangeable Theorem)

In a proportion, the means (or extremes) may be interchanged.

Two possibilities:

$$\text{If } \frac{a}{b} = \frac{c}{d}, \text{ then } \frac{a}{c} = \frac{b}{d} \quad and \quad \text{If } \frac{a}{b} = \frac{c}{d}, \text{ then } \frac{d}{b} = \frac{c}{a}.$$

▶ **Proportion Theorem 4** (The Denominator Addition Theorem)

$$\text{If } \frac{a}{b} = \frac{c}{d}, \text{ then } \frac{a+b}{b} = \frac{c+d}{d}.$$

▶ **Proportion Theorem 5** (The Denominator Subtraction Theorem)

$$\text{If } \frac{a}{b} = \frac{c}{d}, \text{ then } \frac{a-b}{b} = \frac{c-d}{d}.$$

Exercises

1. The following questions refer to this proportion:

$$\frac{7}{11} = \frac{21}{33}.$$

   a) What are the two ratios in this proportion?
   b) Which is the third term of the proportion?
   c) Name the means of the proportion.
   d) Show that the product of the means in this proportion is equal to the product of the extremes.

2. Solve for x in each of the following proportions:

   a) $\dfrac{x}{3} = \dfrac{9}{10}.$

   b) $\dfrac{6}{17} = \dfrac{2x}{51}.$

   c) $\dfrac{x}{x+2} = \dfrac{3}{4}.$

   d) $\dfrac{1}{x} = \dfrac{3}{2x-1}.$

3. Two ratios that are famous in the history of the number π are $\dfrac{22}{7}$ and $\dfrac{355}{113}$.

a) Express $\dfrac{22}{7}$ as a decimal to the nearest hundredth.

b) Express $\dfrac{355}{113}$ as a decimal to the nearest hundredth.

c) Explain, using indirect reasoning and the means-extremes theorem, why, contrary to your results in parts a and b,

$$\frac{22}{7} \neq \frac{355}{113}.$$

© 1968 WALT KELLY. COURTESY OF PUBLISHERS-HALL SYNDICATE

4. Suppose Uncle Albert's arm span is 4 feet and that Alabaster's arm span is 16 inches.

a) Does it seem correct to say that the ratio of their respective arm spans is $\dfrac{4}{16}$ or $\dfrac{1}{4}$?

b) Change Uncle Albert's measurement to inches and find the ratio of their arm spans.

c) Change Alabaster's measurement to feet and find the ratio of their arm spans.

d) Does the ratio of two lengths depend upon the unit of measure if both lengths are in terms of the same unit?

1. What is the missing reason in the following proof of Theorem 1?

    In a proportion, the product of the means is equal to the product of the extremes.

    Hypothesis: $\dfrac{a}{b} = \dfrac{c}{d}$.

    Conclusion: $ad = bc$.

    Proof.

Statements	Reasons
a) $\dfrac{a}{b} = \dfrac{c}{d}$.	Hypothesis.
b) $bd \cdot \dfrac{a}{b} = bd \cdot \dfrac{c}{d}$.	Why?
c) $ad = bc$.	Substitution.

2. Prove Theorem 2. (Hint: Use Theorem 1 and division.)

3. Prove the following possibility of Theorem 3:

    If $\dfrac{a}{b} = \dfrac{c}{d}$, then $\dfrac{a}{c} = \dfrac{b}{d}$.

4. Prove Theorem 4. (Hint: Add 1 to each side of the original proportion.)

5. Prove Theorem 5.

SET III

"As Gold transcends all other Mettals, so doth this Rule all others in Arithmetick. . . . Multiplie the last number by the seconde, and diuide the Product by the first number."*

Although you probably know the "Golden Rule" as "do unto others as you would have them do unto you," the rule stated above was once known by this name instead. It was used by merchants to solve problems such as the following one:

If 100 somethings cost 40 cents, how much money should 15 cost?

The numbers in the problem are 100, 40, and 15. According to the rule, $15 \cdot 40 = 600$ and $600 \div 100 = 6$, so that the answer is 6 cents. Can you explain, using a proportion, why this method works?

---

*This rule appeared in English arithmetic books published in the fifteenth and sixteenth centuries. However, it was known to the Hindus as early as the seventh century.

Lesson 2

More on Proportion

The store in this cartoon evidently lives up to its advertising. We hope that the fellow got what he expected.

Suppose he took in a photograph whose dimensions were 3 inches by 4 inches to be "blown up to poster size." If the store's equipment is limited to making enlargements having dimensions three times those of the original, the picture could be enlarged to 9 inches by 12 inches. Since this hardly qualifies as "poster size," an enlargement of it could be made in turn, blowing the photograph up to 27 inches by 36 inches.

The following proportions indicate the ratios of the corresponding dimensions in going from the original photograph to the first enlargement and from the first enlargement to the second.

$$\frac{3}{9} = \frac{9}{27} \quad \text{and} \quad \frac{4}{12} = \frac{12}{36}.$$

Notice that the means in the first proportion are the same number: 9. As a result, 9 is called the *mean proportional* between 3 and 27. In the second proportion, 12 is the mean proportional between 4 and 36.

▶ Definition

The number b is the *mean proportional* between the numbers a and c iff a, b, and c are positive and $\dfrac{a}{b} = \dfrac{b}{c}$.

On the basis of this definition, it is easy to prove the following theorem.

▶ **Proportion Theorem 6**   (The Mean Proportional Theorem)

If b is the mean proportional between a and c, then $b = \sqrt{ac}$.

In the examples on the preceding page,

$$9 = \sqrt{3 \cdot 27} \quad \text{and} \quad 12 = \sqrt{4 \cdot 36}.$$

One more useful theorem on proportion concerns a series of equal ratios. To illustrate it, we will consider the ratio of each photograph's width to its length. Because all three photographs have the same shape, these ratios are equal.

$$\frac{3}{4} = \frac{9}{12} = \frac{27}{36}.$$

Now if we add their numerators,

$$3 + 9 + 27 = 39,$$

and their denominators,

$$4 + 12 + 36 = 52,$$

and reduce the ratio of the sums,

$$\frac{39}{52} = \frac{3}{4},$$

we find that it is equal to the original ratio! This rather surprising result is true not just for this example, but for any series of equal ratios.

▶ **Proportion Theorem 7**   (The Equal Ratios Theorem)

$$\text{If} \quad \frac{a}{b} = \frac{c}{d} = \frac{e}{f} = k, \quad \text{then} \quad \frac{a + c + e}{b + d + f} = k.$$

1. Complete the following proof of the mean proportional theorem.

If b is the mean proportional between a and c, then $b = \sqrt{ac}$.

Proof.

Statements

a) b is the mean proportional between a and c.

b) a, b, and c are positive and $\dfrac{a}{b} = \dfrac{b}{c}$.

c) $b^2 = ac$.
d) $b = \sqrt{ac}$.

2. Use the mean proportional theorem to find the mean proportional between each of the following pairs of numbers:
   a) 4 and 25. b) $3\sqrt{2}$ and $6\sqrt{2}$. c) 7 and 21.

3. Consider the following proportion: $\dfrac{6}{4} = \dfrac{9}{6}$.

   a) Why does it seem reasonable to speak of an "extreme proportional" in this proportion?
   b) The idea of an "extreme proportional" is not very useful since if $\dfrac{a}{b} = \dfrac{c}{a}$, then $\dfrac{c}{a} = \dfrac{a}{b}$. Which postulate of equality is the reason for this?
   c) Complete the following statement: If a number is the "extreme proportional" between two other numbers, then it is also ▓▓▓▓▓▓.

4. Solve for x and y: a) $\dfrac{x}{4} = \dfrac{1}{y} = 3$. b) $\dfrac{x}{5} = \dfrac{20}{x} = y$.

1. In each of the following exercises, name the theorem on proportion that tells how the second equation follows from the first.

   a) If $\dfrac{a}{b} = \dfrac{8}{5}$, then $5a = 8b$.

   b) If $\dfrac{c}{d} = \dfrac{3}{2}$, then $\dfrac{d}{c} = \dfrac{2}{3}$.

   c) If $\dfrac{e}{7} = \dfrac{4}{9}$, then $\dfrac{9}{7} = \dfrac{4}{e}$.

   d) If $\dfrac{f}{3} = \dfrac{g}{11}$, then $\dfrac{f+3}{3} = \dfrac{g+11}{11}$.

2. Use the theorems on proportion to complete each of the following statements.

   a) If $\dfrac{h}{k} = \dfrac{7}{3}$, then $\dfrac{k}{h} = $ ▒▒▒▒.

   b) If $\dfrac{m}{n} = \dfrac{2}{9}$, then $\dfrac{m}{2} = $ ▒▒▒▒.

   c) If $\dfrac{p}{q} = \dfrac{1}{5}$, then $\dfrac{p+q}{q} = $ ▒▒▒▒.

   d) If $\dfrac{r+s}{s} = \dfrac{11}{6}$, then $\dfrac{r}{s} = $ ▒▒▒▒?

   e) If $\dfrac{1}{t} = \dfrac{t}{13}$, then $t = $ ▒▒▒▒?

3. Complete the following proof of the equal ratios theorem by supplying the reasons.

$$\text{If} \quad \frac{a}{b} = \frac{c}{d} = \frac{e}{f} = k, \quad \text{then} \quad \frac{a+c+e}{b+d+f} = k.$$

Proof.

Statements

a) $\dfrac{a}{b} = \dfrac{c}{d} = \dfrac{e}{f} = k.$

b) $\dfrac{a}{b} = k, \dfrac{c}{d} = k, \dfrac{e}{f} = k.$

c) $a = kb,\ c = kd,\ e = kf.$

d) $a + c + e = kb + kd + kf.$

e) $a + c + e = k(b + d + f).$

f) $\dfrac{a+c+e}{b+d+f} = k.$

4. The use of k in the statement of the equal ratios theorem is convenient in developing the proof, but not necessary in stating it. Which postulate of equality justifies the following restatement of the theorem?

$$\text{If} \quad \frac{a}{b} = \frac{c}{d} = \frac{e}{f}, \quad \text{then} \quad \frac{a+c+e}{b+d+f} = \frac{a}{b}.$$

5. Copy the following statements and use the equal ratios theorem to complete them.

   a) If $\dfrac{x}{y} = \dfrac{1}{2}$, then $\dfrac{x+1}{y+2} = $ ▒▒▒▒.

   b) If $\dfrac{a}{2} = \dfrac{b}{3} = \dfrac{c}{4} = \dfrac{d}{5}$, then $\dfrac{a+b+c+d}{▒▒▒▒} = \dfrac{a}{▒▒▒▒}.$

The mean proportional can be seen in nature in the starfish. Its shape, called a "pentagram," is shown in this figure.

The length ST is the mean proportional between TA and SA: that is,

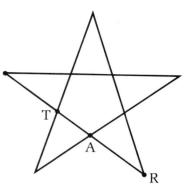

$$\frac{TA}{ST} = \frac{ST}{SA}.$$

On the assumption that ST = AR, it is possible to show that SA is also the mean proportional between two lengths in the figure. Can you do it? (Hint: Apply the denominator addition theorem to the proportion above.)

PHOTOGRAPH BY S. H. ROSENTHAL, JR., RAPHO GUILLUMETTE PICTURES

BROWN BROTHERS

Lesson 3

The Side-Splitter Theorem

In 1889, a dam above Johnstown, Pennsylvania, broke and sent down a torrent of water that destroyed most of the city and killed more than 2,300 people. The photograph above shows an overturned house through which an uprooted tree was hurled!

Before a modern dam is built, a lot of mathematics has to be worked out to insure that it will not collapse. Some of this mathematics is applied geometry.

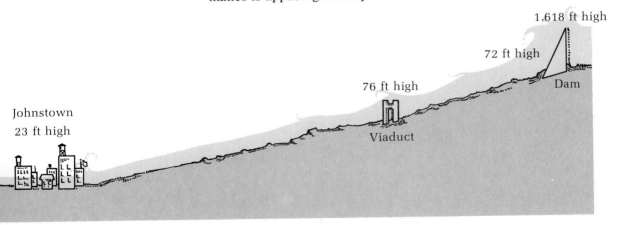

1,618 ft high

72 ft high

76 ft high

Dam

Johnstown
23 ft high

Viaduct

378

The drawing below represents a simplified side view of a dam. The cross section of the dam is shown as $\triangle$DAM; note that the line of the water level behind the dam, $\overleftrightarrow{XY}$, is parallel to the line of the ground, $\overleftrightarrow{AM}$.

The amount of force of the water against the dam depends upon several factors; one of them is the length of $\overline{YM}$, the segment that represents the surface of contact. Can the length of $\overline{YM}$ be determined indirectly from the lengths of other segments in the figure? For instance, suppose DX = 125 feet, XA = 600 feet, and DY = 200 feet; is it possible to find YM from these numbers?

We will show that if $\overleftrightarrow{XY} \parallel \overline{AM}$, then

$$\frac{DX}{XA} = \frac{DY}{YM}.$$

By substituting into this equation,

$$\frac{125}{600} = \frac{200}{YM}.$$

Solving for YM,

$$125 \cdot YM = 200 \cdot 600,$$

$$YM = \frac{200 \cdot 600}{125} = 960,$$

we find that the length of $\overline{YM}$ is 960 feet.

This method depends upon the following fact, which we will call the "Side-Splitter Theorem."

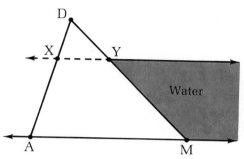

► **Theorem 54**

If a line parallel to one side of a triangle intersects the other two sides in different points, it divides the sides in the same ratio.

Hypothesis: $\triangle$ABC with $\overleftrightarrow{DE} \parallel \overline{BC}$.
Conclusion: $\dfrac{AD}{DB} = \dfrac{AE}{EC}.$

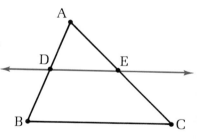

The conclusion of this theorem is that two ratios are equal. We have already proved another theorem about equal ratios: namely, that if two triangles have equal altitudes, the ratio of their areas is equal to the ratio of the lengths of their bases. By adding two line segments to the figure illustrating the Side-Splitter Theorem, we can form two pairs of triangles whose bases are the four segments in the theorem's conclusion. The proof will be based upon showing that

$$\frac{AD}{DB} = \frac{\alpha\triangle AED}{\alpha\triangle DEB} = \frac{\alpha\triangle AED}{\alpha\triangle DEC} = \frac{AE}{EC}.$$

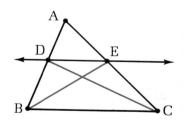

Since we will write it in paragraph form, we will number a couple of the equations for later reference.

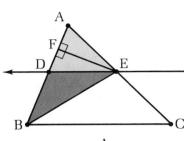

Proof.
Draw $\overline{BE}$ and draw $\overline{EF} \perp \overline{AB}$. Since $\overline{EF}$ is an altitude of both $\triangle AED$ and $\triangle DEB$,

$$\frac{\alpha\triangle AED}{\alpha\triangle DEB} = \frac{AD}{DB}. \qquad (1)$$

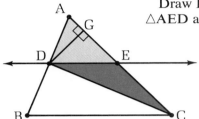

Draw $\overline{DC}$ and draw $\overline{DG} \perp \overline{AC}$. Since $\overline{DG}$ is an altitude of both $\triangle AED$ and $\triangle DEC$,

$$\frac{\alpha\triangle AED}{\alpha\triangle DEC} = \frac{AE}{EC}. \qquad (2)$$

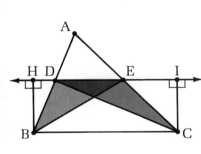

Draw $\overline{BH} \perp \overleftrightarrow{DE}$ and $\overline{CI} \perp \overleftrightarrow{DE}$. We know that $BH = CI$ because $\overleftrightarrow{DE} \parallel \overline{BC}$ and every perpendicular segment joining one of two parallel lines to the other has the same length. Since $\triangle DEB$ and $\triangle DEC$ have equal bases ($DE = DE$) and equal altitudes ($BH = CI$),

$$\alpha\triangle DEB = \alpha\triangle DEC.$$

Substituting this result into equation 2, we have

$$\frac{\alpha\triangle AED}{\alpha\triangle DEB} = \frac{AE}{EC}.$$

Since $\frac{\alpha\triangle AED}{\alpha\triangle DEB} = \frac{AD}{DB}$ [equation (1)], we can substitute again to get

$$\frac{AD}{DB} = \frac{AE}{EC}.$$

A useful fact that follows directly from this theorem is:

► **Corollary**
If a line parallel to one side of a triangle intersects the other two sides in different points, it cuts off segments proportional to the sides.

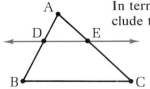

In terms of the adjoining figure, this corollary permits us to conclude that if $\overleftrightarrow{DE} \parallel \overline{BC}$, then

$$\frac{AD}{AB} = \frac{AE}{AC} \quad \text{and} \quad \frac{DB}{AB} = \frac{EC}{AC}.$$

1. Use the Side-Splitter Theorem and its corollary to complete the proportions for the adjoining figures.

Exercises

In △COW, $\overline{RN} \parallel \overline{OW}$.

a) $\dfrac{OR}{RC} = $ ▦.

b) $\dfrac{NW}{CW} = $ ▦.

In △RBL, $\overline{EU} \parallel \overline{LB}$.

c) $\dfrac{EL}{RE} = $ ▦.

d) $\dfrac{RU}{RB} = $ ▦.

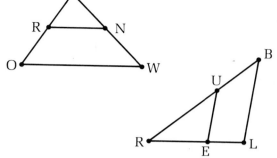

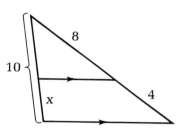

2. To find x in this figure, Obtuse Ollie wrote the proportion

$$\frac{x}{10} = \frac{4}{8}.$$

a) Is this correct?

b) Acute Alice wrote the proportion

$$\frac{x}{x - 10} = \frac{4}{8}.$$

Is she correct?

c) Write a simpler proportion that will give the correct answer.

3. Find x in each of the figures below. (They are not drawn to scale.)

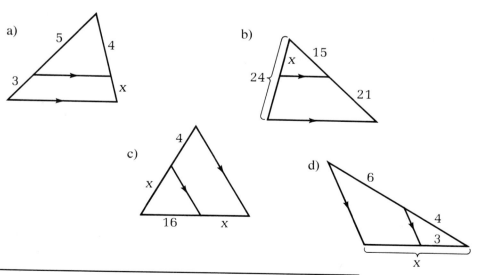

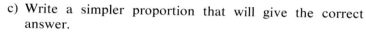

SET II

1. Complete the following proof of the corollary to the Side-Splitter Theorem by supplying the reasons.

   If a line parallel to one side of a triangle intersects the other two sides in different points, it cuts off segments proportional to the sides.

   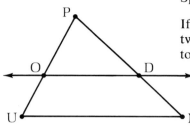

   Hypothesis: △PUN with $\overleftrightarrow{OD} \parallel \overline{UN}$.

   Conclusion: $\dfrac{PO}{PU} = \dfrac{PD}{PN}$ and $\dfrac{OU}{PU} = \dfrac{DN}{PN}$.

   Proof.

   Statements

   a) △PUN with $\overleftrightarrow{OD} \parallel \overline{UN}$.

   b) $\dfrac{PO}{OU} = \dfrac{PD}{DN}$.

   c) $\dfrac{PO + OU}{OU} = \dfrac{PD + DN}{DN}$.

   d) $PO + OU = PU$; $PD + DN = PN$.

   e) $\dfrac{PU}{OU} = \dfrac{PN}{DN}$.

   f) $\dfrac{OU}{PU} = \dfrac{DN}{PN}$.

2. In the previous exercise you have actually proved only one of the two conclusions: that $\dfrac{OU}{PU} = \dfrac{DN}{PN}$. Prove the other conclusion by writing some additional steps. (Hint: Begin by applying the upsidedownable theorem to step b. Then apply the denominator addition theorem to the result.)

   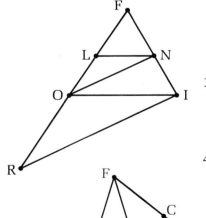

3. Given: △FRI with $\overline{LN} \parallel \overline{OI}$; $\overline{ON} \parallel \overline{RI}$.

   Prove: $\dfrac{FL}{LO} = \dfrac{FO}{OR}$.

   Par 5.

4. Given: △FRN with $\overrightarrow{FA}$ bisecting ∢RFN; $FC = CA$.

   Prove: $\dfrac{FC}{FN} = \dfrac{RA}{RN}$.

   Par 7.

382   *Chapter 10: SIMILARITY*

This photograph shows a device called a *pantograph* being used to enlarge a map. Its four bars are hinged together at A, B, C, and D so that AB = DC and AD = BC. As point D moves around the map, point E moves so that it is always in line with D and P. (Point P is attached to the drawing board and does not move.)

Can you explain why, as the map is enlarged, the ratio of DE to PD never changes?

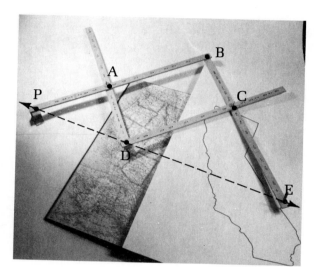

BY PERMISSION OF JOHN HART AND FIELD ENTERPRISES, INC.

Lesson **4**

Similar Triangles

If the moon's angle of elevation is 45°, the shadows it forms are equal in length to the people casting them. Why this is true is illustrated by the left-hand diagram at the top of the next page. The moonlight passing by B.C.'s head at B strikes the ground at A. If ∡A, the moon's angle of elevation, has a measure of 45°, the other acute angle of the right triangle determined by B.C. and his shadow must also have a measure of 45°. The sides opposite these angles are therefore equal so the triangle is isosceles.

The shadow cast by any vertical object will determine, with the object, an isosceles right triangle when the moon is at 45°. The "shadow triangle" for the tree in the right-hand diagram is not congruent to the one for B.C. because the tree is taller than he is. The two triangles, however, have the *same shape*; one might be considered an enlargement of the other. Such triangles are called similar and, to indicate this, we write △TRE ~ △BCA. This notation indicates a special correspondence between the vertices of the two triangles, just as the notation used for congruence does.

384

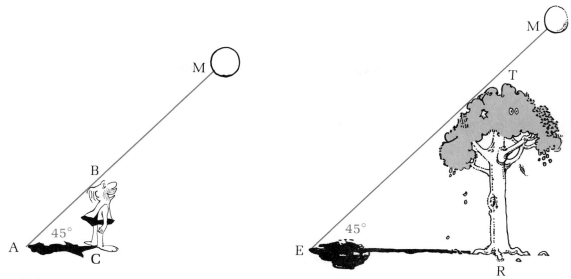

It is evident that the corresponding angles of the two triangles are equal: $\angle T = \angle B$, $\angle R = \angle C$, and $\angle E = \angle A$. Although the corresponding sides of the triangles are not equal, they are *proportional*; in other words, they have the *same ratio.*

$$\frac{TR}{BC} = \frac{ER}{AC} = \frac{TE}{BA}.$$

▶ Definition

Two triangles are **similar** iff there is a correspondence between their vertices such that the corresponding sides of the triangles are proportional and the corresponding angles are equal.

If two triangles are congruent, we can usually write equations for their corresponding parts by looking at the figures and imagining which ones would fit together if they were made to coincide. For similar triangles, however, it is important to base the equations upon the correspondence of their vertices. For example, if we know in the adjoining figure that $\triangle SIM \sim \triangle RLA$, then it follows that

$$\angle S = \angle R, \quad \angle I = \angle L, \quad \angle M = \angle A, \quad \text{and} \quad \frac{SI}{RL} = \frac{IM}{LA} = \frac{SM}{RA}.$$

The definition of similar triangles is sufficient to prove the following theorem.

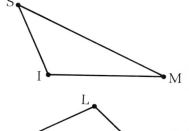

▶ Theorem 55

Any two equilateral triangles are similar.

SET I

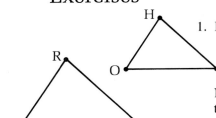

1. If △OHE ~ △NRY, then it is true that

$$\frac{OH}{NR} = \frac{HE}{RY}.$$

Name the theorem on proportion that explains how each of the proportions below follows from this proportion.

a) $\dfrac{OH}{HE} = \dfrac{NR}{RY}.$

b) $\dfrac{NR}{OH} = \dfrac{RY}{HE}.$

c) $\dfrac{OH + NR}{NR} = \dfrac{HE + RY}{RY}.$

d) $\dfrac{OH + HE}{NR + RY} = \dfrac{OH}{NR}.$

2. Complete each of the following statements. No figures are shown since the missing parts can be determined from the correspondences.

If △HEM ~ △ING, then

a) $\dfrac{ME}{GN} = \dfrac{HM}{\text{▨▨▨}}.$

b) ∠E = ∠▨▨▨.

If △ING ~ △WAY, then

c) $\dfrac{WY}{IG} = \dfrac{AW}{\text{▨▨▨}}.$

d) ∠G = ∠▨▨▨.

3. If △CON ~ △RAD, write the appropriate proportions and solve them for x and y.

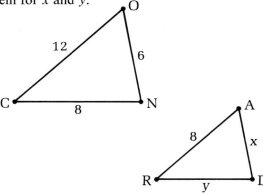

1. Complete the following proof by supplying the reasons.

   Given: △TWI with midsegment $\overline{AN}$.
   Prove: △TWI ~ △NAI.

Proof.

Statements

a) $\overline{AN}$ is a midsegment of △TWI.
b) AN = $\frac{1}{2}$WT.
c) NI = $\frac{1}{2}$TI and AI = $\frac{1}{2}$WI.
d) $\dfrac{AN}{WT} = \dfrac{1}{2}, \dfrac{NI}{TI} = \dfrac{1}{2},$ and $\dfrac{AI}{WI} = \dfrac{1}{2}.$
e) $\dfrac{AN}{WT} = \dfrac{NI}{TI} = \dfrac{AI}{WI}.$
f) $\overline{AN} \parallel \overline{WT}.$
g) ∠T = ∠ANI and ∠W = ∠NAI.
h) ∠I = ∠I.
i) △TWI ~ △NAI.

2. Use the figures shown here to explain why Theorem 55 follows from our definition of similar triangles.

   Any two equilateral triangles are similar.

   Hypothesis: △HUX and △LEY are equilateral.
   Conclusion: △HUX ~ △LEY.

3. Given: △UGH ~ △HGO.
   Prove: HG is the mean
            proportional
            between UG and GO.
   Par 3.

4. Given: △SKA ~ △ASI.
   Prove: △SKA is isosceles.
   Par 4.

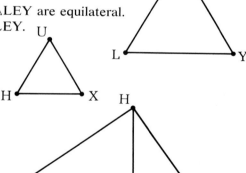

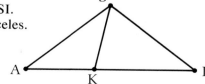

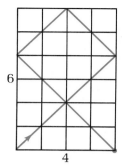

The adjoining figure represents a billiard table on which a ball has been hit from the lower left-hand corner so that it travels at a 45° angle from the sides of the table. The path of the ball has been drawn to show that each time it hits a cushion the ball rebounds from it at the same angle. If it is hit hard enough, the ball will end up in the lower right-hand corner as shown.

1. On graph paper, copy the two tables below and draw the path of a ball hit in the same way from the lower left-hand corner of each.

2. What do you notice?

3. Can you explain?

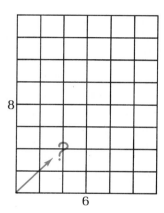

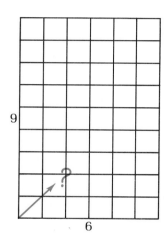

PHOTOGRAPH BY R. KAUFFMAN IN GENTLE WILDERNESS, SIERRA CLUB, © 1967

Lesson 5

The A.A. Similarity Theorem

The tallest trees in the world are the redwoods along the coast of
northern California and southern Oregon. One way to measure
one of these giants is to move some distance from the base of
the tree and place a mirror faceup on the ground so that it is level.
Then move still further away until the top of the tree can be seen
reflected in the mirror.

If you know the height of your eye level above the ground, all
you have to do is to measure your distance to the mirror and the
mirror's distance to the tree in order to find the height of the tree.

389

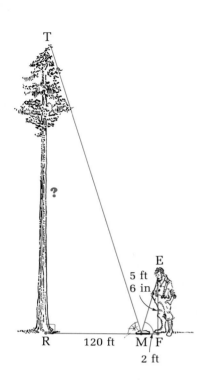

To see how this works, look at the adjoining diagram. Point M represents the position of the mirror so that the light from the top of the tree travels along $\overline{TM}$ and $\overline{ME}$ before reaching the measurer's eyes. The two triangles in the figure can be shown to be similar according to the following correspondence:

$$TRM \leftrightarrow EFM.$$

Therefore, since corresponding sides of similar triangles are proportional,

$$\frac{TR}{EF} = \frac{RM}{FM}, \quad \text{so} \quad TR = \frac{EF \cdot RM}{FM}.$$

Suppose EF = 5 ft 6 in., RM = 120 ft, and FM = 2 ft.

$$TR = \frac{(5.5\,\text{ft})\,(120\,\text{ft})}{2\,\text{ft}} = 330\,\text{ft}.$$

The tree is 330 feet tall.

The solution of this problem is based upon the fact that the two triangles in the diagram are similar. How do we know this? Because *two pairs of their angles are equal:* $\angle F = \angle R$, since they are right angles, and $\angle EMF = \angle TMR$, since they are the angles of incidence and reflection of a beam of light striking a mirror and it is a law of physics that such angles are equal. Finding just two pairs of equal angles such as these is sufficient to prove that two triangles are similar. We will prove this as our next theorem.

▶ **Theorem 56** (The A.A. Similarity Theorem)
If two angles of one triangle are equal to two angles of another triangle, then the triangles are similar.

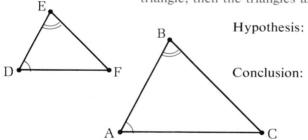

Hypothesis: △ABC and △DEF
with ∠A = ∠D,
and ∠B = ∠E.
Conclusion: △ABC ~ △DEF.

To prove this theorem, we must show that the third pair of angles in the triangles are equal and that all three pairs of corresponding sides are proportional. The angles are easy, so most of the proof will deal with the sides. Our method will be to copy the smaller triangle in one corner of the larger one to form the figure shown at the left at the top of the facing page. We can then prove that we have a triangle with a line segment parallel to one side, permitting us to apply the corollary to the Side-Splitter Theorem to show that corresponding sides of the triangles are proportional.

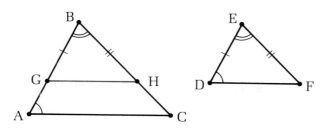

Proof.

Choose G and H on $\overrightarrow{BA}$ and $\overrightarrow{BC}$ such that BG = ED and BH = EF. Draw $\overline{GH}$. Since ∠B = ∠E, △GBH ≅ △DEF (S.A.S.)

Therefore, ∠BGH = ∠D. Since ∠A = ∠D by hypothesis, ∠BGH = ∠A by substitution. So $\overline{GH} \parallel \overline{AC}$.

From the fact that a line parallel to one side of a triangle cuts off segments on the other two sides that are proportional to them, we have

$$\frac{BG}{BA} = \frac{BH}{BC}.$$

Since ED = BG and EF = BH,

$$\frac{ED}{BA} = \frac{EF}{BC}$$

by substitution.

By copying △DEF a second time, as shown in the adjoining figure, and using the same reasoning, we can show that

$$\frac{ED}{BA} = \frac{DF}{AC}.$$

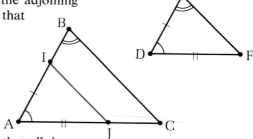

Hence, $\frac{ED}{BA} = \frac{EF}{BC} = \frac{DF}{AC}$, and so we have shown that all three pairs of corresponding sides of the triangles are proportional.

Since all three pairs of corresponding angles in the triangles are equal (∠C = ∠F because if two angles of one triangle are equal to two angles of another, the third pair of angles are equal), the triangles are similar by definition.

Although we could have proved the following theorem directly from the definition of similar triangles, the A.A. Similarity Theorem makes its proof especially easy.

► **Corollary**

Two triangles similar to a third triangle are similar to each other.

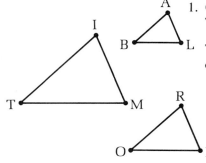

Exercises

SET I

1. Complete the following proof of the corollary to the A.A. Theorem by supplying the reasons.

   Two triangles similar to a third triangle are similar to each other.

   Hypothesis: △BAL ~ △ORE and △TIM ~ △ORE.
   Conclusion: △BAL ~ △TIM.

   Proof.

   Statements

   a) △BAL ~ △ORE and △TIM ~ △ORE.
   b) ∠B = ∠O and ∠A = ∠R; ∠T = ∠O and ∠I = ∠R.
   c) ∠B = ∠T and ∠A = ∠I.
   d) △BAL ~ △TIM.

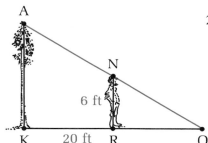

2. Dilcue thinks he has figured out another way to measure the height of a tree—this one using its shadow. He walks away from the tree along its shadow until his head is in line with the top of the tree and the tip of the shadow. Suppose he is 6 feet tall and his distance to the base of the tree is 20 feet.
   a) Why is △NRO ~ △AKO?
   b) What other distance would Dilcue need to know in order to determine the tree's height?
   c) If this distance is 10 feet, how tall is the tree?

3. To find the height of a bridge between two buildings, Dilcue stood at one end and looked down to the ground at the other end.
   a) In the diagram, why is △FLI ~ △TNI?
   b) If LI = 10 ft, IN = 25 ft, and FL = 6 ft, find NT.

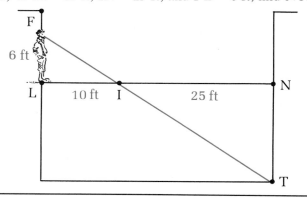

1. Given: △MBI with altitudes $\overline{BL}$ and $\overline{IO}$.
   Prove: △BOE ~△ILE.
   Par 6.

2. Given: SALE is a trapezoid with bases $\overline{SE}$
   and $\overline{AL}$ and diagonals $\overline{SL}$ and $\overline{EA}$.
   Find and prove that a pair of triangles
   in the figure are similar.
   Par 4.

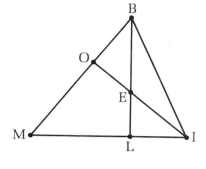

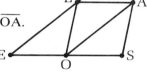

3. Given: △EPS with midsegments $\overline{LA}$, $\overline{LO}$, and $\overline{OA}$.
   Prove: △AOL ~ △EPS.
   Par 5.

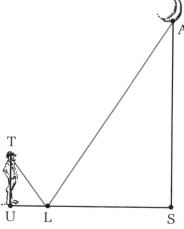

Inspired by his previous successes, Dilcue has decided to try to measure the distance to the moon by the mirror method described in this lesson. He plans to mark the point on the ground at which the moon is directly overhead, place the mirror at a convenient distance from this point, and then move away from the mirror until he can see the moon's reflection in it.

Do you think the method will work? This painting by the Belgian artist René Magritte contains a hint.

COURTESY OF L. ARNOLD WEISSBERGER

Lesson 5: The A.A. Similarity Theorem 393

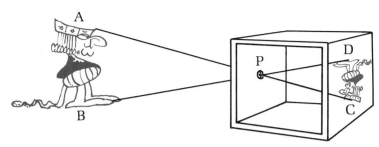

Lesson **6**

Proportional Line Segments

Did you know that a camera capable of taking recognizable pictures can be made from an ordinary box simply by poking a pinhole in the center of one wall? The diagram above shows such a "pinhole camera" and its interior.

Notice that the camera produces an image that is upside-down with respect to the object being photographed. The size of the image depends upon the distance of the object from the pinhole. In the somewhat simplified version below, $\overline{AB}$ represents the object, $\overline{CD}$ represents the image, and P represents the pinhole. If $\overline{AB} \parallel \overline{DC}$, it is easy to show that $\triangle ABP \sim \triangle CDP$.

The distances from the pinhole of the object and image are the lengths of altitudes $\overline{PE}$ and $\overline{PF}$ in these two triangles. Altitudes in two triangles such as these are called *corresponding altitudes* because they are drawn from corresponding vertices of the triangles. We know that corresponding sides of similar triangles are proportional, so that

$$\frac{AB}{CD} = \frac{BP}{DP} = \frac{AP}{CP}.$$

What about corresponding altitudes of the triangles? It is easy to show that they have the same ratio as these pairs of corresponding sides, so that, for example,

$$\frac{PE}{PF} = \frac{AB}{CD}.$$

Applying the means-extremes theorem to this proportion, we can write

$$PE \cdot CD = PF \cdot AB.$$

Dividing by PE gives the equation,

$$CD = \frac{PF \cdot AB}{PE}.$$

We can use this equation to predict the size of the image produced by the camera, CD, if we know the size of the object, AB, the distance of the image from the pinhole (which is the length of the camera box), PF, and the distance of the object from the pinhole, PE. For example, if the object is 24 inches tall and the camera is 5 inches long and 30 inches from the object,

$$CD = \frac{5 \cdot 24}{30} = 4 \text{ in.}$$

The image will be 4 inches tall.

This method depends upon the following fact.

► **Theorem 57**

Corresponding altitudes of similar triangles have the same ratio as the corresponding sides.

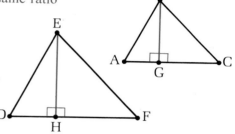

Hypothesis: $\overline{BG}$ and $\overline{EH}$ are altitudes of △ABC and △DEF, respectively; △ABC ~ △DEF.

Conclusion: $\dfrac{BG}{EH} = \dfrac{AB}{DE} = \dfrac{BC}{EF} = \dfrac{AC}{DF}.$

Proof.

Statements	Reasons
1. $\overline{BG}$ and $\overline{EH}$ are altitudes of △ABC and △DEF.	Hypothesis.
2. $\overline{BG} \perp \overline{AC}; \overline{EH} \perp \overline{DF}.$	An altitude of a triangle is perpendicular to the line of the opposite side.
3. ∡AGB and ∡DHE are right angles.	Perpendicular lines form right angles.
4. ∠AGB = ∠DHE.	All right angles are equal.
5. △ABC ~ △DEF.	Hypothesis.
6. ∠A = ∠D.	Corresponding angles of similar triangles are equal.
7. △BGA ~ △EHD.	A.A. Similarity Theorem.
8. $\dfrac{BG}{EH} = \dfrac{AB}{DE}.$	Corresponding sides of similar triangles are proportional.
9. Also, $\dfrac{AB}{DE} = \dfrac{BC}{EF} = \dfrac{AC}{DF}.$	Same reason as step 8 (from step 5).
10. $\dfrac{BG}{EH} = \dfrac{AB}{DE} = \dfrac{BC}{EF} = \dfrac{AC}{DF}.$	Transitive.

Exercises

SET I

1. Use Theorem 57 to solve for x in each of the following exercises.

a)

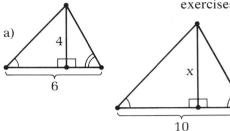

b)

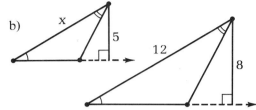

2. We assumed in our proof of Theorem 57 that the altitudes were inside the triangles, yet we have just applied it to a case (Exercise 1, part b) in which they are not. Read the proof of the theorem again, on page 395, thinking in terms of the figure shown here, in which the pair of corresponding altitudes shown lie outside the triangles.

   a) One of the statements in the proof is now ambiguous. Which is it and how should it be restated?

   b) If you have restated it correctly, it is evident that some steps should be inserted between it and the following step. What should they say?

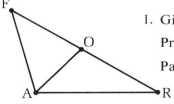

SET II

1. Given: $\angle F = \angle OAR.$

   Prove: $\dfrac{FR}{AR} = \dfrac{AR}{OR}.$

   Par 4.

2. Given: $\square BRID$; $\overrightarrow{RG}$ bisects $\angle BRI.$

   Prove: $\dfrac{BE}{ER} = \dfrac{IG}{GR}.$

   Par 7.

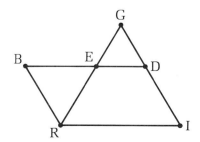

3. Given: $\triangle$WSH with $\overline{\text{TI}} \parallel \overline{\text{WH}}$;
   $\angle$WHT $= \angle$S.

   Prove: $\dfrac{\text{WT}}{\text{HI}} = \dfrac{\text{WH}}{\text{HT}}$.

   Par 9.

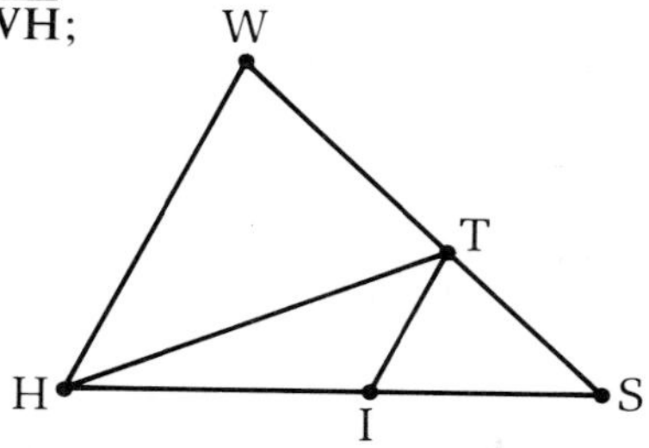

SET III

Experiment: The Pinhole Camera. The following experiment will enable you to see an image made by a pinhole camera. All you need is a stiff card (a file card is convenient), a candle, and a dark room.

Poke a hole through the center of the card with your compass point. The hole should have a diameter of about $\frac{1}{16}$ inch. Light the candle and hold it about 6 inches from a wall. Then hold the card between the candle and the wall so that the hole, flame, and wall are in line. You should observe an upside-down image of the flame on the wall.

1. Hold the card close to the wall and move it toward the candle. What happens to the size of the flame's image?

2. Can you use the diagram and the equation for the size of the image in a pinhole camera on page 394 to explain this?

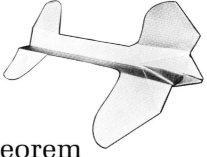

Lesson *7*

The Angle Bisector Theorem

Three of the winners among more than eleven thousand entries in a paper airplane competition held by *Scientific American** are shown at the top of this page and the next. Which one do you suppose flew the farthest? Which one stayed in the air the longest?

So that they will fly straight, each of these planes is symmetrical; that is, their wings match in length and form equal angles with the center line of the plane. The figure below illustrates a lopsided airplane in which the wing angles are equal (∠CBD = ∠DBA), but the lengths of their side edges are not (BC = 7 in. and BA = 8 in.). As a result of this, the back edges of the wings, $\overline{\text{CD}}$ and $\overline{\text{DA}}$, evidently do not have equal lengths either. If we assume that C, D, and A are collinear so that the two wings are coplanar (an especially appropriate word in this particular case!), then it is possible to prove that the back edges must have the same length ratio as the side edges; that is,

$$\frac{\text{CD}}{\text{DA}} = \frac{\text{CB}}{\text{BA}} = \frac{7}{8}.$$

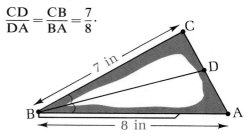

*For more details, see *The Great International Paper Airplane Book*, by Jerry Mander, George Dippel, and Howard Gossage (Simon & Schuster, 1967). Photographs: Copyright © 1967 by Shade Tree Corp. Reprinted by permission of Simon & Schuster.

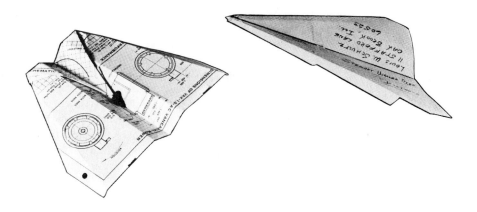

   This result depends upon the fact that, since $\angle CBD = \angle DBA$, $\overrightarrow{BD}$ bisects $\angle CBA$. Stated more generally, if a line bisects one angle of a triangle, it divides the opposite side into segments that have the same ratio as the other two sides. We will prove this and call it the "Angle Bisector Theorem."

▶ **Theorem 58**

An angle bisector in a triangle divides the opposite side into segments that have the same ratio as the other two sides.

> Hypothesis: In $\triangle ABC$, $\overrightarrow{BD}$ bisects $\angle ABC$.
> Conclusion: $\dfrac{AD}{DC} = \dfrac{AB}{BC}$.

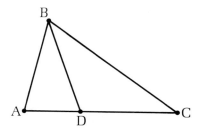

   One way to prove that a set of line segments are proportional is to show that they are corresponding sides of similar triangles. The four segments in the conclusion of the theorem are sides of $\triangle ABD$ and $\triangle DBC$. These triangles, however, are not necessarily similar since we know only that *one* pair of angles in them are equal.

Another way to prove that segments are proportional is to use the Side-Splitter Theorem. It, however, applies only to a triangle in which there is a line parallel to one side that intersects the other two sides. The angle bisector in the figure illustrating the theorem is obviously not parallel to any side of the triangle. However, by carefully choosing certain lines to add to the figure, we can form a triangle to which the Side-Splitter Theorem *will* apply. By extending $\overline{BC}$ and drawing, through A, a line parallel to $\overline{BD}$, a triangle is produced in which a line is parallel to one side: $\triangle AEC$. At the same time, we can prove that an isosceles triangle has also been formed: $\triangle ABE$. The conclusion of the theorem can easily be derived from these results.

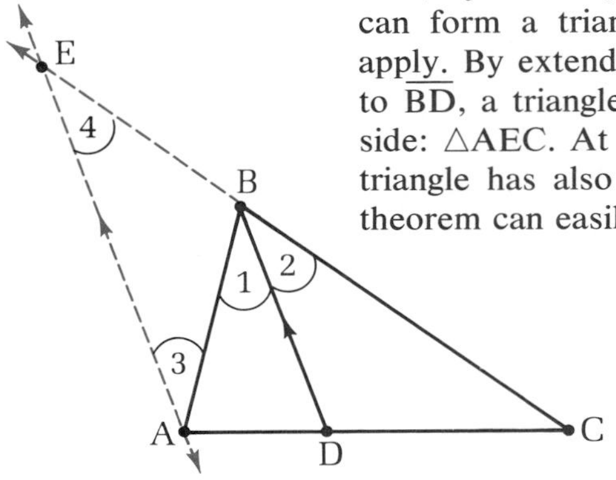

Proof.

Statements	Reasons
1. Draw $\overleftrightarrow{BC}$.	Two points determine a line.
2. Through A, draw $\overleftrightarrow{AE} \parallel \overline{BD}$.	Through a point not on a line, there is exactly one parallel to the line.
3. In $\triangle AEC$, $\dfrac{AD}{DC} = \dfrac{EB}{BC}$.	The Side-Splitter Theorem.
4. $\angle 3 = \angle 1$.	Parallel lines form equal alternate interior angles with a transversal.
5. $\angle 4 = \angle 2$.	Parallel lines form equal corresponding angles with a transversal.
6. $\overrightarrow{BD}$ bisects $\angle ABC$.	Hypothesis.
7. $\angle 1 = \angle 2$.	A bisected angle is divided into two equal angles.
8. $\angle 3 = \angle 4$.	Substitution.
9. $AB = EB$.	If two angles of a triangle are equal, the sides opposite them are equal.
10. $\dfrac{AD}{DC} = \dfrac{AB}{BC}$.	Substitution.

Of the three paper airplanes at the beginning of this lesson, the second stayed in the air the longest and the third flew the farthest.

1. Which of the following proportions follow from the fact that $\overrightarrow{AE}$ bisects ∡WAV in △WAV?

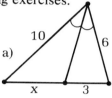

a) $\dfrac{WE}{EV} = \dfrac{WA}{AV}$.

b) $\dfrac{WE}{EV} = \dfrac{VA}{AW}$.

c) $\dfrac{WE}{WA} = \dfrac{EV}{AV}$.

d) $\dfrac{AV}{AW} = \dfrac{VE}{EW}$.

2. Use the Angle Bisector Theorem to solve for x in each of the following exercises.

a)

b)

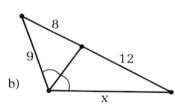

Study the following example before doing exercises c and d.

If the length of the side cut by the angle bisector is 4 and one of its segments has length x, the other segment must have the length $4 - x$.

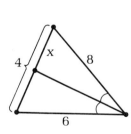

$$\frac{x}{4 - x} = \frac{8}{6}.$$

$$6x = 8(4 - x)$$

$$6x = 32 - 8x$$

$$14x = 32$$

$$x = \frac{32}{14} = 2\frac{2}{7}.$$

c)

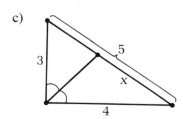

d)

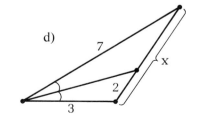

Lesson 7: The Angle Bisector Theorem   401

3. Supply the reasons for the following proof of the converse of the Angle Bisector Theorem.

If a ray from the vertex of one angle of a triangle divides the opposite side into segments that have the same ratio as the other two sides, then it bisects the angle.

Hypothesis: $\triangle OCA$ with $\overrightarrow{CN}$ such that $\dfrac{ON}{NA} = \dfrac{OC}{CA}$.

Conclusion: $\overrightarrow{CN}$ bisects $\measuredangle OCA$.

Proof.

<u>Statements</u>

a) Draw $\overleftrightarrow{OC}$.
b) Through A, draw $\overleftrightarrow{AE} \parallel \overline{NC}$.
c) $\dfrac{ON}{NA} = \dfrac{OC}{CE}$.
d) $\dfrac{ON}{NA} = \dfrac{OC}{CA}$.
e) $\dfrac{OC}{CE} = \dfrac{OC}{CA}$.
f) $\dfrac{CE}{OC} = \dfrac{CA}{OC}$.
g) $CE = CA$.
h) $\angle 3 = \angle 4$.
i) $\angle 1 = \angle 4$.
j) $\angle 2 = \angle 3$.
k) $\angle 1 = \angle 2$.
l) $\overrightarrow{CN}$ bisects $\measuredangle OCA$.

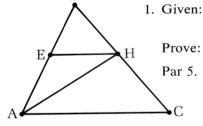

SET II

1. Given: $\overrightarrow{AH}$ bisects $\measuredangle BAC$ in $\triangle BAC$; $\overline{EH} \parallel \overline{AC}$.

   Prove: $\dfrac{BE}{EA} = \dfrac{BA}{AC}$.

   Par 5.

2. Given: $\triangle SND$ with $\angle 1 = \angle 2$; $SA = AN$.

   Prove: $SD = DN$.

   Par 6.

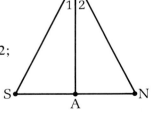

The winning origami entry in the *Scientific American* paper airplane competition is shown in the photograph below.* It was
created by Professor James M. Sakoda of Brown University,
Rhode Island.

The plane is made from a square sheet of paper, first folded as
shown in the diagram so that two consecutive edges meet along a
diagonal. As a result, the four numbered angles are equal.

Can you figure out the ratio of the area of the part folded over
(shaded in the diagram) to the area of the rest of the square?

Directions for folding the plane are given on the next page.

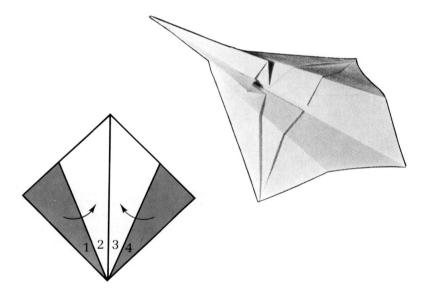

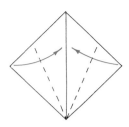

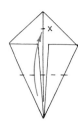

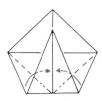

1 Take a square piece of paper.
   White bond paper 8½″ × 8½″ is
   suitable. Make a crease along
   one diagonal and fold two
   sides to this diagonal line
   to form the nose and wings
   of the plane. See next figure
   for desired result.

2 Fold the nose of the plane
   back to the point marked X.

3 Fold over leading edges
   of the wings; then tuck in
   between wing and nose.

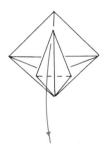

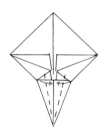

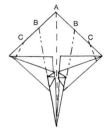

4 Pull nose out forward.

5 Narrow down nose by
   folding edge to center line.

6 Fold up plane in half along
   center line A. Fold down
   wings along B, and then
   spread out horizontally.
   Turn up trailing edge of
   wings at slight angle at C.
   Adjust this angle to provide
   proper amount of lift.
   If plane drops rapidly, lift
   flap up more. If it rises too
   rapidly and stalls, flatten
   the flap down more.

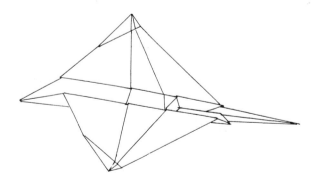

KEY

— — — — — Fold up

— ·· — ·· — ·· Fold down

Lesson **8**

Perimeters and Areas
of Similar Triangles

Suppose that B.C. has two small fields on which he grows "ratabagas." Also, that the fields are rectangular in shape and each side of one is three times as long as the corresponding side of the other, as shown here.

If B.C. decides to build a fence around each field, how would the amount of fencing needed to surround them compare? The figure shows that the larger field requires *three* times as much fencing. Its area, however, is *not* three times that of the smaller field: it is *nine* times as great.

Since the corresponding sides of B.C.'s fields are proportional,

$$\left(\frac{AB}{EF} = \frac{BC}{FG} = \frac{CD}{GH} = \frac{DA}{HE} = \frac{1}{3}\right),$$

and their corresponding angles are equal (they are right angles), the rectangles, like triangles, are said to be *similar*. We have observed that the perimeters of these similar figures have the same ratio as a pair of corresponding sides. The ratio of their areas, however, is equal to the *square* of the ratio of a pair of corresponding sides.

We will prove that for pairs of similar triangles, these relationships are always true.

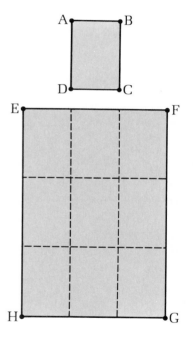

► **Theorem 59**

The ratio of the perimeters of two similar triangles is equal to the ratio of the corresponding sides.

► **Theorem 60**

The ratio of the areas of two similar triangles is equal to the square of the ratio of the corresponding sides.

Exercises

SET I

Supply the reasons for the following proofs of the theorems in this lesson.

1. *Theorem 59.* The ratio of the perimeters of two similar triangles is equal to the ratio of the corresponding sides.

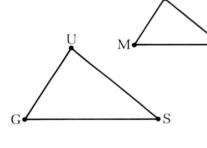

Hypothesis: $\triangle MIN \sim \triangle GUS$.

Conclusion: $\dfrac{\rho\triangle MIN}{\rho\triangle GUS} = \dfrac{MI}{GU}$.*

Proof.

Statements
a) $\triangle MIN \sim \triangle GUS$.
b) $\dfrac{MI}{GU} = \dfrac{IN}{US} = \dfrac{NM}{SG}$.
c) $\dfrac{MI + IN + NM}{GU + US + SG} = \dfrac{MI}{GU}$.
d) $\rho\triangle MIN = MI + IN + NM$, $\rho\triangle GUS = GU + US + SG$.
e) $\dfrac{\rho\triangle MIN}{\rho\triangle GUS} = \dfrac{MI}{GU}$.

2. *Theorem 60.* The ratio of the areas of two similar triangles is equal to the square of the ratio of the corresponding sides.

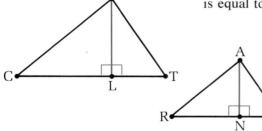

Hypothesis: $\triangle COT \sim \triangle RAE$.

Conclusion: $\dfrac{\alpha\triangle COT}{\alpha\triangle RAE} = \left(\dfrac{CT}{RE}\right)^2$.

*Remember that the symbol ρ represents perimeter.

Proof.

Statements

a) $\triangle COT \sim \triangle RAE$.

b) Through O, draw $\overline{OL} \perp \overline{CT}$; through A, draw $\overline{AN} \perp \overline{RE}$.

c) $\overline{OL}$ and $\overline{AN}$ are altitudes of $\triangle COT$ and $\triangle RAE$.

d) $\dfrac{OL}{AN} = \dfrac{CT}{RE}$.

e) $\alpha\triangle COT = \tfrac{1}{2}CT \cdot OL$,
   $\alpha\triangle RAE = \tfrac{1}{2}RE \cdot AN$.

f) $\dfrac{\alpha\triangle COT}{\alpha\triangle RAE} = \dfrac{\tfrac{1}{2}CT \cdot OL}{\tfrac{1}{2}RE \cdot AN}$.

g) $\dfrac{\alpha\triangle COT}{\alpha\triangle RAE} = \dfrac{CT}{RE} \cdot \dfrac{OL}{AN} = \dfrac{CT}{RE} \cdot \dfrac{CT}{RE} = \left(\dfrac{CT}{RE}\right)^2$.

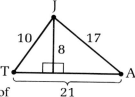

3. In the figure shown here, $\triangle TJA \sim \triangle DER$.

   a) Comparing the triangles in this order, what is the ratio of their corresponding sides?

   b) Find the perimeter of each triangle and verify that the ratio of the perimeters is equal to the ratio of the corresponding sides.

   c) Find the area of each triangle and verify that the ratio of the areas is equal to the *square* of the ratio of the corresponding sides.

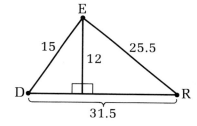

SET II

Before solving the following problems, study the example below.

Example.
The areas of two similar triangles are 16 and 9, respectively. If the perimeter of the first triangle is 12, find the perimeter of the second triangle.

Solution.
Let a and b be the lengths of a pair of corresponding sides of the two triangles.

$\dfrac{16}{9} = \left(\dfrac{a}{b}\right)^2$ (Theorem 60), so $\dfrac{4}{3} = \dfrac{a}{b}$ (square roots postulate).

Let x be the perimeter of the second triangle.

$\dfrac{12}{x} = \dfrac{a}{b}$ (Theorem 59), so $\dfrac{12}{x} = \dfrac{4}{3}$, $4x = 36$, $x = 9$.

The perimeter of the second triangle is 9.

1. a) The perimeters of two similar triangles are 25 and 49, respectively. Find the ratio of their sides.

   b) A pair of corresponding sides of two similar triangles have lengths of $\sqrt{5}$ and 5, respectively. Find the ratio of their areas.

   c) Two similar triangles have areas of 100 and 144, respectively. If one side of the first triangle is 20 units long, how long is the corresponding side of the second triangle?

   d) Two similar triangles have corresponding sides of lengths 4 and 6, respectively. If the area of the first triangle is 32, find the area of the second triangle.

2. Given: $\triangle KUP$ with $\overline{RA} \parallel \overline{UP}$.

   Prove: $\dfrac{\alpha\triangle KRA}{\alpha\triangle KUP} = \left(\dfrac{RA}{UP}\right)^2$

   Par 4.

3. Given: $\triangle DAV$ and $\triangle VIS$ are equilateral.

   Prove: $\dfrac{\rho\triangle DAV}{\rho\triangle VIS} = \dfrac{DS}{SV} + 1.$

   Par 5.

SET III

Some Sphinx Puzzles. This figure looks somewhat like a "sphinx." Make nine copies of it on stiff paper and cut them out.

1. Can you fit four of them together to form a larger sphinx? (Hint: Like that of similar triangles, the ratio of the areas of two similar sphinxes is equal to the square of the ratio of the corresponding sides.)

2. Can you fit all nine together to form an even larger sphinx?

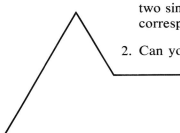

Chapter 10/Summary and Review

Basic Ideas

Mean proportional   374
Proportion   369
Ratio   369
Similar triangles   385

Theorems on Proportion   370, 374

1. Means-Extremes
2. Upsidedownable
3. Interchangeable
4. Denominator Addition
5. Denominator Subtraction
6. Mean Proportional
7. Equal Ratios

Theorems

54. *The Side-Splitter Theorem.*   If a line parallel to one side of a triangle intersects the other two sides in different points, it divides the sides in the same ratio.   379
    Corollary.   If a line parallel to one side of a triangle intersects the other two sides in different points, it cuts off segments proportional to the sides.   380
55. Any two equilateral triangles are similar.   385
56. *The A.A. Similarity Theorem.*   If two angles of one triangle are equal to two angles of another triangle, then the triangles are similar.   390
    Corollary.   Two triangles similar to a third triangle are similar to each other.   391

Exercises

SET I

1. Which of the following conditions are sufficient to prove that △ORA and △NGE are similar?
   a) △ORA and △NGE are equilateral.
   b) ∠R = ∠N = 45°, ∠A = 75°, ∠G = 60°.
   c) △ORA and △NGE are isosceles right triangles.
   d) OR = OA = EN = EG.

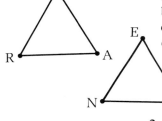

2. In this figure, $\overrightarrow{LM}$ bisects ∡PLU of △PLU.
   a) Find PL.
   b) Find ρ△PLU.
   c) Is △PLU a right triangle?

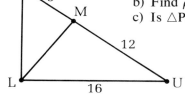

3. In △AGE, $\overline{RP} \parallel \overline{GE}$.
   a) Find $\dfrac{RP}{GE}$.
   b) If ρ△ARP = 24, find ρ△AGE.
   c) If α△ARP = 32, find α△AGE.

SET II

1. During the Renaissance, artists became interested in geometry as a key to understanding perspective. Leonardo da Vinci, in his book *Treatise on Painting*, went so far as to write, "Let no one who is not a mathematician read my works."

   A couple of basic rules of perspective are illustrated in the figure below. Vertical lines such as $\overleftrightarrow{EH}$ and $\overleftrightarrow{AC}$ are drawn vertical and parallel. Horizontal lines that are parallel to each other and that intersect the horizon line ℓ, such as $\overleftrightarrow{EA}$ and $\overleftrightarrow{HC}$, are drawn so that they intersect it at the same point.

   Prove that, according to these rules, the following lengths must be proportional.

   Given: $\overleftrightarrow{EH} \parallel \overleftrightarrow{AC}$.

   Prove: $\dfrac{EH}{AC} = \dfrac{PH}{PC}$.

   Par 4.

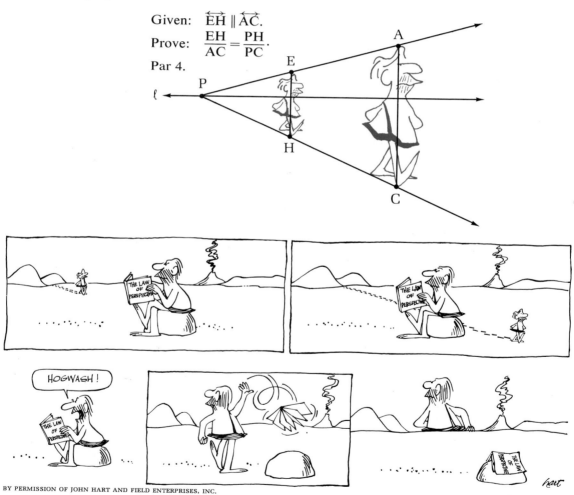

2. Given: $\overline{AP} \parallel \overline{OR} \parallel \overline{CI}$.
   Prove: $\dfrac{AO}{PR} = \dfrac{OC}{RI}$.
   Par 5. (Hint: draw $\overline{AI}$ and label
   the point in which it intersects
   $\overline{OR}$ point T.)

3. Given: QUIN is a rectangle.
   Prove: $CE \cdot EU = NE \cdot EI$.
   Par 7.

SET III

According to Pythagoras, the area of the square on the hypotenuse of a right triangle is equal to the sum of the areas of the squares on the two legs.

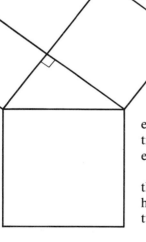

Would this still be true if the squares were replaced with equilateral triangles? In other words, is the area of an equilateral triangle on the hypotenuse equal to the sum of the areas of two equilateral triangles on the legs?

What if similar pictures of Pythagoras himself were placed on the three sides of a right triangle. Would the Pythagoras on the hypotenuse be equal to the sum of the Pythagorases on the two legs?

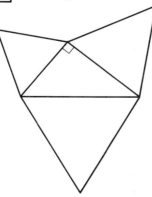

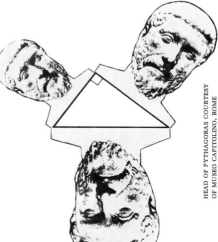

Chapter 11

THE RIGHT TRIANGLE

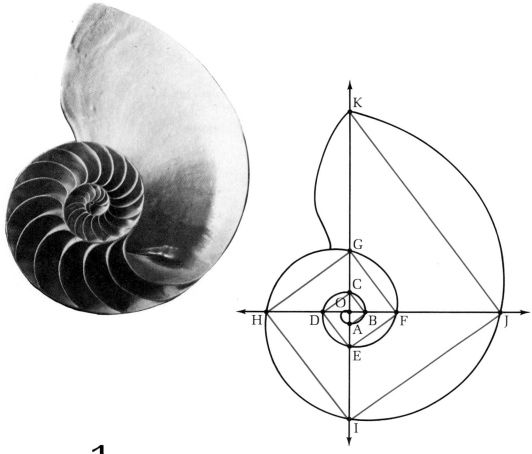

Lesson 1

Proportions in a Right Triangle

An attractive example of proportion in nature is the shell of the chambered nautilus. This sea creature moves through a series of successively larger compartments as it grows. The photograph above shows a shell that has been cut in half to reveal these compartments.

Part of the beauty of the shell of the nautilus is due to the fact that as the chambers increase in size their shape always remains the same. The drawing at the right of the photograph reveals the way in which the shell grows. The curve is called a spiral, and through its center, point O, two perpendicular lines have been drawn. As the spiral winds around point O, it intersects these lines in the series of points labeled A, B, C, ... If these points are joined in order by the series of line segments $\overline{AB}$, $\overline{BC}$, $\overline{CD}$, . . . , each of these segments is perpendicular to the next: $\overline{AB} \perp \overline{BC} \perp \overline{CD}$ . . . . Because of this, the consecutive pairs of segments are the legs of a series of progressively larger right triangles: $\triangle ABC$, $\triangle BCD$, $\triangle CDE$, ...

It is easy to prove that these triangles are similar, and, as a result, that

$$\frac{OA}{OB} = \frac{OB}{OC} = \frac{OC}{OD} = \cdots$$

Since $\overline{OB}$, $\overline{OC}$, $\overline{OD}$, . . . are the altitudes to the hypotenuses of these triangles, we see that each successive altitude is the mean proportional between the altitude before and after it as we move about the spiral. In terms of one of the right triangles, say $\triangle ABC$ for example, this is equivalent to saying that the altitude to the hypotenuse is the mean proportional between the segments into which it divides the hypotenuse.

Before proving this, we will show that the altitude separates the triangle into two triangles that are similar both to it and to each other.

▶ **Theorem 61**

The altitude to the hypotenuse of a right triangle forms two triangles that are similar to each other and to the original triangle.

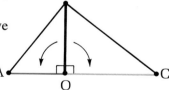

Hypothesis: Right $\triangle ABC$ with altitude $\overline{BO}$ to hypotenuse $\overline{AC}$.

Conclusion: $\triangle AOB \sim \triangle BOC \sim \triangle ABC$.

Proof.

Since $\triangle AOB$ and $\triangle ABC$ are right triangles with a common acute angle at A, $\triangle AOB \sim \triangle ABC$. Also, since $\triangle BOC$ and $\triangle ABC$ have a common acute angle at C, $\triangle BOC \sim \triangle ABC$. Finally, $\triangle AOB \sim \triangle BOC$ since, if two triangles are similar to a third triangle, they are similar to each other.

From the similarity correspondence, $\triangle AOB \sim \triangle BOC$, we can write the proportion:

$$\frac{AO}{BO} = \frac{BO}{CO}.$$

BO is the mean proportional between AO and CO.

▶ **Corollary 1**

The altitude to the hypotenuse of a right triangle is the mean proportional between the segments into which it divides the hypotenuse.

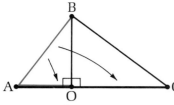

AB is the mean
proportional between
AO and AC.

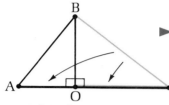

BC is the mean
proportional between
OC and AC.

From the correspondences $\triangle AOB \sim \triangle ABC$ and $\triangle BOC \sim \triangle ABC$, follow the proportions:

$$\frac{AO}{AB} = \frac{AB}{AC} \quad \text{and} \quad \frac{OC}{BC} = \frac{BC}{AC}.$$

The first equation shows that leg $\overline{AB}$ of the triangle is the mean proportional between the hypotenuse $\overline{AC}$ and the segment $\overline{AO}$ of the hypotenuse. The second equation reveals a comparable relationship for leg $\overline{BC}$. We will refer to the segments $\overline{AO}$ and $\overline{OC}$ as the *projections* of sides $\overline{AB}$ and $\overline{BC}$ on the hypotenuse, respectively.

▶ **Corollary 2**

Each leg of a right triangle is the mean proportional between the hypotenuse and its projection on the hypotenuse.

Exercises

SET I

1. In this figure, right $\triangle DOV$ has been reflected through the line containing its hypotenuse and $\overline{OO'}$ has been drawn.
   a) What relationship does $\overline{OO'}$ have to $\overline{DV}$?
   b) What is $\overline{OE}$ called with respect to $\triangle DOV$?
   c) What are $\overline{DE}$ and $\overline{EV}$ called with respect to $\overline{DO}$ and $\overline{OV}$?

2. The altitude to the hypotenuse of right $\triangle IDL$ is $\overline{DA}$. We have proved that the three triangles in the figure are similar, but to write an appropriate correspondence between their vertices can be confusing. A convenient way to do it is shown in the second figure, in which the two smaller triangles have been reflected through the legs of the original triangle.
   a) By naming the vertices of the triangles in the order indicated in the third figure, complete the following similarity correspondence: $\triangle IAD \sim \triangle \text{▨▨▨} \sim \triangle \text{▨▨▨}$.

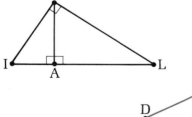

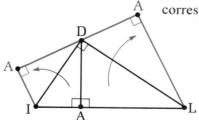

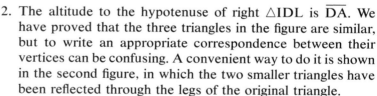

Use your answer to part a to complete the following proportions.

b) $\dfrac{IA}{DA} = \dfrac{AD}{\text{▓▓▓}}$.

c) $\dfrac{IL}{ID} = \dfrac{ID}{\text{▓▓▓}}$.

d) $\dfrac{IL}{DL} = \dfrac{DL}{\text{▓▓▓}}$.

The proportions in parts b, c, and d illustrate the corollaries to Theorem 61.

e) Look back at the first figure in this exercise to identify them.

3. In right $\triangle ZES$, $\overline{ET} \perp \overline{ZS}$.
   a) Between what two lengths is ET the mean proportional?
   b) Which segment is the projection of $\overline{ZE}$ on $\overline{ZS}$?
   c) Between what two lengths is ZE the mean proportional?
   d) What length is the mean proportional between ZS and TS?

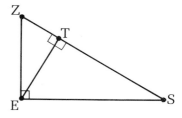

4. Solve for x in each figure below.

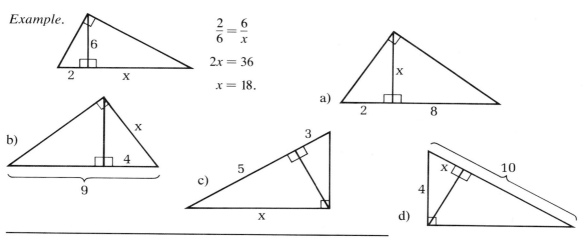

Example.

$\dfrac{2}{6} = \dfrac{6}{x}$

$2x = 36$

$x = 18.$

SET II

1. We have proved that each leg of a right triangle is the mean proportional between the hypotenuse and its *projection* on the hypotenuse. The word "projection" has a more general meaning than that indicated by its use in this corollary. In fact, there are several different types of projections. We will consider the properties of one type, called "orthogonal," in this exercise.

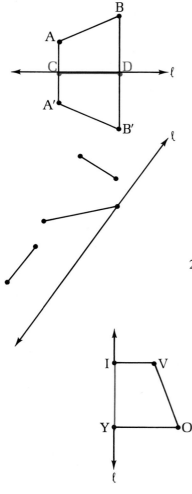

In this figure, $\overline{A'B'}$ is the reflection of $\overline{AB}$ through line ℓ, and $\overline{CD}$ is the *orthogonal projection* of $\overline{AB}$ on line ℓ. The prefix "ortho" comes from a Greek word meaning "right." An orthodontist sets teeth "right." An orthodox doctrine is one that is "right."

a) Explain why the projection illustrated in this figure is called "orthogonal."

b) Copy the second adjoining figure and find the projection of each segment on line ℓ.

What relationship does a line segment have to a line if

c) the line segment is equal in length to its projection on the line?

d) the projection of the line segment on the line is a point?

2. In this figure, $\overline{IY}$ is the orthogonal projection of $\overline{VO}$ on line ℓ. Hence, $\overline{VI} \perp \ell$ and $\overline{OY} \perp \ell$. On the assumption that $\overline{VO}$ is not parallel to line ℓ, answer the following questions to explain why $\overline{IY}$ must be shorter than $\overline{VO}$.

a) Through V, draw $\overline{VR} \parallel \overline{IY}$. What permits us to do this?
b) Why is $\overline{VR} \perp \overline{YO}$?
c) Why is $VR < VO$?
d) Why is $\overline{IV} \parallel \overline{YO}$?
e) Why is $IY = VR$?
f) Why is $IY < VO$?

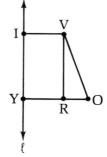

3. Suppose that the sun is directly overhead. How is the length of the shadow of a pole on the ground affected by the position of the pole?

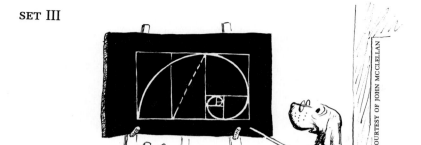

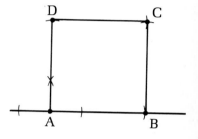

The spiral of the chambered nautilus shell is closely related to a figure called the *golden rectangle*. Discovered by the Greeks in the fifth century B.C., the golden rectangle is thought to have an especially pleasing shape. A simple method for constructing it is illustrated in the drawing above and described below.

First construct a square with sides several inches long and label it ABCD as shown in the first figure. Bisect the base of the square and label the midpoint E. With E as center and EC as radius, draw an arc intersecting $\overleftrightarrow{AB}$ in point F. Construct a line perpendicular to $\overleftrightarrow{AB}$ at F and label the point in which it intersects $\overleftrightarrow{DC}$ point G.

Quadrilateral ADGF is a golden rectangle and is divided by $\overline{BC}$ into a square and another golden rectangle! Because both ADGF and CGFB are golden rectangles, their corresponding dimensions are proportional.

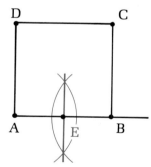

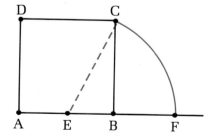

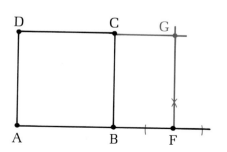

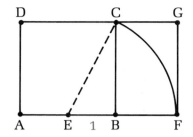

1. To show why this is so, let EB = 1 unit and find each of the following lengths.
   a) BC.
   b) EC.
   c) EF.
   d) BF.
   e) AF.

2. Use your answers to Exercise 1 to show that the corresponding dimensions of rectangles ADGF and CGFB are proportional.

3. Use your straightedge and compass to subdivide CGFB into five squares and a small rectangle as shown here. If you draw a quarter circle in each of the six squares in the figure as shown, the result is a good approximation of the nautilus spiral.

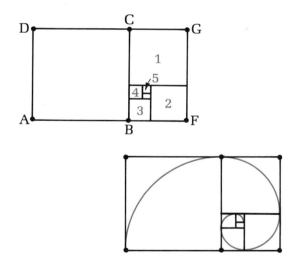

Lesson 2

The Pythagorean Theorem Revisited

Three Navaho women sit side by side on the ground. The first woman, who is sitting on a goatskin, has a son who weighs 140 pounds. The second woman, who is sitting on a deerskin, has a son who weighs 160 pounds. The third woman, who weighs 300 pounds, is sitting on a hippopotamus skin. Therefore, the squaw on the hippopotamus is equal to the sons of the squaws on the other two hides!*

There are many different ways of proving that the square on the hypotenuse is equal to the sum of the squares on the other two sides. You have studied several proofs of the Pythagorean Theorem that are based upon area. In this lesson, we will consider a proof based upon similarity. It depends upon the second corollary to Theorem 61—the fact that each leg of a right triangle is the mean proportional between the hypotenuse and its projection upon the hypotenuse—and is presented on the next page.

*From *The Unexpected Hanging*, by Martin Gardner (Simon & Schuster, 1969).

421

▶ The Pythagorean Theorem

In a right triangle, the square of the hypotenuse is equal to the sum of the squares of the legs.

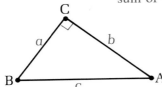

Hypothesis: Right $\triangle ABC$ with legs of lengths a and b and hypotenuse of length c.

Conclusion: $a^2 + b^2 = c^2$.

Proof.

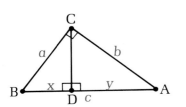

Project the legs of $\triangle ABC$ on the hypotenuse by drawing $\overline{CD}$ $\perp \overline{BA}$; let the lengths of the two projections be called x and y.

Since either leg of a right triangle is the mean proportional between the hypotenuse and its projection upon the hypotenuse,

$$\frac{c}{a} = \frac{a}{x} \quad \text{and} \quad \frac{c}{b} = \frac{b}{y}.$$

Hence,

$$a^2 = cx \quad \text{and} \quad b^2 = cy \text{ (the means-extremes theorem)}.$$

Adding,

$$a^2 + b^2 = cx + cy$$
$$= c(x + y).$$

Since $x + y = c$ (betweenness of points: D is between B and A),

$$a^2 + b^2 = c \cdot c = c^2.$$

Exercises

SET I

A set of three integers that can be the lengths of the sides of a right triangle is called a *Pythagorean triple*. We will call a right triangle all of whose sides have lengths that are integers a *Pythagorean triangle*.

The simplest Pythagorean triple is the set "3, 4, 5." These numbers are the lengths of the sides of a "3-4-5" Pythagorean right triangle. The list below contains all of the Pythagorean triples in which no number is more than 50.

3, 4, 5	9, 12, 15	14, 48, 50	20, 21, 29
5, 12, 13	9, 40, 41	15, 20, 25	21, 28, 35
6, 8, 10	10, 24, 26	15, 36, 39	24, 32, 40
7, 24, 25	12, 16, 20	16, 30, 34	27, 36, 45
8, 15, 17	12, 35, 37	18, 24, 30	30, 40, 50

1. Verify that the set "7, 24, 25" is a Pythagorean triple.

2. Can all three numbers of a Pythagorean triple be even?

3. Does it appear that all three numbers can be odd?

4. Some of the Pythagorean triples in the list above seem to be related to each other. For example, 3, 4, 5 and 6, 8, 10 and some of the other triples are related in a special way. Rewrite the list, separating it into sets of related triples.

5. Supply the reasons for the following proof that if a, b, and c are the lengths of the legs and hypotenuse of a right triangle, and k is a positive number, then a triangle with sides of lengths ka, kb, and kc is also a right triangle.

Proof.

Statements
_____

a) a, b, and c are the lengths of the legs and hypotenuse of a right triangle.
b) $a^2 + b^2 = c^2$.
c) $k^2a^2 + k^2b^2 = k^2c^2$.
d) $(ka)^2 + (kb)^2 = (kc)^2$.
e) A triangle with sides of lengths ka, kb, and kc is a right triangle.

6. Use the following two facts to explain why it is impossible for all three numbers of a Pythagorean triple to be odd.

   A number is odd iff its square is odd.
   The sum of two odd numbers is even.

_____

SET II

1. Find the lettered lengths in this figure.

2. Given: △ALP with $\overline{LS} \perp \overline{AP}$.
   Prove: $a^2 - x^2 = b^2 - y^2$.
   Par 6.

3. As an introduction to the Triangle Inequality Theorem, we considered the shortest path that an anteater might take in walking to a stream and then to an anthill. We assumed that the anteater and anthill are 45 feet and 30 feet from the stream, respectively, and showed that the shortest path is the one for which $\angle 1 = \angle 2$.

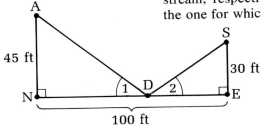

In this exercise, we will use the Pythagorean Theorem to find the length of this path.

a) Why is $\triangle AND \sim \triangle SED$?

b) Why is $\dfrac{AN}{SE} = \dfrac{ND}{ED}$?

c) If we let $ND = x$, what is ED?

d) Use the proportion in step b to find ND and ED.

e) Find AD and DS.

f) What is the length of the anteater's path?

SET III

The following proof exercise appears in a Greek geometry book. Can you figure it out?

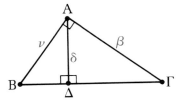

Αν ΑΔ εἶναί ἡ ἀπόστασις τῆς κορυφῆς Α ὀρθογωνίου τριγώνου ἀπὸ τὴν ὑποτείνουσαν ΒΓ, νὰ ἀποδείξητε ὅτι: $\dfrac{1}{\nu^2} + \dfrac{1}{\beta^2} = \dfrac{1}{\delta^2}$

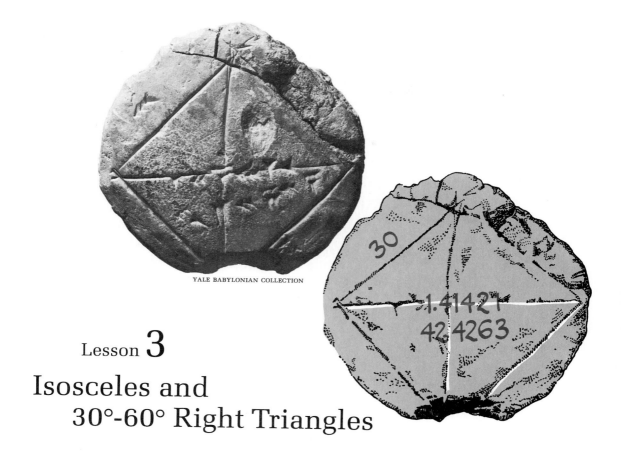

Lesson 3

Isosceles and
30°-60° Right Triangles

Archeologists digging in the land between the Tigris and Euphrates rivers in the late nineteenth century found thousands of clay tablets with writing on them. Some of these tablets, which date back to about 1700 B.C., reveal what the ancient Babylonians knew about mathematics. The one in the photograph above, for instance, shows that they knew the relationship between the length of a diagonal of a square and the length of one of its sides.* The three wedge-shaped symbols at the upper left indicate that the side of the square is 30 units long. The symbols along the diagonal represent the number 1.41421 and the symbols below them indicate that the diagonal is 42.4263 units long.

These numbers imply that the Babylonians knew of the Pythagorean Theorem more than a thousand years before Pythagoras was born! So it was not first discovered by the man for whom it is named.

---

*Asger Aaboe, *Episodes from the Early History of Mathematics* (Random House, 1964), pp. 25–27.

425

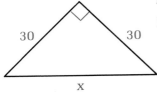

To understand how the Pythagorean Theorem applies, notice that each diagonal of a square is the hypotenuse of two isosceles right triangles. One of these is shown here.

By the Pythagorean Theorem,

$$x^2 = 30^2 + 30^2$$
$$= 2 \cdot 30^2$$
$$x = \sqrt{2 \cdot 30^2} = 30\sqrt{2}.$$

The square root of 2 is approximately 1.41421, so

$$x = 30(1.41421)$$
$$= 42.4263 \text{ (approximately)}.$$

The Babylonians recognized that to solve this problem, they could multiply the length of one leg of the triangle by $\sqrt{2}$ to find the length of the hypotenuse. It is this number, in fact, that is written along the diagonal on the tablet.

▶ **Theorem 62**
In an isosceles right triangle, the hypotenuse is $\sqrt{2}$ times the length of one leg.

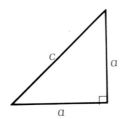

Hypothesis: An isosceles right triangle with legs of length a and hypotenuse c.

Conclusion: $c = a\sqrt{2}$.

Proof.
By the Pythagorean Theorem,

$$c^2 = a^2 + a^2 = 2a^2.$$
$$c = \sqrt{2a^2} = \sqrt{2} \cdot \sqrt{a^2} = \sqrt{2}a = a\sqrt{2}.$$

It is convenient to restate this theorem in terms of the diagonal of a square.

▶ **Corollary**
Each diagonal of a square is $\sqrt{2}$ times the length of one side.

Since each acute angle of an isosceles right triangle has a measure of 45°, we might call it a "45°-45° right triangle." Another important triangle in geometry is the "30°-60° right triangle."

▶ **Theorem 63**
In a 30°-60° right triangle, the hypotenuse is twice the length of the shorter leg and the longer leg is $\sqrt{3}$ times the length of the shorter leg.

Hypothesis: Right $\triangle ABC$ with right $\measuredangle C$
and $\angle A = 30°$, $\angle B = 60°$;
$BC = a$, $AC = b$, and $AB = c$.
Conclusion: $c = 2a$ and $b = a\sqrt{3}$.

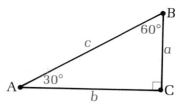

Proof.
Reflect $\triangle ABC$ through $\overleftrightarrow{AC}$. Since $\triangle AB'C \cong \triangle ABC$ (a triangle and its reflection through a line are congruent), $\angle B' = \angle B = 60°$ and $\angle CAB' = \angle CAB = 30°$; so $\angle BAB' = 60°$. Therefore, $\triangle ABB'$ is equiangular and, hence, equilateral.

Since $BB' = a + a = 2a$, $AB = c$, and $AB = BB'$, $c = 2a$.
By the Pythagorean Theorem, $a^2 + b^2 = c^2$. Substituting, we get:

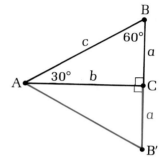

$$a^2 + b^2 = (2a)^2,$$

or

$$a^2 + b^2 = 4a^2.$$

$$b^2 = 3a^2 \text{ (subtraction)}.$$

Since $a > 0$ and $b > 0$,

$$b = \sqrt{3}a \text{ (square roots postulate)}$$

or

$$b = a\sqrt{3}.$$

Because $\overline{AC}$ is an altitude of $\triangle ABB'$, we can use the theorem we have just proved to derive a formula for the length of an altitude of an equilateral triangle.

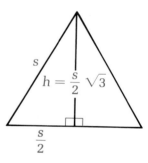

▶ **Corollary**
An altitude of an equilateral triangle is $\dfrac{\sqrt{3}}{2}$ times the length of one side.

In the exercises for this lesson, express all irrational answers in simple radical form.

Exercises

SET I

1. In an isosceles right triangle, the ratio of the length of the hypotenuse to the length of a leg is $\dfrac{\sqrt{2}}{1}$.

   a) What is the ratio of the length of a leg to the length of the hypotenuse?
   b) What is the ratio of the measure of one of the acute angles to the measure of the right angle?

2. Determine the following ratios for a 30°-60° right triangle.

a) $\dfrac{\text{hypotenuse}}{\text{shorter leg}}$.

b) $\dfrac{\text{longer leg}}{\text{shorter leg}}$.

c) $\dfrac{\text{hypotenuse}}{\text{longer leg}}$.

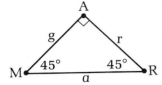

3. Find the following lengths in △MAR and △VEN.

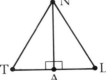

a) If $g = 5$, find a.
b) If $a = 7\sqrt{2}$, find r.
c) If $r = \sqrt{8}$, find a.
d) If $a = 6$, find g.
e) If $n = 4$, find e and d.
f) If $e = 10$, find n and d.
g) If $d = 3$, find n and e.

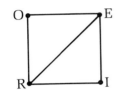

4. Find the following lengths in square ORIE and equilateral △NTL.

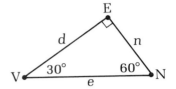

a) If OR = 8, find RE.
b) If RE = 8, find OR.
c) If NT = 9, find NA.
d) If NA = 9, find NT.

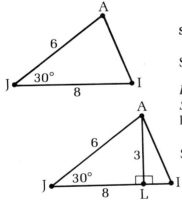

SET II

Study the following:

Example. Find the area of △JAI.
Solution. Draw altitude $\overline{AL}$, forming △JAL as shown at the left. Since △JAL is a 30°-60° right triangle,

$$AL = \tfrac{1}{2}JA = 3.$$

So

$$\alpha \triangle JAI = \tfrac{1}{2}(JI \cdot AL) = \tfrac{1}{2}(8 \cdot 3) = 12.$$

1. Find h if the base angles of this triangle each have a measure of
   a) 30°.
   b) 45°.
   c) 60°.

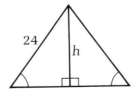

2. Find the area of each triangle below.

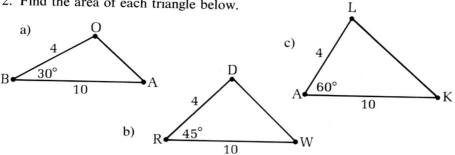

a)

c)

b)

3. Find the perimeter and area of each rhombus below.

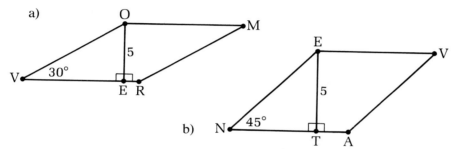

a)

b)

4. Find the area of each triangle below. (Hint: A wise choice of an altitude is important for part b.)

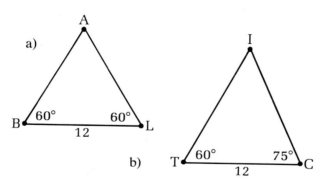

a)

b)

5. We have proved, by means of Heron's Theorem, that the area of an equilateral triangle with sides of length a is

$$\frac{a^2}{4} \sqrt{3}.$$

Use the corollary about the length of the altitude of an equilateral triangle to derive the same area formula in a different way.

This article about how to cut down a tree appeared in a news-paper several years ago. You know enough geometry now to be able to recognize that something in the article is incorrect. How would you change it so that it makes more sense?

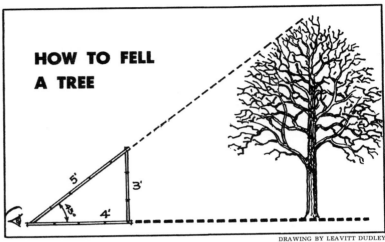

DRAWING BY LEAVITT DUDLEY

Tree removal used to be a hard, hazardous job, one better left to tree removal companies. But along came the lightweight chain saw and now it's easy to cut a 30-foot-tall tree into neat fireplace-length logs in a few hours. There's just one problem. Felling the tree is still hazardous, although it need not be. If you have analyzed its fall correctly and have made the proper cuts, it will fall exactly where you plan. But if you haven't, no amount of guy wires or ropes can be trusted to guide its crash to the ground.

The first thing you have to do is find how much room it needs when it falls. You can do this by using the Boy Scout method of measuring heights. A triangle with sides of three, four and five feet is placed so that you sight along the five-foot side while you move the triangle to a point where the top of the tree comes into sight, as shown above. That's how far the tree will reach when it's felled.

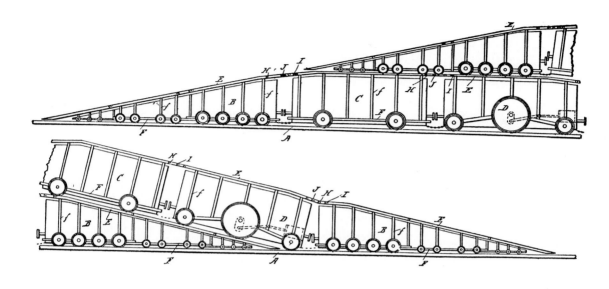

Lesson 4

The Tangent Ratio

Back in 1895, an unusual invention for preventing train collisions was registered in the U.S. Patent Office. It is illustrated in the figures above.* If a train meets another one approaching on the same track, it can run up the slanted rails of a special end car, along rails on top of the rest of the cars, and down the other end car. Instead of colliding, one train merely passes over the other! If two trains are traveling in the same direction, one can overtake and pass the other by the same procedure.

The figures reveal that the end cars are approximately right triangular in shape. If the box cars on the train are 14 feet high and the track on one of these end cars rises at a 30° angle with the horizontal, how long must the car be?

---

*From *Absolutely Mad Inventions* by A. E. Brown and H. A. Jeffcott, Jr. (Dover).

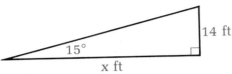

Looking at the adjoining figure, we see that the problem is to find the length of the longer leg of a 30°-60° right triangle. Since we know the length of the shorter leg and the longer leg is $\sqrt{3}$ times as long, we can write:

$$x = 14\sqrt{3} \approx 14(1.73) \approx 24.2 \text{ ft.}$$

The end car should be about 24 feet long.

If the track on the end car actually rose at a 30° angle with the horizontal, it would be too steep to be very safe. A smaller angle would be more practical.

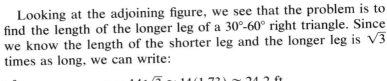

Suppose the track rose at a 15° angle instead. How long would the end car have to be? Since we have no theorem about the relationships of the lengths of the sides of a 15°-75° right triangle, we cannot immediately answer this question. In fact, there are only two right triangles for which, given the measures of the angles, we know the relative lengths of the sides. They are the 45°-45° and 30°-60° right triangles that we studied in the last lesson. It would be very inconvenient to have to prove and learn a long list of theorems for other right triangles. Instead, we will develop a method that will work for every possibility.

Look at the figure below in which $\triangle ABC$ and $\triangle DEF$ are right triangles with $\angle A = \angle D$. Since $\triangle ABC \sim \triangle DEF$, $\dfrac{BC}{EF} = \dfrac{AB}{DE}$. By the interchangeable theorem, $\dfrac{BC}{AB} = \dfrac{EF}{DE}$.

We have shown that if two right triangles have an acute angle of one equal to an acute angle of the other, then *the ratio of the length of the leg opposite that angle to the length of the other leg*, called the adjacent leg, *is the same* for both triangles. This ratio, then, depends only upon the measure of the acute angle; it does not depend upon the size of the triangle. It is called the *tangent* of the angle.

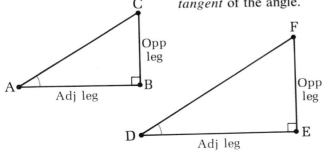

► Definition

The *tangent* of an acute angle of a right triangle is the ratio of the length of the opposite leg to the length of the adjacent leg.

In the figure shown here, the tangent of ∡A is $\frac{a}{b}$ and the tangent of ∡B is $\frac{b}{a}$. These relationships are customarily abbreviated as:

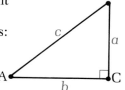

$$\tan A = \frac{a}{b} \quad \text{and} \quad \tan B = \frac{b}{a}.$$

The tangent is one of the basic ratios in the branch of mathematics called *trigonometry*. The word "trigonometry" is derived from two Greek words meaning "triangle" and "measurement." To use the tangent ratio to measure the legs of a right triangle, we need to know its values for angles of different measures. Some of these values are listed in the table shown here. Most of them have been rounded off, so they are not exact.

Angle	Tangent	Angle	Tangent
5°	.087	50°	1.192
10°	.176	55°	1.428
15°	.268	60°	1.732
20°	.364	65°	2.145
25°	.466	70°	2.747
30°	.577	75°	3.732
35°	.700	80°	5.671
40°	.839	85°	11.430
45°	1.000		

We can now use the tangent ratio to find the length of the end car if the track on it rises at a 15° angle with the horizontal.

$$\tan 15° = \frac{14}{x}$$

Multiplying by x, $x \tan 15° = 14$, so $x = \frac{14}{\tan 15°}$. From the table, $\tan 15° \approx .268$, so

$$x \approx \frac{14}{.268} \approx 52.2 \text{ ft.}$$

For a rise of 15°, the end car would be about 52 feet long.

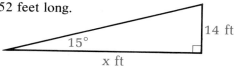

1. Find the tangent of the lettered angle in each figure below.

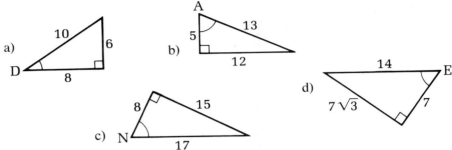

a) b) c) N d)

2. In this figure, $\overline{SE} \perp \overline{LE}$ and each of the five small angles at L has a measure of 10°. Use a ruler and protractor to make an accurate enlargement of this figure so that LE = 100 mm.

In your enlargement, the length of EI should be about 18 mm. For △LIE, we can write

$$\tan 10° = \frac{EI}{LE} = \frac{18 \text{ mm}}{100 \text{ mm}} = .18$$

which is the same as the value in the table on page 433 after it has been rounded off.

a) Measure the following lengths in your enlargement, each to the nearest millimeter: EN, EA, EP, and ES.

b) Using the method illustrated above, find approximate values of the following tangents and see how close they come to those listed in the table: tan 20°, tan 30°, tan 40°, and tan 50°.

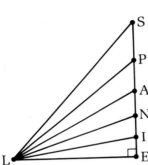

3. Solve for the length marked x in each figure. Use the table of tangents on page 433 and express each length to the nearest tenth.

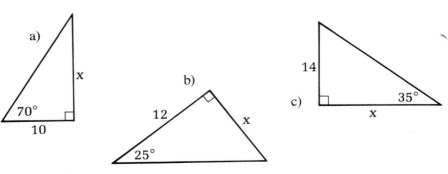

a) b) c)

4. If we know the lengths of both legs of a right triangle, we can use the tangent ratio to find the measure of either of its acute angles. For example, in the figure at the right,

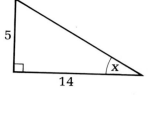

$$\tan x = \frac{5}{14} \approx .36$$

so

$$x \approx 20°.$$

Solve for the angle marked x in each figure below. Round each tangent to the nearest hundredth and use the table on page 433.

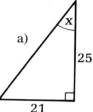

a)

b)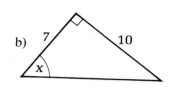

c)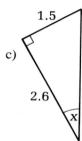

SET II

1. In this figure $\triangle HUK$ is a right triangle.
   a) Suppose $\angle H = 35°$ and $\angle U = 55°$. Find the tangents of these angles in the table and multiply. In other words, find $(\tan 35°)(\tan 55°)$.
   b) If you remember that most of the tangents in the table are not exact, you will realize that your answer to part a is not exact either. To what number is it almost equal?
   c) Use $\triangle HUK$ to explain why, for *any* right triangle,

   $$(\tan H)(\tan U) = 1.$$

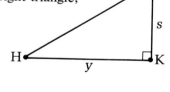

2. In this figure, $\overline{HW} \perp \overline{CW}$.
   a) Explain why $\angle HCW > \angle OCW$.
   b) Explain why $\tan \angle HCW > \tan \angle OCW$.

3. Given: In right $\triangle PUG$, $\angle P < \angle U$.
   Prove: $\tan P < 1$.
   Par 5.

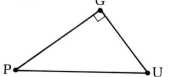
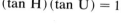

A palm tree 21 feet tall casts a horizontal shadow 30 feet long. If the sun rose at 6:00 A.M. and will be directly overhead at noon, can you figure out what time it is?

Lesson **5**

The Sine and Cosine Ratios

Having given up on basketball, the king of Id has found a sport he likes. Suppose that after the serve shown in this cartoon the king returns the ball so that it hits the table at a 20° angle. If he hits it from a height of 3 feet, how far does the ball travel before hitting the table?

In the figure, we see that the problem is to find the length of the hypotenuse of a right triangle, given the measure of an acute angle and the length of the opposite leg. One way to solve this would be first to use the tangent ratio to find the length of the other leg and then to use the Pythagorean Theorem to find the length of the hypotenuse. This method, however, requires a lot of computation. A simpler method would be to find the length of the hypotenuse directly. We can do this by means of a trigonometric ratio called the *sine* ratio.

437

► Definition

The *sine* of an acute angle of a right triangle is the ratio of the length of the opposite leg to the length of the hypotenuse.

In this figure, the sine of ⊀A is $\frac{a}{c}$ and the sine of ⊀B is $\frac{b}{c}$. These relationships are abbreviated as:

$$\sin A = \frac{a}{c} \quad \text{and} \quad \sin B = \frac{b}{c}.$$

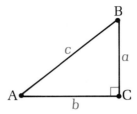

The sines of angles of different measures, like their tangents, can be found in tables. A table listing these ratios for angles in 1° intervals appears on the next page.

We will now use the sine ratio to solve the problem of how far the ball would travel if it hit the table at a 20° angle.

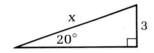

$$\sin 20° = \frac{3}{x}.$$

Multiplying by x, $x \sin 20° = 3$, so $x = \dfrac{3}{\sin 20°}$. We see from the table that $\sin 20° \approx .342$, so

$$x \approx \frac{3}{.342} \approx 8.77 \text{ ft.}$$

Another useful trigonometric ratio is the *cosine*. It relates, for a given acute angle in a right triangle, the lengths of the adjacent leg and the hypotenuse.

► Definition

The *cosine* of an acute angle of a right triangle is the ratio of the length of the adjacent leg to the length of the hypotenuse.

In the figure above, the cosine of ⊀A is $\frac{b}{c}$ and the cosine of ⊀B is $\frac{a}{c}$. These relationships are abbreviated as:

$$\cos A = \frac{b}{c} \quad \text{and} \quad \cos B = \frac{a}{c}.$$

Table of Trigonometric Ratios

A	sin A	cos A	tan A	A	sin A	cos A	tan A
1°	.017	1.000	.017	46°	.719	.695	1.035
2°	.035	.999	.035	47°	.731	.682	1.072
3°	.052	.999	.052	48°	.743	.669	1.111
4°	.070	.998	.070	49°	.755	.656	1.150
5°	.087	.996	.087	50°	.766	.643	1.192
6°	.105	.995	.105	51°	.777	.629	1.235
7°	.122	.993	.123	52°	.788	.616	1.280
8°	.139	.990	.141	53°	.799	.602	1.327
9°	.156	.988	.158	54°	.809	.588	1.376
10°	.174	.985	.176	55°	.819	.574	1.428
11°	.191	.982	.194	56°	.829	.559	1.483
12°	.208	.978	.213	57°	.839	.545	1.540
13°	.225	.974	.231	58°	.848	.530	1.600
14°	.242	.970	.249	59°	.857	.515	1.664
15°	.259	.966	.268	60°	.866	.500	1.732
16°	.276	.961	.287	61°	.875	.485	1.804
17°	.292	.956	.306	62°	.883	.469	1.881
18°	.309	.951	.325	63°	.891	.454	1.963
19°	.326	.946	.344	64°	.899	.438	2.050
20°	.342	.940	.364	65°	.906	.423	2.145
21°	.358	.934	.384	66°	.914	.407	2.246
22°	.375	.927	.404	67°	.921	.391	2.356
23°	.391	.921	.424	68°	.927	.375	2.475
24°	.407	.914	.445	69°	.934	.358	2.605
25°	.423	.906	.466	70°	.940	.342	2.747
26°	.438	.899	.488	71°	.946	.326	2.904
27°	.454	.891	.510	72°	.951	.309	3.078
28°	.469	.883	.532	73°	.956	.292	3.271
29°	.485	.875	.554	74°	.961	.276	3.487
30°	.500	.866	.577	75°	.966	.259	3.732
31°	.515	.857	.601	76°	.970	.242	4.011
32°	.530	.848	.625	77°	.974	.225	4.331
33°	.545	.839	.649	78°	.978	.208	4.705
34°	.559	.829	.675	79°	.982	.191	5.145
35°	.574	.819	.700	80°	.985	.174	5.671
36°	.588	.809	.727	81°	.988	.156	6.314
37°	.602	.799	.754	82°	.990	.139	7.115
38°	.616	.788	.781	83°	.993	.122	8.144
39°	.629	.777	.810	84°	.995	.105	9.514
40°	.643	.766	.839	85°	.996	.087	11.430
41°	.656	.755	.869	86°	.998	.070	14.301
42°	.669	.743	.900	87°	.999	.052	19.081
43°	.682	.731	.933	88°	.999	.035	28.636
44°	.695	.719	.966	89°	1.000	.017	57.290
45°	.707	.707	1.000				

Exercises

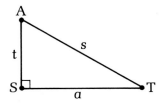

1. Express the following ratios for △SAT and △IRE in terms of the lengths of their sides.
   a) sin A.
   b) cos A.
   c) tan A.
   d) sin I.
   e) cos I.
   f) tan I.

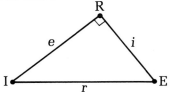

2. Use the table on page 439 to find each of the following angles. If the ratio does not appear in the table, choose the angle whose ratio is closest to it.
   a) ∠P if sin P = .731.
   b) ∠A if cos A = .927.
   c) ∠R if tan R = .810.
   d) ∠O if sin O = .340.
   e) ∠D if tan D = 1.615.
   f) ∠Y if cos Y = .095.

3. Which ratio would you use to solve for the part labeled x in each of the following figures, if the only parts you know are the ones labeled a and b?

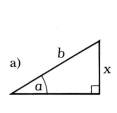

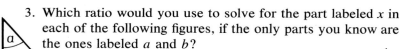

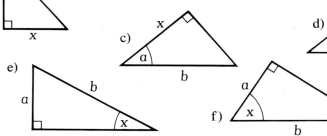

4. Solve for x in each figure. Express each length to the nearest tenth.

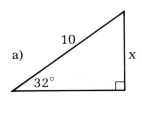

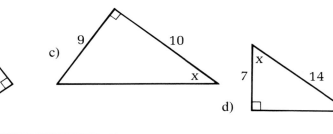

Write and solve an equation, using an appropriate trigonometric ratio, for each of the following exercises. Express each length to the nearest unit. The figures are not drawn accurately.

1. How tall is a totem pole if the line of sight from a point 100 feet from its base makes an angle of 39° with the horizontal?

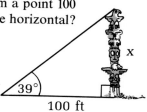

100 ft

2. A crow sitting on a tree branch 25 feet above the ground sees something to eat on the ground and swoops down upon it at an angle of 20° with the horizontal. How long is the crow's flight?

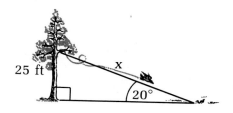

3. In teeing off, a golfer hooks his drive at an angle of 14° from the line to the hole. The hole is 150 yards from the tee. If the ball ends up due left of the hole,
   a) how far is it from the hole?
   b) how far did the ball go?

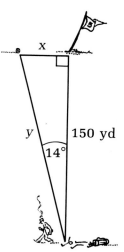

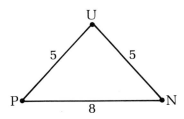

4. Copy the figure shown here. The problem is to figure out the sine of ∡P.

 a) Obtuse Ollie says that sin P $= \frac{5}{8}$. What is wrong with this conclusion?

 b) After adding something to the figure, Acute Alice decides that sin P $= \frac{3}{5}$. Explain.

5. If $\overline{EO}$ is an altitude of △JKE, show that

$$\alpha\triangle JKE = \tfrac{1}{2}ab \sin J.$$

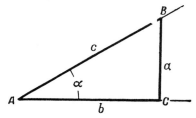

SET III

In this passage from a Russian trigonometry book, a relationship is derived that is so useful that it is considered a theorem in the subject. Can you explain what the passage is about?

Возьмем произвольный острый угол α и построим прямоугольный треугольник *ABC*, у которого один из острых углов равен α.   Пусть *BC* = *a*, *CA* = *b* и *AB* = *c*.

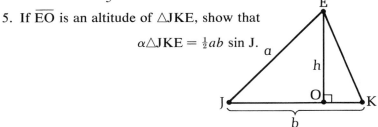

Из прямоугольного треугольника *ABC* по теореме Пифагора находим:

$$a^2 + b^2 = c^2.$$

Разделив обе части этого равенства на *c²*, получим:

$$\left(\frac{a}{c}\right)^2 + \left(\frac{b}{c}\right)^2 = 1,$$

или

$$\sin^2 \alpha + \cos^2 \alpha = 1,$$

так как $\frac{a}{c} = \sin \alpha$ и $\frac{b}{c} = \cos \alpha.$

Chapter 11 / Summary and Review

Basic Ideas

Cosine 438
Projection 417
Pythagorean triple 422
Sine 438
Tangent 433

Theorems

61. The altitude to the hypotenuse of a right triangle forms two triangles that are similar to each other and to the original triangle. 415
    Corollary 1. The altitude to the hypotenuse of a right triangle is the mean proportional between the segments into which it divides the hypotenuse. 415
    Corollary 2. Each leg of a right triangle is the mean proportional between the hypotenuse and its projection on the hypotenuse. 416
    The Pythagorean Theorem. In a right triangle, the square of the hypotenuse is equal to the sum of the squares of the legs. 422
62. In an isosceles right triangle, the hypotenuse is $\sqrt{2}$ times the length of one leg. 426
    Corollary. Each diagonal of a square is $\sqrt{2}$ times the length of one side. 426
63. In a 30°-60° right triangle, the hypotenuse is twice the length of the shorter leg and the longer leg is $\sqrt{3}$ times the length of the shorter leg. 426
    Corollary. An altitude of an equilateral triangle is $\frac{\sqrt{3}}{2}$ times the length of one side. 427

Exercises

1. Without referring to a table, solve for x in each figure.

a)

b)

c)

d)

e)

f)

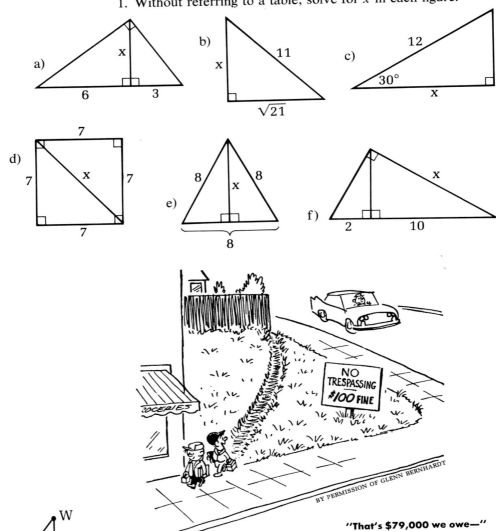

BY PERMISSION OF GLENN BERNHARDT

"That's $79,000 we owe—"

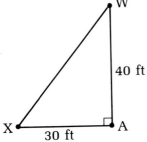

2. Suppose, for simplicity, that the shortcut taken by the kids in this cartoon is the hypotenuse of a right triangle along whose legs they are supposed to walk.
   a) If WA = 40 ft and AX = 30 ft, find WX.
   b) What distance do they save by taking the shortcut?
   c) Find $\angle W$ and $\angle X$, each to the nearest degree.

3. Draw a figure to illustrate each part of the following exercise. Find, without referring to a table, the exact length of the projection of $\overline{AB}$ on line ℓ if AB = 10 and
   a) $\angle 1 = 30°$.
   b) $\angle 1 = 45°$.
   c) $\angle 1 = 60°$.
   d) $\angle 1 = 90°$.

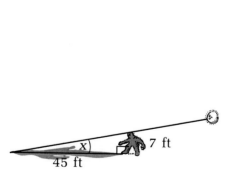

   e) Do you notice a pattern in your answers to parts a–d?

SET II

Use the table of trigonometric ratios on page 439 to solve the following problems. Express each answer to the nearest unit.

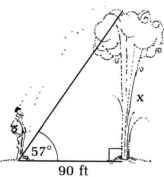

1. How high is an eruption of Old Facefull if the line of sight from a point 90 feet from its base makes an angle of 57° with the horizontal?

2. A 25-foot ladder leans against a burning building. If it just reaches a window 18 feet above the ground, what angle does it make with the horizontal?

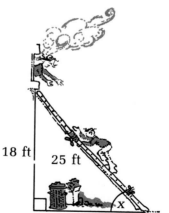

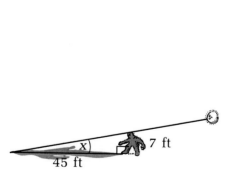

3. What is the angle of elevation of the moon if a 7-foot creature that walks in the night casts a shadow 45 feet long?

4. Explain why, in a right triangle, the sine of one acute angle is equal to the cosine of the other. In other words, why, in right $\triangle GUM$, sin G = cos U.

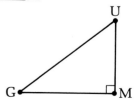

5. In $\square$PAST, $\overline{AE} \perp \overline{PT}$, PA = 6, AS = 7, and sin P = $\dfrac{\sqrt{5}}{3}$.
   a) Find AE.
   b) Find $\alpha\square$PAST.
   c) Find PE.
   d) Find $\alpha\triangle$PEA.

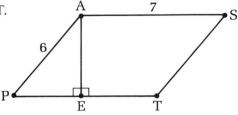

SET III

Because of the structure of their eyes, certain kinds of insects do not see straight ahead. As a result, they do not move directly toward a light but travel in a complex path around it. The path is, in fact, a spiral comparable to that of the shell of the chambered nautilus with which we began this chapter.

The insect moves so that it always sees the light at the same angle. An approximation of part of its path is shown in this figure in which the rays represent rays of light from a candle at point C. The successive segments of the path form equal angles with the successive rays of light—in this case, right angles.

If the small angles formed by the light rays each have a measure of 30° and the insect starts at point A, 64 centimeters from the candle, can you figure out how far from the candle it is when it has gone halfway around it and is at point B?

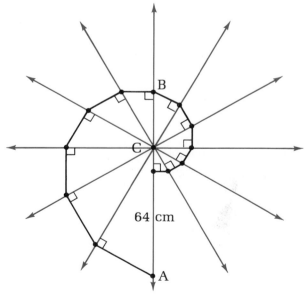

CIRCLES

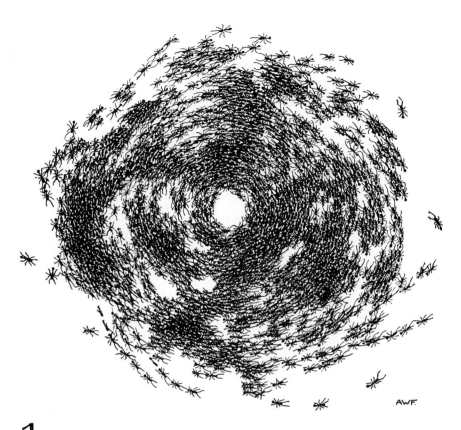

Lesson **1**

Circles, Radii, and Chords

Army ants are blind and find their way by following the scent trails left by other ants. When a group of these ants is prevented from traveling along its usual path, the ants sometimes begin milling around in circles. The ants shown in this drawing walked around like this for more than thirty hours, stopping only when all of them were dead.*

The path in which the ants are walking is the most symmetric of all geometric figures. In fact, it was in our study of symmetry that we first defined the term "circle." In this chapter, we will consider more of the properties of this figure; we begin by reviewing a few familiar terms and introducing some new ones.

---

*T. C. Schneirla, *Army Ants*, edited by H. R. Topoff (W. H. Freeman and Company, 1971). Drawing courtesy of the American Museum of Natural History.

▶ Definition

A *circle* is the set of all points in a plane that are at a given distance from a given point in the plane.

The given point is called the *center* of the circle. Two or more circles that lie in the same plane and that have the same center are said to be *concentric*. The ants in the picture seem to be walking in concentric circles.

The given distance in the definition of a circle is called its *radius*. A point is *inside*, *on*, or *outside* a circle depending upon whether its distance from the center is *less than*, *equal to*, or *more than* the circle's radius.

A circle is ordinarily named for its center, so that the adjoining figure illustrates circle O with radius OP. Point A is inside circle O, so OA < OP. Point B is outside, so OB > OP.

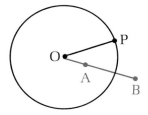

We have defined the radius of a circle to be a *number*. In circle O, the radius is OP, the *distance* between O and P. The word "radius" is also used to name the *line segment*, $\overline{OP}$.

▶ Definition

A *radius* of a circle is a line segment that joins the center of the circle to a point of the circle.

It would be better not to use the term "radius" in two different ways, but this is what is ordinarily done. The context in which the word is used, however, should make it clear whether we mean a *line segment* or a *number*.

A useful fact that follows directly from the definitions of "circle" and "radius" is:

▶ Corollary

All radii of a circle are equal.

The adjoining figure illustrates two more line segments related to circles that are of interest; $\overline{AB}$ is called a *chord* of circle O and $\overline{CD}$ is called a *diameter*.

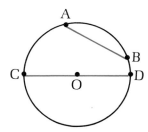

▶ Definitions

A *chord* of a circle is a line segment that joins two points of the circle.

A *diameter* of a circle is a chord that contains the center.

The word "diameter," like "radius," is used in two ways. It refers to both a *line segment* and the *number* that is its length. Thinking in terms of numbers, if we represent a circle's diameter by d and its radius by r, then

$$d = 2r.$$

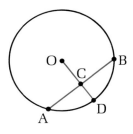

The adjoining figure shows a circle in which a radius intersects a chord. If the radius is perpendicular to the chord, we can prove that it bisects it. Conversely, if the radius bisects the chord, we can show that it must be perpendicular to it. Symbolically,

and

$$\overline{OD} \perp \overline{AB} \rightarrow AC = CB$$

$$AC = CB \rightarrow \overline{OD} \perp \overline{AB}.$$

We will prove these relationships, along with a third one, as theorems.

▶ **Theorem 64**
If a line through the center of a circle is perpendicular to a chord, it also bisects it.

▶ **Theorem 65**
If a line through the center of a circle bisects a chord that is not a diameter, it is also perpendicular to it.

▶ **Theorem 66**
The perpendicular bisector of a chord (in the plane of a circle) passes through the center of the circle.

Exercises

SET I

1. Which do you think is the better illustration of a circle: a perfectly round pizza or a bicycle tire? Explain your answer.

2. In the figure below, one vertex of △JIG is the center of the circle.
   a) What is the name of the circle?
   b) How many vertices of the triangle are on the circle?
   c) What are $\overline{JI}$ and $\overline{JG}$ called?
   d) What is $\overline{IG}$ called?
   e) What kind of triangle must △JIG be?
   f) Why?
   g) If $\angle I = 60°$, does it follow that △JIG must be equilateral?
   h) Explain.

3. Complete the following proof of Theorem 64 by supplying the reasons.

If a line through the center of a circle is perpendicular to a chord, it also bisects it.

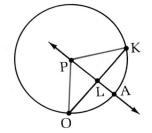

Hypothesis: Circle P with $\overleftrightarrow{PA} \perp \overline{OK}$.
Conclusion: $\overleftrightarrow{PA}$ bisects $\overline{OK}$.

Proof.

<u>Statements</u>

a) Circle P with $\overleftrightarrow{PA} \perp \overline{OK}$.
b) ∡PLO and ∡PLK are right angles.
c) Draw $\overline{PO}$ and $\overline{PK}$.
d) △PLO and △PLK are right triangles.
e) PO = PK.
f) PL = PL.
g) △PLO ≅ △PLK.
h) OL = LK.
i) $\overleftrightarrow{PA}$ bisects $\overline{OK}$.

SET II

1. Explain why the reflection of a circle through its center is the circle itself. (Reflection through a point is defined on page 219.)

2. Use the adjoining figure to prove Theorem 65.

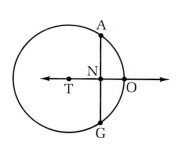

If a line through the center of a circle bisects a chord that is not a diameter, it is also perpendicular to it.

Hypothesis: Circle T, $\overleftrightarrow{TO}$ bisects $\overline{AG}$.
Conclusion: $\overrightarrow{TO} \perp \overline{AG}$.
Par 5.   (Hint: Use Theorem 12, p. 179.)

3. Obtuse Ollie says that, if a line through the center of a circle bisects a chord, it is also perpendicular to it. Acute Alice thinks that this isn't necessarily true. Whom do you agree with?

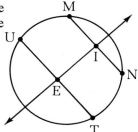

4. Given: $\overleftrightarrow{IE}$ is the perpendicular bisector of $\overline{MN}$; $\overline{MN} \parallel \overline{UT}$.
   Prove: $\overleftrightarrow{IE}$ bisects $\overline{UT}$ *without* adding anything to the figure.
   Par 5.

5. Supply the missing reasons for the following indirect proof of Theorem 66.

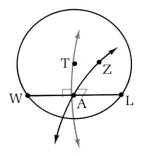

The perpendicular bisector of a chord (in the plane of a circle) passes through the center of the circle.

Hypothesis: Circle T with $\overleftrightarrow{ZA}$ the perpendicular bisector of $\overline{WL}$.
Conclusion: $\overleftrightarrow{ZA}$ contains point T.

Proof.
Line $\overleftrightarrow{ZA}$ is the perpendicular bisector of $\overline{WL}$ in circle T. Suppose $\overleftrightarrow{ZA}$ does not contain point T. Draw $\overleftrightarrow{TA}$.
a) What permits us to do this?

Then $\overleftrightarrow{TA} \perp \overline{WL}$.
b) Why?

So through point A there are two lines perpendicular to $\overline{WL}$: $\overleftrightarrow{TA}$ and $\overleftrightarrow{ZA}$. But this is impossible.
c) Why?

Therefore, the assumption that $\overleftrightarrow{ZA}$ does not contain point T is wrong. So $\overleftrightarrow{ZA}$ contains point T.

SET III

Instead of doing his geometry assignment, Dilcue decided to blow some soap bubbles. One of them turned out to look like the one shown here.

Suppose that Dilcue's "double bubble" consists of two spheres whose centers are 14 centimeters apart. (A sphere is the set of all points in space at a given distance from a given point.) If the two spheres intersect in a circle whose radius is 12 centimeters and the radius of one sphere is 13 centimeters, can you figure out the radius of the other?

Lesson 2

Tangents

One way to propel an object with great force into space is to whirl it around a circular path several times before releasing it. The diagram at the left below represents a heavy object that is tied to one end of a rope at P and spun around in a circle whose center is the other end of the rope at O. If the rope is let go while the object is traveling in this circular path, it will fly off along a path that appears to be a straight line.

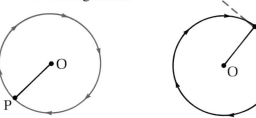

The second diagram shows the relationship of this line to the circle: it intersects it in exactly one point, the point that is the position of the object when the rope is released. Such a line is called a *tangent* to the circle.

453

► Definition

A *tangent* to a circle is a line in the plane of the circle that intersects it in exactly one point.

You probably know that one use of the word "tangent" in everyday conversation refers to someone "going off on a tangent," meaning that they are changing suddenly from one train of thought or course of action to another. Our illustration of the change in path of the whirling object reveals where this expression came from.

In circle O on the preceding page, what relationship does the tangent line seem to have to the radius drawn to the point of contact, P? They look as if they are perpendicular, and we will prove this as our next theorem.

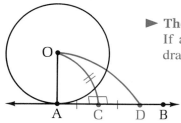

► **Theorem 67**

If a line is tangent to a circle, it is perpendicular to the radius drawn to the point of contact.

Hypothesis: $\overleftrightarrow{AB}$ is tangent to circle O at point A.

Conclusion: $\overleftrightarrow{AB} \perp \overline{OA}$.

We will prove this theorem by the indirect method.

Proof.
Line $\overrightarrow{AB}$ is tangent to circle O at point A. Suppose $\overleftrightarrow{AB}$ not $\perp \overline{OA}$ (i.e., $\overline{OA}$ not $\perp \overleftrightarrow{AB}$). We can draw $\overline{OC} \perp \overleftrightarrow{AB}$ since, through a point not on a line, there is exactly one perpendicular to the line. Next we choose point D on the ray opposite $\overrightarrow{CA}$ such that CD = CA, and draw $\overline{OD}$.

Now $\triangle OCA \cong \triangle OCD$ (S.A.S.), so OD = OA. Since OA is the radius of the circle, this means that point D is also at this distance from the center. Therefore, D must be a point on the circle since a circle is the set of *all* points in the plane at the distance of the radius from the center.

So $\overleftrightarrow{AB}$ intersects the circle in *two* points: A and D. But this contradicts the hypothesis that AB is tangent to the circle, since a tangent to a circle intersects it in exactly one point. This means that our assumption that $\overleftrightarrow{AB}$ not $\perp \overline{OA}$ is wrong, and so $\overleftrightarrow{AB} \perp \overline{OA}$.

The converse of this theorem is also true:

► **Theorem 68**

If a line is perpendicular to a radius at its outer endpoint, then it is tangent to the circle.

1. How many tangents do you think can be drawn to a circle
   a) at a point on the circle?
   b) from a point outside it?
   c) through a point inside it?

Exercises

2. Supply the missing reasons for the following indirect proof of Theorem 68.

   If a line is perpendicular to a radius at its outer endpoint, then it is tangent to the circle.

   Hypothesis: Circle J with $\overleftrightarrow{DE} \perp \overline{JD}$.
   Conclusion: $\overleftrightarrow{DE}$ is tangent to circle J.

   Proof.
   Suppose that $\overleftrightarrow{DE}$ is not tangent to circle J and that it intersects the circle in a second point, U. Now, $\overleftrightarrow{DE} \perp \overline{JD}$, so ∡JDU is a right angle and $\angle JDU = 90°$.
   We also know that JD = JU.
   a) Why?

   So $\angle JDU = \angle JUD$.
   b) Why?

   It follows that $\angle JUD = 90°$, so that ∡JUD is a right angle and $\overleftrightarrow{UE} \perp \overline{JU}$. But this result contradicts several theorems that we have previously proved.
   c) What is one of them?

   So the assumption that $\overleftrightarrow{DE}$ intersects circle J in more than one point is wrong. Therefore, $\overleftrightarrow{DE}$ is tangent to circle J.

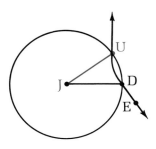

3. **Construction 9**
   To construct the tangent to a circle at a given point on the circle.

   Draw a circle, mark a point on it, and then use your straightedge and compass to construct a line tangent to the circle at the given point. Base your construction on Theorem 68.

4. In the adjoining figure, there are exactly three lines that are tangent to both circles. These lines are called *common tangents* of the circles.
   Can a pair of circles be drawn so that they have
   a) four common tangents?
   b) exactly two common tangents?
   c) only one common tangent?
   d) no common tangents?
   e) more than four common tangents?

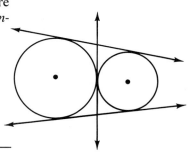

1. In the figure below, $\overline{HO}$ and $\overline{HN}$ are tangents to circle J and $\overline{JO}$ and $\overline{JN}$ are radii.
   Explain why ∡J and ∡H must be supplementary.

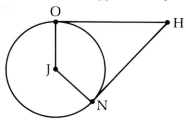

2. Given: $\overline{AE}$ is tangent to circle J at A and to circle S at E, and $\overline{JS}$ joins the centers of the two circles.
   Prove: ∠J = ∠S.
   Par 4.

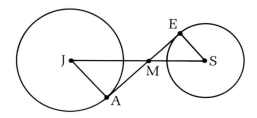

3. Given: M is the center of both circles and $\overline{AK}$ is tangent to the smaller circle at R.
   Prove: AR = RK.
   Par 5.

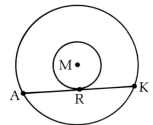

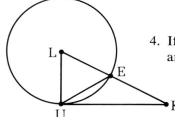

4. If $\overline{UK}$ is tangent to circle L and UE = LU, present a logical argument to explain why $\overline{UE}$ bisects $\overline{LK}$.

An Exercise in Op Art. Draw a circle on your paper that has the same radius as the circle shown here, place it over it and trace the set of 18 points lettered A through R.

Now draw the set of chords $\overline{AI}$, $\overline{BJ}$, $\overline{CK}$, and so forth, around the circle. (The last chord will be $\overline{RH}$.)

Then draw the following sets of chords, each set in a different color:

$$\overline{AH}, \overline{BI}, \overline{CJ}, \ldots, \overline{RG}.$$

$$\overline{AG}, \overline{BH}, \overline{CI}, \ldots, \overline{RF}.$$

$$\overline{AF}, \overline{BG}, \overline{CH}, \ldots, \overline{RE}.$$

$$\overline{AE}, \overline{BF}, \overline{CG}, \ldots, \overline{RD}.$$

What do you notice in the finished figure?

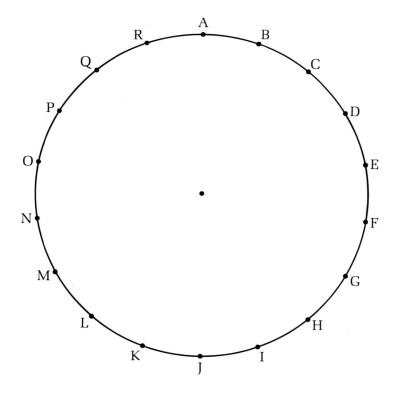

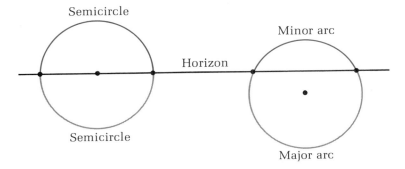

Semicircle

Minor arc

Horizon

Semicircle

Major arc

Lesson **3**

Central Angles and Arcs

A rainbow is produced when sunlight hits a bank of raindrops in either a cloud or falling rain. The drops act as tiny prisms, the apparent color of each drop being determined by the angle at which it is viewed. The rainbow itself consists of a set of circular arcs that have a common center. If the sun is on the horizon, exactly half of each of these circles, called a *semicircle*, can be seen. For positions of the sun above the horizon, the center of the rainbow is below the horizon. As a result, the visible part of the rainbow is less than a semicircle. It is called a *minor arc*.

458

To indicate how much of the rainbow can be seen, it is convenient to measure its arc by means of an angle at its center. For instance, if ∢AOB in the diagram has a measure of 120°, then we say that arc AB (written $\overset{\frown}{AB}$) also has a measure of 120°. Angle AOB is called a *central angle* of circle O.

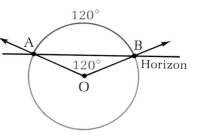

► Definitions
A **central angle** of a circle is an angle whose vertex is the center of the circle.

A **minor arc** of a circle is the set of the points of the circle that lie on or inside a central angle.

A **major arc** of a circle is the set of the points of the circle that lie on or outside a central angle.

A **semicircle** is the set of the points of the circle that lie on or on one side of a line containing a diameter.

Every pair of points on a circle determines two arcs. A minor arc is usually named by just these two points, called its endpoints. A major arc or a semicircle is named with three letters, the middle letter naming a third point on the arc. In the circle at the right, for example, the symbol $\overset{\frown}{AB}$ refers to the minor arc (in color) and the symbol $\overset{\frown}{ACB}$ refers to the major arc (in black).

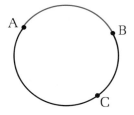

We have seen that an arc can be measured by the central angle associated with it; we will call this the "degree measure" of the arc and represent it by the letter *m*. If ∠AOB = 120°, then $m\overset{\frown}{AB} = 120°$ and $m\overset{\frown}{ACB} = 240°$.

► Definitions
The **degree measure** of a
   minor arc is the measure of its central angle;
   semicircle is 180°;
   major arc is 360° minus the measure of the corresponding minor arc;
   circle is 360°.

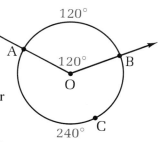

In the circle above, it appears that $m\overset{\frown}{AC} + m\overset{\frown}{CB} = m\overset{\frown}{ACB}$. Although this equation is similar to the one for betweenness of points, we cannot say that C is between A and B because the three points are not collinear. We will assume that this measure relation between arcs of a circle is true without proving it.

► **Postulate 20** (The Arc Addition Postulate)
If C is on $\overset{\frown}{AB}$, then $m\overset{\frown}{AC} + m\overset{\frown}{CB} = m\overset{\frown}{ACB}$.

For simplicity, we will refer to arcs that have equal degree measures as "equal arcs." It is obvious from the definition of the degree measure of an arc that the following theorem is true.

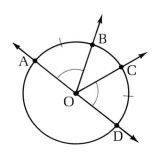

▶ **Theorem 69**

In a circle, two central angles are equal iff their minor arcs are equal.

This is illustrated in the first figure shown here. The second figure suggests a comparable relationship between chords and their minor arcs.

▶ **Theorem 70**

In a circle, two chords are equal iff their minor arcs are equal.

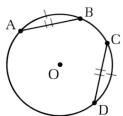

Exercises

SET I

1. The two circles in this figure are concentric; $\overline{DN}$ is a diameter of the larger circle and ∢DRE is acute.
   a) Name the minor arcs of the larger circle.
   b) Name two major arcs of the smaller circle.
   c) What kind of arc is $\overset{\frown}{DEN}$?
   d) Which arc has the greater measure: $\overset{\frown}{DE}$ or $\overset{\frown}{MK}$?

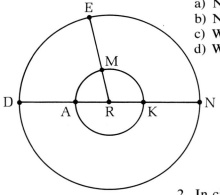

2. In circle Y, $\overline{YA} \perp \overline{TL}$ and $m\overset{\frown}{IT} = 40°$.
   a) Find $m\overset{\frown}{TA}$.
   b) Find $m\overset{\frown}{IA}$.
   c) Find $m\overset{\frown}{ILA}$.
   d) Find $m\overset{\frown}{ILT}$.

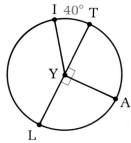

3. The statement of Theorem 69 actually contains two theorems:

   A. In a circle, if two central angles are equal, their minor arcs are equal.

   B. In a circle, if two minor arcs are equal, their central angles are equal.

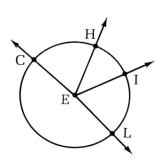

Supply the reasons for the following proof of theorem A.

Proof.

Statements

a) $\angle CEH = \angle IEL$.
b) $m\widehat{CH} = \angle CEH$ and $m\widehat{IL} = \angle IEL$.
c) $m\widehat{CH} = m\widehat{IL}$.
d) What change could be made in the proof above in order to prove theorem B instead?

4. Theorem 70 also contains two theorems:

   A. In a circle, if two chords are equal, their minor arcs are equal.

   B. In a circle, if two minor arcs are equal, their chords are equal.

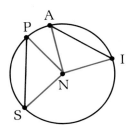

Supply the reasons for the following proof of theorem A.

Proof.

Statements

a) Draw $\overline{NS}$, $\overline{NP}$, $\overline{NA}$, and $\overline{NI}$.
b) $NS = NA$ and $NP = NI$.
c) $SP = AI$.
d) $\triangle SNP \cong \triangle ANI$.
e) $\angle SNP = \angle ANI$.
f) $m\widehat{SP} = m\widehat{AI}$.

SET II

1. Prove Theorem 70B.

   In a circle, if two minor arcs are equal, their chords are equal.

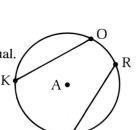

      Hypothesis: Circle A with $m\widehat{KO} = m\widehat{RE}$.
      Conclusion: $KO = RE$.
      Par 6.

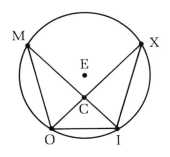

2. Given: Circle E with MO = XI.
   Prove: MI = XO.
   Par 6.

3. Given: Circle H with diameter $\overline{CI}$;
            $\overline{CA} \parallel \overline{HN}$.
   Prove: $m\overset{\frown}{AN} = m\overset{\frown}{NI}$.
   Par 8.

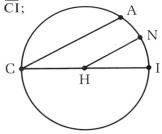

SET III

This photograph of the night sky was taken in Australia and shows stars close to the South Pole. A radar antenna used to study them appears in the foreground.

1. Why do the stars appear as arcs of concentric circles rather than as points of light?

2. How do the arcs of the different stars compare in measure as their distances from the center of the circles (the point above the South Pole) increase?

FRITZ GORO—TIME-LIFE PICTURE AGENCY

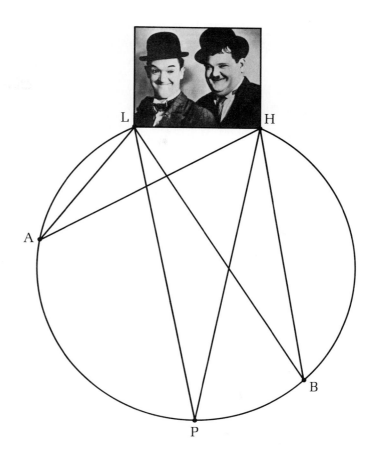

Lesson 4

Inscribed Angles

If you have ever watched a wide-screen movie from a seat near the front of a theater, you know that it is almost impossible to see everything on the screen at once. This is because the angle that the side edges of the screen make with your eyes is very large. At the back of the theater, this angle is much smaller. It is tempting to conclude from this that the farther someone is from the screen, the smaller his viewing angle of it will be. But such is not the case!

In the diagram, $\overline{LH}$ represents the screen and point P represents the center seat of the back row, directly below the projection booth. For a person seated at P, the screen angle to which we have been referring is ∡P. Where else in the theater is the screen angle equal to that at P?

The other locations having the same screen angle are on the circle that contains points L, H, and P. Every angle whose vertex is on this circle and whose sides pass through points L and H is equal to the angle at P. You can verify this for the angles at A and B by measuring both of them and the angle at P with a protractor.

Such angles are called *inscribed angles*, and to prove that they are equal we will derive a formula for their measure in terms of the measure of the common arc they intercept, $\widehat{LH}$ ($\widehat{LH}$ is covered by the screen in the diagram).

▶ Definition
An *inscribed angle* is an angle whose vertex is on a circle and each of whose sides intersects it in another point.

In the adjoining figure, ∡A is an inscribed angle of circle O. It is said to *intercept* $\widehat{BC}$ and to be *inscribed* in $\widehat{BAC}$. Note that these two arcs make up the entire circle. This is true of every inscribed angle in a circle. It divides the circle into two arcs, one of which it intercepts and the other in which it is inscribed.

By definition, a central angle is equal in measure to the arc of the circle that it intercepts: its minor arc. We will prove that an inscribed angle is equal in measure to *half* its intercepted arc.

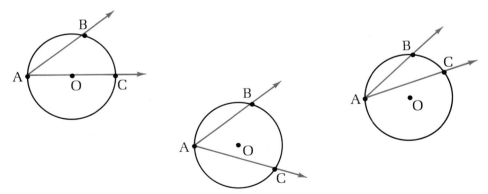

In each of the figures above, then, $\angle A = \frac{1}{2}m\widehat{BC}$. To prove this, we have three possibilities to consider: that the center of the circle lies *on a side* of the inscribed angle, that it lies *inside* the angle, and that it lies *outside* the angle.

▶ **Theorem 71**

An inscribed angle is equal in measure to half its intercepted arc.

Hypothesis: ∡A is inscribed in circle O.
Conclusion: $\angle A = \frac{1}{2}m\widehat{BC}$.

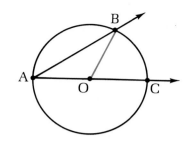

The adjoining figure illustrates the first possibility: that O lies on a side of ∡A. To prove the theorem for this case, we draw radius $\overline{OB}$ to form a central angle, ∡BOC, that intercepts the same arc as ∡A: $\widehat{BC}$. We can then derive an equation relating ∡A and ∡BOC from the fact that they are interior and exterior angles of △AOB.

Proof.

Statements	Reasons
1. ∡A is inscribed in circle O.	Hypothesis.
2. Draw $\overline{OB}$.	Two points determine a line.
3. OA = OB.	All radii of a circle are equal.
4. ∠ABO = ∠A.	If two sides of a triangle are equal, the opposite angles are equal.
5. ∠BOC = ∠A + ∠ABO.	An exterior angle of a triangle is equal in measure to the sum of the measures of the two remote interior angles.
6. ∠BOC = ∠A + ∠A = 2∠A.	Substitution.
7. $\angle A = \frac{1}{2}\angle BOC$.	Division.
8. $m\widehat{BC} = \angle BOC$.	A minor arc is equal in measure to its central angle.
9. $\angle A = \frac{1}{2}m\widehat{BC}$.	Substitution.

The other two possibilities, in which O lies inside or outside ∡A, can now easily be proved and one of them is left as an exercise.

This theorem has a couple of useful corollaries. The first of them justifies the statement that the screen angles at every point on the circle are equal.

▶ **Corollary 1**

Inscribed angles that intercept the same arc or equal arcs are equal.

▶ **Corollary 2**

An angle inscribed in a semicircle is a right angle.

1. In the figure below, △JAD is said to be "inscribed" in circle E because each of its angles is an inscribed angle of the circle.
   a) What arc does ∡J intercept?
   b) In what arc is ∡J inscribed?
   c) Write an equation relating ∠D to the measure of an arc of the circle.
   d) Write an equation relating the measures of two angles in the figure, if $m\widehat{JD} = m\widehat{AD}$.

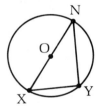

2. Use the figure below to explain why an angle inscribed in a semicircle is a right angle.

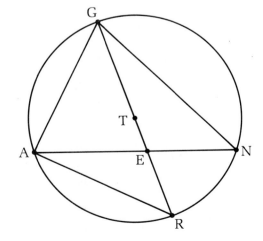

3. In circle T, $\overline{GR}$ is a diameter, $m\widehat{AG} = 70°$, and $\angle NAR = 30°$. Copy the figure, determine the following measures, and write each in the appropriate place:
   a) ∠N.
   b) ∠R.
   c) $m\widehat{NR}$.
   d) $m\widehat{GN}$.
   e) ∠GAN.
   f) ∠GAR.

4. Theorem 71, in which the center of the circle lies outside the angle, is proved below. Supply reasons for the lettered statements.

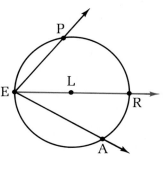

Proof.
Angle PEA is inscribed in circle L. (Hypothesis.)
a) Draw $\overrightarrow{EL}$.

$\angle PER = \frac{1}{2}m\widehat{PR}$ and $\angle REA = \frac{1}{2}m\widehat{RA}$. (Proved in first case.)

b) $\angle PER + \angle REA = \frac{1}{2}m\widehat{PR} + \frac{1}{2}m\widehat{RA} = \frac{1}{2}(m\widehat{PR} + m\widehat{RA})$.
c) $\angle PEA = \angle PER + \angle REA$.
d) $m\widehat{PR} + m\widehat{RA} = m\widehat{PA}$.
e) $\angle PEA = \frac{1}{2}m\widehat{PA}$.

5. Look again at the three figures on page 464 illustrating Theorem 71. In the previous exercise, you used the result of the proof for the first figure to explain the proof for the second. How do you think the theorem would be proved for the third figure?

SET II

1. Given: Circle O with diameter $\overline{TA}$; $m\widehat{PA} = m\widehat{AZ}$.
   Prove: $\triangle PAT \cong \triangle ZAT$.
   Par 8.

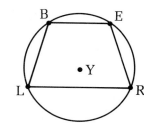

2. Given: Circle O, $\square DIAN$, I-A-M.
   Prove: $DM = DI$.
   Par 5.

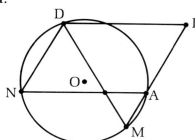

3. Given: In circle Y, $BL = ER$.
   Prove: $\overline{BE} \parallel \overline{LR}$.
   Par 5.

The diagram below shows a top view of Acute Alice's new hat after Obtuse Ollie sat on it. On the assumption that $\angle ATH$ is inscribed in the larger circle, Ollie says that $m\overset{\frown}{AH} > m\overset{\frown}{MY}$. Alice says that $m\overset{\frown}{AH} = m\overset{\frown}{MY}$.

What's your opinion?

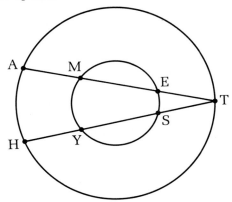

Lesson 5

Secant Angles

In this circular woodcut by Escher, pictures of angels and devils are fitted together like pieces in a jigsaw puzzle. But, unlike those of a jigsaw puzzle, the parts become progressively smaller and more numerous as they approach the edge. By this device, the artist has managed to convey an impression of the infinite within a finite region.

It is apparent from the intricate and orderly design of the picture that Escher has been strongly influenced by ideas from geometry. For example, if the wing tips of the large devils and angels that meet in the center are joined to it and to each other by line segments, six equilateral triangles are formed.

In the first adjoining figure, one of these triangles is shown. Since it is equilateral, the angle at the center has a measure of 60°. If we extend its sides so that they intersect the circle, the intercepted arc must also have a measure of 60°.

If the vertex of this angle were on the circle rather than at its center, as shown in the second adjoining figure, it would be an inscribed angle and would intercept a larger arc with a measure of 120°.

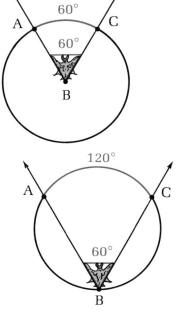

469

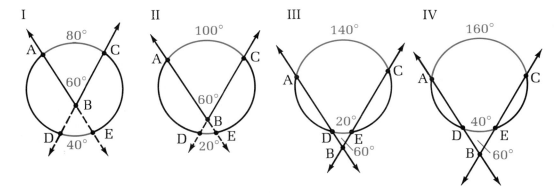

We know the relationship between the measures of central and inscribed angles and the arcs they intercept. What about other angles, such as those shown above?

In figures I and II, the vertex of the angle is inside the circle. Although the size of the arc intercepted by the angle differs in each picture, the *sum of the measures of it and the arc intercepted by its vertical angle does not change*. A comparable result holds for figures III and IV, where the vertex of the angle is outside the circle. Here, the *difference of the measures of the two intercepted arcs does not change*.

It would be convenient to have a name for such angles. We will call them *secant angles* and define them so that central angles and inscribed angles are included as special cases.

In every case, the sides of the angles are contained in lines that intersect the circle in two points. Such lines are called *secants*.

► Definition
A *secant* is a line that intersects a circle in two points.

To include angles such as ∡1, ∡2, and ∡3 as secant angles but to exclude angles such as ∡4, ∡5, and ∡6 from being considered so, we make the following definition.

► Definition
A *secant angle* is an angle whose sides are contained in two secants of a circle so that each side intersects the circle in at least one point other than the angle's vertex.

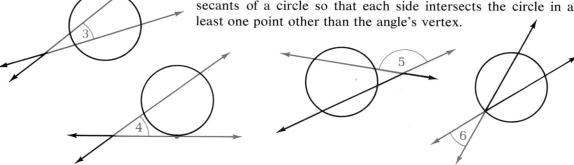

Figures I–IV on the facing page suggest the following relationships between secant angles and the arcs they intercept on a circle.

▶ **Theorem 72**
A secant angle whose vertex is inside a circle is equal in measure to half the sum of the arcs intercepted by it and its vertical angle.

▶ **Theorem 73**
A secant angle whose vertex is outside a circle is equal in measure to half the positive difference of its intercepted arcs.

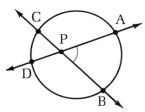

$\angle APB = \frac{1}{2}(m\widehat{AB} + m\widehat{CD})$

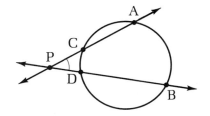

$\angle APB = \frac{1}{2}(m\widehat{AB} - m\widehat{CD})$

SET I

Exercises

1. Supply the reasons for the following proof of Theorem 72.

   A secant angle whose vertex is inside a circle is equal in measure to half the sum of the arcs intercepted by it and its vertical angle.

   Hypothesis: Secant ∡TIU with I inside circle N.
   Conclusion: $\angle TIU = \frac{1}{2}(m\widehat{TU} + m\widehat{PR})$.

   Proof.

   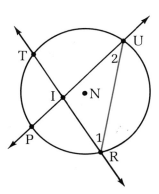

   Statements

   a) Secant ∡TIU with I inside circle N.
   b) Draw $\overline{UR}$.
   c) $\angle TIU = \angle 1 + \angle 2$.
   d) $\angle 1 = \frac{1}{2}m\widehat{TU}$ and $\angle 2 = \frac{1}{2}m\widehat{PR}$.
   e) $\angle TIU = \frac{1}{2}m\widehat{TU} + \frac{1}{2}m\widehat{PR} = \frac{1}{2}(m\widehat{TU} + m\widehat{PR})$.

2. Use the figure at the top of the next page to prove Theorem 73. The proof is similar to the one for Theorem 72.

   A secant angle whose vertex is outside a circle is equal in measure to half the positive difference of its intercepted arcs.

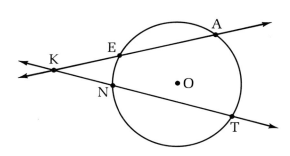

Hypothesis: Secant ⊀ AKT with K outside circle O.

Conclusion: ∠AKT = ½(m$\widehat{AT}$ − m$\widehat{EN}$).

Par 6. (Hint: Draw either $\overline{ET}$ or $\overline{NA}$.)

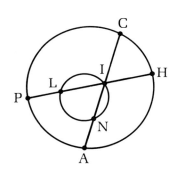

3. In the adjoining circle, m$\widehat{DL}$ = 80°, m$\widehat{LA}$ = 130°, and m$\widehat{AG}$ = 20°. Find each of the following measures:

      a) ∠ALG.   c) ∠N.

      b) ∠AOG.   d) ∠LDG.

SET II

1. Central angles and inscribed angles are special cases of secant angles. A central angle is a secant angle whose vertex is the center of the circle.

    a) Does Theorem 72 apply to central angles? Explain.

    b) What kind of secant angle is an *inscribed angle*?

    c) Both Theorems 72 and 73 might be considered to apply to inscribed angles. Explain.

2. Given: $\overline{CA}$ and $\overline{PH}$ are chords of the larger of the two circles.

   Prove: m$\widehat{LN}$ > m$\widehat{CH}$.

   Par 6.

3. Given: ⊀C is a secant angle of the circle; m$\widehat{AS}$ = 2m$\widehat{HE}$.

   Prove: △ACE is isosceles.

   Par 8.

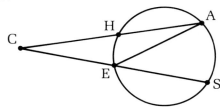

The King of Id has commissioned a statue of himself for the village square. Unfortunately, what the royal sculptor has turned out so far doesn't seem to be what he has in mind.

The figure below shows three positions of the king as he walks toward the original statue. The measure of the angle formed by the king's lines of sight to the top and bottom of the statue depends upon where the king is standing. Can you describe what happens to it as the king walks from a great distance away to the base of the statue?

BY PERMISSION OF JOHN HART AND FIELD ENTERPRISES, INC.

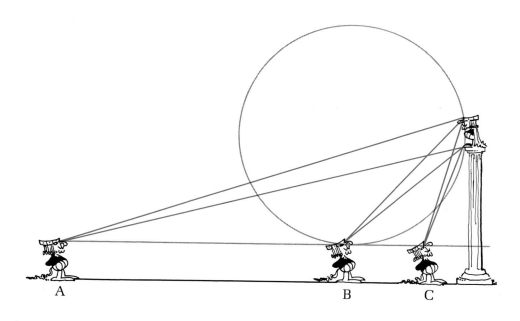

Lesson 5: Secant Angles   473

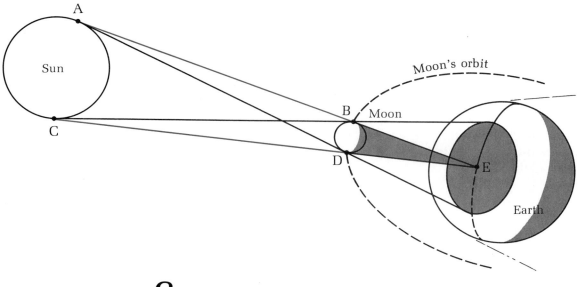

Tangent Segments

A total eclipse of the sun is an impressive sight. The sky suddenly changes from sunlight to darkness, the stars come out, flowers close up and birds become confused and think it is night. The darkness lasts for just a few minutes (about seven at the most) and then the light reappears as swiftly as it had gone.

Such an eclipse occurs when the moon passes between the earth and the sun so that its shadow touches the earth. This shadow is in the shape of a cone and, because of the relative sizes of the moon and the sun and their relative distances from the earth, the tip of the cone often misses the earth's surface. It is only when the cone touches the earth that a total eclipse can be seen by the people within it. The tip of the cone moves across the earth's surface as the earth rotates and is shown as the dark band in the diagram.

At the same time, a partial eclipse of the sun can be seen from a much larger region, shown shaded in the diagram. Within this region, the disk of the sun is only partially covered by the moon.

The lines of the shadows in the figure are tangent to the circles that represent the sun and moon. Two of them, $\overleftrightarrow{AB}$ and $\overleftrightarrow{CD}$, meet in point E. The segments $\overline{EA}$ and $\overline{EC}$ are called *tangent segments* from point E to the circle of the sun; segments $\overline{EB}$ and $\overline{ED}$ are tangent segments from point E to the circle of the moon.

► Definition

If a line is tangent to a circle, then any segment of the line having the point of tangency as one of its endpoints is a *tangent segment* to the circle.

From the diagram, it looks as if one of the two tangent segments from an external point to a circle might be longer than the other. It is easy to prove that, no matter what the "relative positions" of the point and circle, they must be equal.

► Theorem 74

The tangent segments to a circle from an external point are equal.

Hypothesis: Circle O with tangent segments $\overline{PA}$ and $\overline{PB}$.

Conclusion: PA = PB.

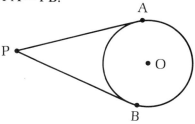

Proof.

Draw $\overline{PO}$, $\overline{OA}$, and $\overline{OB}$. Since $\overline{PA}$ and $\overline{PB}$ are tangent segments to circle O, $\overline{PA} \perp \overline{OA}$ and $\overline{PB} \perp \overline{OB}$ (if a line is tangent to a circle, it is perpendicular to the radius drawn to the point of contact). Therefore, ⦨PAO and ⦨PBO are right angles and so △PAO and △PBO are right triangles. Since OA = OB (all radii of a circle are equal) and PO = PO, △PAO ≅ △PBO (H.L.). Therefore, PA = PB.

SET I

Exercises

1. Tangent segments have been drawn to the circles in the figures below. Find x in each figure.

a)

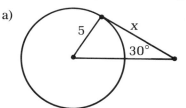

b)

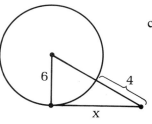

c)

7

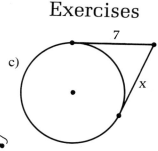

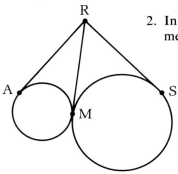

2. In the adjoining figure, $\overline{RA}$, $\overline{RM}$, and $\overline{RS}$ are tangent segments from point R to the circles. Explain why RA = RS.

3. **Construction 10**

   To construct the tangents to a circle from a given external point.

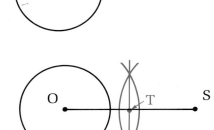

• S

   Draw a figure similar to the adjoining one and use the following method to construct the tangents from point S to circle O.

   Draw $\overline{OS}$.

   Bisect $\overline{OS}$ and label its midpoint T.

   Draw a circle with T as center and TO as radius.

   Label the two points in which it intersects circle O points C and L.

   Draw $\overline{SC}$ and $\overline{SL}$.

   Segments $\overline{SC}$ and $\overline{SL}$ are the tangents from S to circle O. Draw $\overline{OC}$ and $\overline{OL}$ and answer the following questions about this construction.

   a) Why are $\angle OCS$ and $\angle OLS$ right angles?

   b) Why is $\overline{SC} \perp \overline{OC}$ and $\overline{SL} \perp \overline{OL}$?

   c) Why are $\overline{SC}$ and $\overline{SL}$ tangent to circle O?

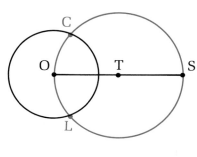

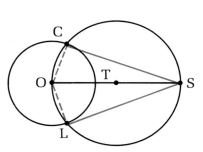

1. Given: $\overline{JE}$ and $\overline{JS}$ are tangent segments
   from point J to circle T.
   Prove: $\overrightarrow{JT}$ bisects ∡EJS.
   Par 8.

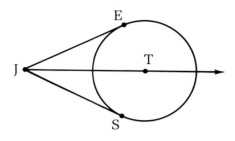

2. Given: △OBN is isosceles (BO = BN)
   and its sides are tangent to the circle
   at R, S, and W.
   Prove: OW = WN.
   Par 7.

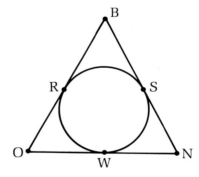

3. Given: Circle S with diameter $\overline{CF}$
   and tangents $\overline{CH}$, $\overline{HE}$, and $\overline{EF}$.
   Prove: $\alpha CHEF = \frac{1}{2}CF \cdot HE$.
   Par 10.

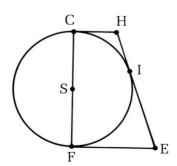

Obtuse Ollie, having had enough of being shown up by Acute Alice, got Dilcue to make up the following puzzle for her.

In △PAC, AP = 36 mm and PC = 52 mm. In quadrilateral KERS, KS = 24 mm, SR = 14 mm, and RE = 40 mm. Ollie has placed a quarter inside each figure so that it touches each side in exactly one point and tells Alice that she can have the quarters if she can figure out the lengths of *both* $\overline{AC}$ and $\overline{KE}$ without doing any additional measuring.

Ollie is safe in making this offer because, although there is an easy way to find one of the lengths, the other part of the puzzle is impossible! Can you figure out which is which?

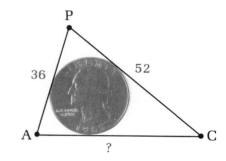

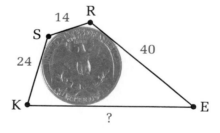

Lesson 7

Chord and Secant Segments

Although the wheel has been used in transportation for more than five thousand years, the first practical self-propelled vehicle to use the wheel, the bicycle, was invented only a little more than one hundred years ago. The modern bicycle is one of the most efficient means of transportation with respect to the amount of energy needed to carry a given amount of weight.

In order to prevent the frame of a bicycle from wobbling up and down, the hub of each wheel must be located in the wheel's exact center. This is because the center of a circle is the only point in the plane of the circle that is equidistant from every point of it. As a result, the sum of the lengths of the two segments

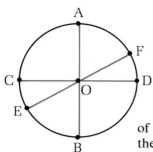

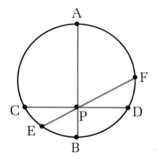

of each chord through the center is the same. In the first figure at the left, AO + OB = CO + OD = EO + OF.

The center of the circle is the *only* point for which this is true. If the hub of the wheel were placed "off center," such as at point P in the second figure at the left, then the sum of the lengths of the two segments of each chord through point P would not be the same. For example, AP + PB > CP + PD.

Is there any point in a circle for which the *product* of the lengths of the two segments of each chord through it is the same? The center is such a point because, in the first figure, AO · OB = CO · OD = EO · OF = r^2. But the center is not the only point. Amazing as it may seem, *any* point within the circle will work! In the second figure, for instance, AP · PB = CP · PD = EP · PF.

Although this relationship is quite surprising, it is very easy to prove.

▶ **Theorem 75** (The Intersecting Chords Theorem)
If two chords intersect in a circle, the product of the lengths of the segments of one chord is equal to the product of the lengths of the segments of the other.

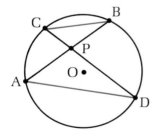

Hypothesis: Circle O with chords $\overline{AB}$ and $\overline{CD}$ intersecting at point P.

Conclusion: AP · PB = CP · PD.

Proof.

Statements	Reasons
1. Chords $\overline{AB}$ and $\overline{CD}$ intersect at point P in circle O.	Hypothesis.
2. Draw $\overline{CB}$ and $\overline{AD}$.	Two points determine a line.
3. ∠A = ∠C, ∠D = ∠B.	Inscribed angles that intercept the same arc are equal.
4. △APD ∼ △CPB.	A.A.
5. $\dfrac{AP}{CP} = \dfrac{PD}{PB}$.	Corresponding sides of similar triangles are proportional.
6. AP · PB = CP · PD.	Means-extremes theorem.

Perhaps even more remarkable is the fact that this relationship holds true for chords that "intersect outside" a circle. In the adjoining figure, the lines containing chords $\overline{AB}$ and $\overline{CD}$ have been extended to meet at point P. For any such pair of chords $\overline{AB}$ and $\overline{CD}$, it is true that

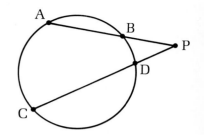

$$AP \cdot PB = CP \cdot PD.$$

We will call segments $\overline{AP}$ and $\overline{CP}$ *secant segments* since they intersect the circle in more than one point.

► Definition
If a segment intersects a circle in two points, exactly one of which is an endpoint of the segment, then the segment is a *secant segment* to the circle.

► Theorem 76 (The Secant Segments Theorem)
If two secant segments are drawn to a circle from an external point, the product of the lengths of one secant segment and its external part is equal to the product of the lengths of the other secant segment and its external part.

SET I

Exercises

1. Supply the reasons for the following proof of the Secant Segments Theorem.

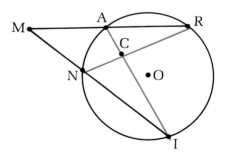

Hypothesis: Circle O with secant segments $\overline{MR}$ and $\overline{MI}$.
Conclusion: $MR \cdot MA = MI \cdot MN.$

Proof.

Statements

a) $\overline{MR}$ and $\overline{MI}$ are secant segments to circle O.
b) Draw $\overline{AI}$ and $\overline{NR}$.
c) $\angle M = \angle M$.
d) $\angle R = \angle I$.
e) $\triangle MNR \sim \triangle MAI$.
f) $\dfrac{MN}{MR} = \dfrac{MA}{MI}$.
g) $MR \cdot MA = MI \cdot MN.$

2. Since the Intersecting Chords Theorem and the Secant Segments Theorem express essentially the same relationship, they might be stated as a single theorem. Copy and complete the following statement of it.

If two secants contain chords $\overline{AB}$ and $\overline{CD}$ in a circle and intersect in a point P that is not on the circle, then ▓▓▓▓.

3. Solve for x in each of the figures below.

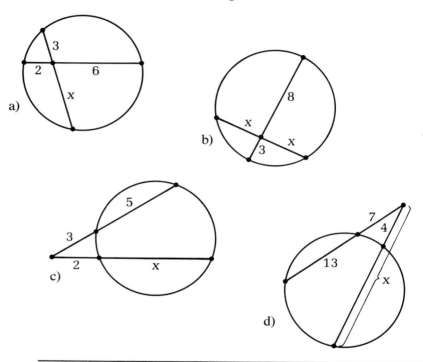

a)

b)

c)

d)

SET II

1. Given: $\overline{IE}$ and $\overline{IN}$ are secants
   to circle O; IE = IN.
   Prove: ID = IS.
   Par 4.

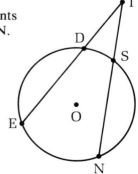

2. **Given:** Chords $\overline{WS}$ and $\overline{IH}$ intersect at G;
   $\overleftrightarrow{RT}$ bisects $\measuredangle WGI$ and $\measuredangle HGS$.

   Prove: $\dfrac{WR}{RI} = \dfrac{HT}{TS}$.

   Par 6.

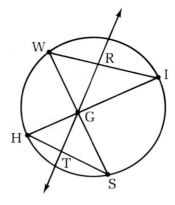

This rather challenging problem from a Korean geometry book consists of three consecutive parts. How many of them can you prove?

△ABC 에서 ∠A 의 이등분선이 BC 와 외접원과의 교점을 각각 D, E 라고 하면

① △ABE∽△ADC

② AB · AC = AD · AE

③ AD² = AB · AC − BD · DC

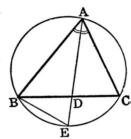

Lesson 8

The Inverse of a Point

This strange-looking oil painting was created by an eighteenth-century artist who intended that it be viewed through its reflection in the cylindrical mirror standing on its surface. The mirror eliminates the distortion of the picture to reveal a group of people being serenaded by a musician.*

The picture illustrates a kind of geometric transformation more complex than those we have studied previously. Reflections, rotations, and translations are sometimes called *isometry* transformations because they preserve distance. ("Isometry" is derived from two Greek words meaning "equal measure.") Since the transformation in the cylindrical mirror alters distance, it is not an isometry. If we replace the mirror with a circle and define a way to locate the image of each point of the plane with respect to the circle, we will have a new transformation, called an *inversion*.

How shall we locate the image of a point with respect to a circle? For example, where is the image of point P in the adjoining figure? It seems logical to look for it on the line determined by P and O, the center of the circle, and, since the point is outside the circle, its image might be inside it. But just exactly where the image will be depends upon our definition of how to find it. A convenient definition is based upon the radius of the circle.

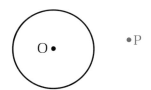

*Another example of deliberate distortion in art, a style quite popular for a time, is the portrait of Edward VI on page 239 of this book.

484

▶ Definition

The *inverse of a point P* with respect to a given circle O with radius r is the point P' on $\overrightarrow{OP}$ such that $OP \cdot OP' = r^2$.

This definition is convenient because it, in effect, turns the plane inside out with respect to the circle. Every point *outside* the circle has an inverse *inside* the circle and, except for the center, every point inside the circle has an inverse outside. Furthermore, every point *on* the circle is *its own inverse*.

For example, consider the figure below, in which the radius of circle O is 6 units and the distances of points A, B, and C from O are 9, 3, and 6 units, respectively. To find the inverse of point A, we write

$$OA \cdot OA' = r^2$$
$$9 \cdot OA' = 6^2 = 36$$
$$OA' = 4.$$

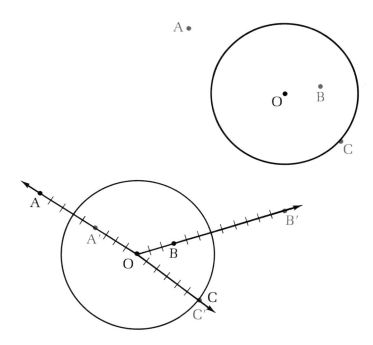

Point A', the inverse of A, is the point on $\overrightarrow{OA}$ 4 units from O. For the inverses of points B and C, we have

$$OB \cdot OB' = r^2 \qquad OC \cdot OC' = r^2$$
$$3 \cdot OB' = 36 \qquad 6 \cdot OC' = 36$$
$$OB' = 12 \qquad OC' = 6$$

Point B' is 12 units from O on $\overrightarrow{OB}$, and, since there is only one point on $\overrightarrow{OC}$ that is 6 units from O, C' is C itself.

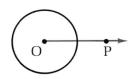

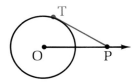

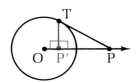

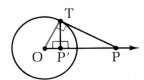

The inverse of a point can also be found geometrically. The method for finding the inverse of a point that is outside a circle is illustrated in the adjoining figures.

First, draw $\overrightarrow{OP}$. Next, draw a tangent segment from P to the circle, $\overline{PT}$.* Then draw, through T, a line perpendicular to $\overrightarrow{OP}$. The point in which this line intersects $\overrightarrow{OP}$ is the inverse of P.

To prove that P′ is the inverse of P, we must show that $OP \cdot OP' = r^2$. First, draw $\overline{OT}$. Since $\overline{PT}$ is tangent to circle O, $\overline{PT} \perp \overline{OT}$ (a tangent to a circle is perpendicular to the radius drawn to the point of contact). Therefore, $\triangle OTP$ is a right triangle and, since $\overline{TP'} \perp \overline{OP}$, $\overline{TP'}$ is the altitude to the hypotenuse. Now, we know that either leg of a right triangle is the mean proportional between the hypotenuse and its projection on the hypotenuse. So, OT is the mean proportional between OP and OP′.

$$\frac{OP}{OT} = \frac{OT}{OP'}$$

Since $\overline{OT}$ is also a radius of the circle, $OT = r$.

$$\frac{OP}{r} = \frac{r}{OP'}$$
$$OP \cdot OP' = r^2.$$

If the point whose inverse we want to find is inside the circle, we simply reverse the steps as shown in the figures below.

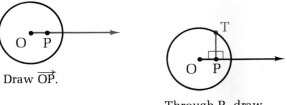

Draw $\overrightarrow{OP}$.

Through P, draw $\overrightarrow{PT} \perp \overrightarrow{OP}$.

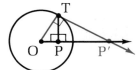

Through T, draw $\overrightarrow{TP'}$ tangent to circle O $(\overrightarrow{TP'} \perp \overline{OT})$.

*The method for doing this was described in Construction 10 on page 476.

1. Trace the figure below and use the equation $OP \cdot OP' = r^2$ to locate the inverse of each of the points S, A, B, L, and E. Note that $r = 6$.

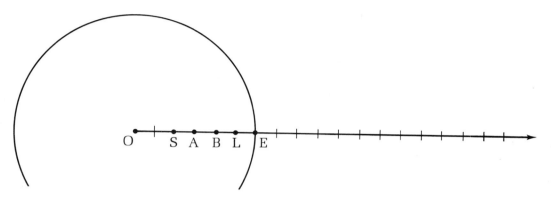

2. Use your results for Exercise 1 as a basis for answering the following questions.
   a) As a point moves away from the center of the circle through which it is being inverted, in what direction does its inverse move?
   b) Where is the inverse of a point that is on the inversion circle?
   c) Which point inside the circle does not have an inverse?
   d) Use the equation $OP \cdot OP' = r^2$, to explain why this point has no inverse.

3. The figure for Exercise 1 is shown, reduced in size, below. The geometric method described in this lesson for finding the inverse of a point has been used to locate the inverse of L. Three segments were drawn: $\overline{OT}$, $\overline{LT}$, and $\overline{L'T}$.
   a) Which one of these three segments was drawn first?
   b) How was it drawn?
   c) Which one was drawn second?
   d) How?
   e) Which one was drawn last?
   f) How?

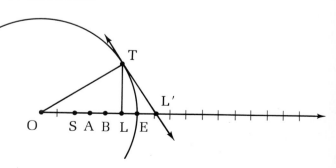

4. Use the figure you drew for Exercise 1 for the following exercise.
   a) Locate the inverses of S, A, and B by means of the geometric method illustrated in Exercise 3. Compare the results with those you got before.
   b) What would happen if you tried to use this geometric method to find the inverse of O? Try it.

SET II

In each of the exercises in this section, use the geometric method for finding the inverses of the points.

1. Copy the figure below.
   a) Locate the inverses of points F and X.
   b) Does inversion of a set of points through a circle preserve distance?

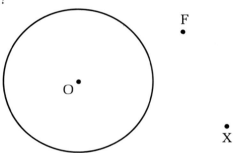

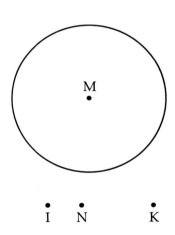

2. Copy the figure at the left.
   a) Locate the inverses of points I, N, and K.
   b) Does inversion of a set of points through a circle preserve collinearity?
   c) Does it preserve betweenness?

To find the inverse of an infinite set of points such as a line, we can locate the inverses of a limited number of the points and guess the rest of the figure from them. Copy the following figures and use this procedure to find the inverses of the sets of points in each.

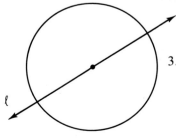

3. What is the inverse of line ℓ?

4. What is the inverse of $\overline{AB}$?

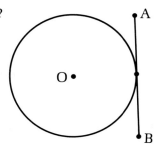

SET III

What would a "happy face" look like if his features were inverted with respect to the circle of his head?

Lesson 9

Inverses of Lines and Circles

This picture of a checkerboard with a circle in its center was drawn by a mathematician to illustrate some of the properties of inversive geometry.* The circle separates the checkerboard into two regions: the region within is the result of "turning" the region outside "inside out." To do this, the inverse of each point outside the circle was found as a point inside it.

Part of the picture is shown on the facing page. The inverse of the brown square labeled A is the small region labeled A', the inverse of the white square B is B', and so on.

It is evident that squares outside the circle do not invert into squares inside. In fact, their inverses do not even all have the same size or shape. The ring of squares directly surrounding the inversion circle have the largest inverses, whereas those around

*Adapted from "Mathematical Games," by Martin Gardner. Copyright © 1965 by Scientific American, Inc. All rights reserved.

490

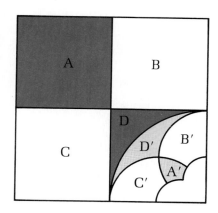

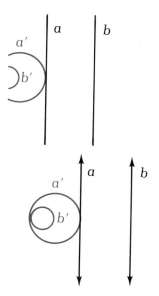

the border of the checkerboard have the smallest ones. Further-more, the *entire* plane outside the checkerboard inverts into the small white region in the center of the figure!

The inverted checkerboard seems to consist of arcs of many small circles of varying size. These arcs are the inverses of the line segments separating the rows of squares of the original board. In the first adjoining figure, for example, arc *a'* is the inverse of segment *a* and arc *b'* is the inverse of segment *b*. If the segments were longer, then so would their inverse arcs be. In fact, if they were *lines*, their inverses would seem to be *circles*.

We will prove this, with the exception of one point, for a line that does not intersect the inversion circle.

▶ **Theorem 77**

The inverse of a line that does not intersect the inversion circle, O, is a circle through O except for the point O itself.

> Hypothesis: Line ℓ does not intersect
> the inversion circle, O.
> Conclusion: The inverse of ℓ is a circle
> through O, except for point
> O itself.

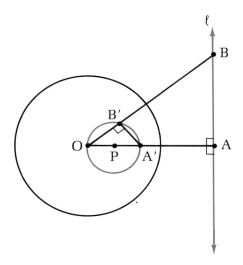

Proof.
Through O, draw $\overline{OA} \perp \ell$. Let A' be the inverse of point A. Draw the circle with P, the midpoint of $\overline{OA'}$, as its center and OA' as its diameter. We will show that, except for O, the points on this circle are the inverses of the points on line ℓ.

Let B be any other point on ℓ and let B' be the point in which $\overline{OB}$ intersects circle P. Draw $\overline{A'B'}$. Angle OB'A' is a right angle since it is inscribed in a semicircle, so $\angle OB'A' = \angle OAB$; also $\angle O = \angle O$. So $\triangle OB'A' \sim \triangle OAB$, and hence

$$\frac{OB'}{OA} = \frac{OA'}{OB}.$$

By the means-extremes theorem,

$$OB \cdot OB' = OA \cdot OA'.$$

But $OA \cdot OA' = r^2$, where r is the radius of the inversion circle, since A' is the inverse of A. Therefore,

$$OB \cdot OB' = r^2$$

and so B' is the inverse of B. Except for point O, we can match up the rest of the points on circle P with inverse points on line ℓ in the same way. So the inverse of ℓ is a circle through O, except for point O itself.

1. In the figure below, the small circle (except for point O) is the inverse of the line ℓ through circle O. Several points on the line and their inverse points on the circle have been labeled.
   a) Point N on line ℓ is the endpoint of the opposite rays $\overrightarrow{NR}$ and $\overrightarrow{NS}$. What is the inverse of $\overrightarrow{NR}$?
   b) As a point moves upward along $\overrightarrow{NR}$, how does its inverse move?
   c) Where would you expect to find the inverse of a point on $\overrightarrow{NR}$ that was a great distance, say one mile, above the top of this page?

Exercises

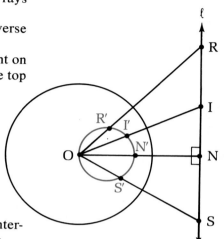

2. We have proved that the inverse of a line that does not intersect the inversion circle is, except for one point, a *circle*.
   a) What does the figure at the left below suggest about the inverse of a line that is *tangent* to the circle of inversion?
   b) What does the figure at the right suggest about the inverse of a line that intersects the circle of inversion in *two* points?
   c) Do you think the conclusion you have drawn in the previous exercise is always true? Look back at Exercise 3 of Lesson 8, Set II (page 488).
   d) What do you think happens to the inverse of a line as the line moves toward the center of the inversion circle?

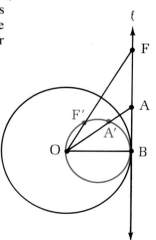

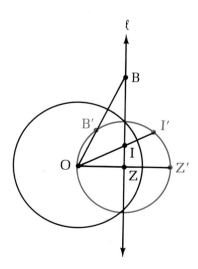

a)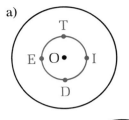

1. You now have an idea of what the inverse of a line is. What does the inverse of a *circle* look like? Find the inverse of the small circle with respect to the large circle O in each of the figures shown here by using the following method. Make an accurate copy of each figure and then locate, except for O, the inverses of the lettered points. Guess the position of the rest of the figure.

b)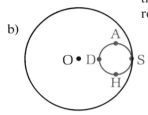

c)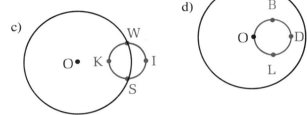

d)

2. What conclusion seems reasonable on the basis of these exercises?

SET III

The device in this photograph can be used to draw a straight line without a straightedge. It is called a Peaucellier linkage and is based in principle upon Theorem 77 in this lesson. The linkage consists of seven rods hinged to each other and to the drawing board. The two hinge points labeled X and Y are fixed, and the four labeled A, B, C, and D are free to move. Links XB and XD are equal in length, as are AB, BC, CD, and DA.

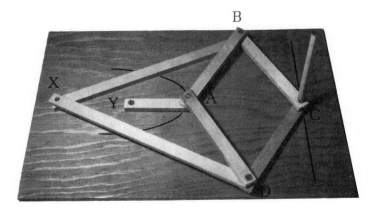

It can be proved that A and C are inverse points with respect to a circle with its center at X. Since Y is fixed, A moves around a circle (having radius YA) that passes through X, the center of this inversion circle (YX = YA). The inverse of the path of A (a circle passing through the center of the inversion circle) is a line, as Theorem 77 suggests. Therefore, as A moves around a circle, C, its inverse, moves along a line.

Part of the linkage is shown in this figure. Can you answer the questions in the following explanation of why A and C are inverse points?

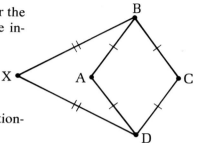

1. Why are points X, A, and C collinear? (Hint: What relationship do they have to points B and D?)

2. Why does $\overline{BD}$ bisect $\overline{AC}$?

We will prove that C is the inverse of A with respect to a circle with its center at X by showing that the product XA · XC never changes.

$$XA \cdot XC = (XO - OA)(XO + OC).$$

Since OA = OC,

$$XA \cdot XC = (XO - OA)(XO + OA) = XO^2 - OA^2.$$

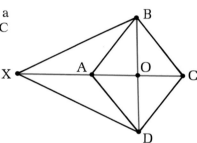

3. Why is $\overline{BD} \perp \overline{AC}$?

4. Why is $XO^2 + OB^2 = XB^2$ and $OB^2 + OA^2 = AB^2$?

5. Why is $XO^2 = XB^2 - OB^2$ and $OA^2 = AB^2 - OB^2$?

We have already shown that

$$XA \cdot XC = XO^2 - OA^2.$$

By substitution,

$$XA \cdot XC = (XB^2 - OB^2) - (AB^2 - OB^2)$$
$$= XB^2 - AB^2.$$

Since XB and AB are constants (they are the lengths of two of the links), $XB^2 - AB^2$ is a constant. So XA · XC is a constant and never changes.

Since XA · XC > 0, let $r = \sqrt{XA \cdot XC}$. Then

$$XA \cdot XC = r^2.$$

Hence C is the inverse of A with respect to a circle with its center at X and radius r where $r = \sqrt{XA \cdot XC}$.

Chapter 12 / Summary and Review

Basic Ideas

Central angle 459
Chord 449
Circle 449
Degree measure of an arc 459
Diameter 449
Inscribed angle 464
Inverse of a point 485
Major arc 459
Minor arc 459
Radius 449
Secant 470
Secant angle 470
Secant segment 481
Semicircle 459
Tangent 454
Tangent segment 475

Constructions

9. To construct the tangent to a circle at a given point on the circle. 455
10. To construct the tangents to a circle from a given external point. 476

Postulate

20. *The Arc Addition Postulate.* If C is on $\widehat{AB}$, then $m\widehat{AC} + m\widehat{CB} = m\widehat{ACB}$. 459

Theorems

64. If a line through the center of a circle is perpendicular to a chord, it also bisects it.   450
65. If a line through the center of a circle bisects a chord that is not a diameter, it is also perpendicular to it.   450
66. The perpendicular bisector of a chord (in the plane of a circle) passes through the center of the circle.   450
67. If a line is tangent to a circle, it is perpendicular to the radius drawn to the point of contact.   454
68. If a line is perpendicular to a radius at its outer endpoint, then it is tangent to the circle.   454
69. In a circle, two central angles are equal iff their minor arcs are equal.   460
70. In a circle, two chords are equal iff their minor arcs are equal.   460
71. An inscribed angle is equal in measure to half its intercepted arc.   465

    Corollary 1.   Inscribed angles that intercept the same arc or equal arcs are equal.   465

    Corollary 2.   An angle inscribed in a semicircle is a right angle.   465
72. An internal secant angle is equal in measure to half the sum of the arcs intercepted by it and its vertical angle.   471
73. An external secant angle is equal in measure to half the positive difference of its intercepted arcs.   471
74. The tangent segments to a circle from an external point are equal.   475
75. *The Intersecting Chords Theorem.*   If two chords intersect in a circle, the product of the lengths of the segments of one chord is equal to the product of the lengths of the segments of the other.   480
76. *The Secant Segments Theorem.*   If two secant segments are drawn to a circle from an external point, the product of the lengths of one secant segment and its external part is equal to the product of the lengths of the other secant segment and its external part.   481
77. The inverse of a line that does not intersect the inversion circle, O, is a circle through O except for the point O itself.   491

Exercises

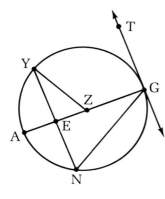

1. In the adjoining figure, $\overline{AG}$ is a diameter of circle Z, $\overline{ZY}$ is a radius, $\overline{YN}$ and $\overline{NG}$ are chords, and $\overleftrightarrow{TG}$ is a tangent. State a reason to justify each of the following conclusions.
   a) ZY = ZA.
   b) $m\widehat{YA} + m\widehat{AN} = m\widehat{YN}$.
   c) If $\overline{AG} \perp \overline{YN}$, then YE = EN.
   d) $\angle YZG = m\widehat{YG}$.
   e) $\overleftrightarrow{TG} \perp \overline{ZG}$.
   f) If $m\widehat{YN} = m\widehat{GN}$, then YN = GN.
   g) $\angle AGN = \frac{1}{2}m\widehat{AN}$.
   h) YE · EN = AE · EG.

2. Solve for x in each of the figures below. (The figures are not drawn according to the measurements shown.)

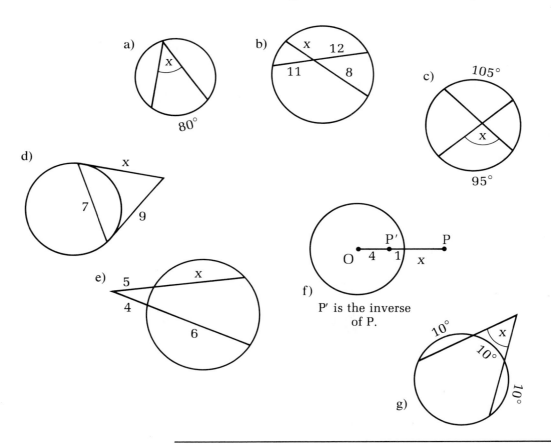

a) 80°
  x

b) x 12
   11   8

c) 105°
   x
   95°

d) x
   7
   9

f) P' P
   O 4 1
         x
   P' is the inverse of P.

e) 5
   4
   x
   6

g) 10°
   x
   10°
   10°

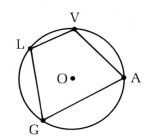

1. If all four vertices of a quadrilateral lie on a circle, then its opposite angles must be supplementary. Use the adjoining figure to explain why.

2. Given: Circle G with $\overline{NI} \parallel \overline{ER}$.
   Prove: $m\widehat{NE} = m\widehat{IR}$.
   Par 6.

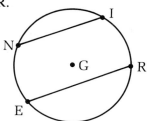

3. Given: $\overline{RI}$ is a diameter of circle H and $\overline{EI}$ is a chord of circle R.
   Prove: $IN = NE$.
   Par 7.

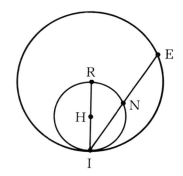

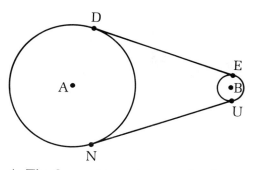

4. The figure above shows a circular turntable being driven by a belt. The belt is tangent to the turntable at D and N and tangent to a circular drive wheel at E and U. On the assumption that $\overline{DE}$ and $\overline{NU}$ are not parallel, can you explain why $DE = NU$?

5. The sides of square NILE are tangent to circle O in the adjoining figure. Trace it and sketch the inverse of the square with respect to the circle.

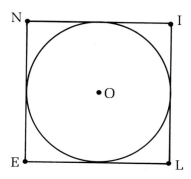

SET III

A small astronaut starts at the edge of his flying saucer, which is circular in shape, and walks due north until he reaches the edge again. He then walks due west as far as he can go. If he has traveled 14 centimeters and 48 centimeters, respectively, can you figure out the radius of the saucer?

"Sorry I'm late. I boarded a
Frisbee by mistake."

S. GROSS
BY PERMISSION OF S. GROSS

Chapter **13**

THE CONCURRENCE
THEOREMS

Lesson 1

Concyclic Points

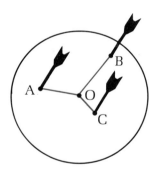

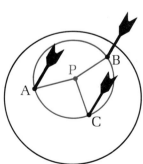

If an archer hits a target with several arrows, it is probable that they will land at varying distances from the center of the bull's-eye. In the first figure at the left, if O represents the center of the target and A, B, and C represent the landing points of three arrows, it is unlikely that OA, OB, and OC will be equal. On the other hand, as long as A, B, and C are not collinear, it is possible to prove that they are equidistant from *some* point in the plane they determine. In the second figure, this point is represented by P, so that AP = BP = CP. This means that a circle can be drawn with P as center that contains all three points.

▶ Definition
Three or more points that lie on the same circle are called *concyclic points*.

Three noncollinear points determine a triangle. Since it can be proved that there is a point equidistant from any three non-collinear points, this means that for every triangle there exists a circle that contains all of its vertices.

▶ Definition
A triangle or quadrilateral is *cyclic* iff there exists a circle that contains all of its vertices.

502

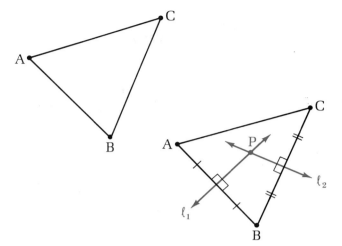

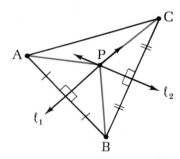

► Theorem 78
Every triangle is cyclic.

Let $\triangle ABC$ represent any triangle. To prove that it is cyclic, we must show that there is a point equidistant from A, B, and C. Let ℓ_1 and ℓ_2 be the perpendicular bisectors of $\overline{AB}$ and $\overline{BC}$, respectively, and let their point of intersection be called P. Draw $\overline{PA}$, $\overline{PB}$, and $\overline{PC}$.

Now point B is the reflection through line ℓ_1 of point A and point C is the reflection through line ℓ_2 of point B. (One point is the reflection through a line of another point if the line is the perpendicular bisector of the segment that joins the two points.) Furthermore, P is its own reflection through ℓ_1 and ℓ_2 since it lies on each line. Because reflection through a line preserves distance, PA = PB and PB = PC. Since PA = PB = PC, P is equidistant from A, B, and C. Therefore, a circle can be drawn with P as center and PA as radius that contains all three points.

The triangle is said to be *inscribed* in the circle that contains its vertices and the circle is said to be *circumscribed* about the triangle. The circle is called the *circumcircle* of the triangle and its center is called the *circumcenter* of the triangle.

The theorem we have just proved might be restated as the following corollary.

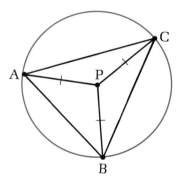

► Corollary 1
Three noncollinear points determine a circle.

To prove Theorem 78, we drew the perpendicular bisectors of two sides of the triangle and showed that the point in which they intersect is equidistant from the three vertices. It is easy to prove that the perpendicular bisector of the third side passes through this same point. You may recall that lines that contain the same point are called *concurrent*.

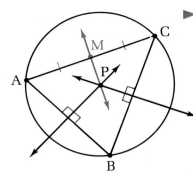

► **Corollary 2**

The perpendicular bisectors of the sides of a triangle are concurrent.

Let P be the center of the circumcircle of △ABC (the point in which the perpendicular bisectors of sides $\overline{AB}$ and $\overline{BC}$ intersect) and let M be the midpoint of side $\overline{AC}$. Draw $\overleftrightarrow{PM}$. Now, $\overline{AC}$ is a chord of circle P and $\overleftrightarrow{PM}$ is a line through the center of the circle that bisects $\overline{AC}$. Hence, $\overleftrightarrow{PM} \perp \overline{AC}$. Therefore, $\overleftrightarrow{PM}$ is the perpendicular bisector of $\overline{AC}$ and so the perpendicular bisectors of the sides of the triangle are concurrent.

Exercises

SET I

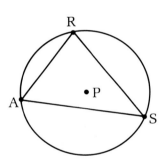

1. In the adjoining figure, circle P is circumscribed about △RAS.
   a) What relationship do points R, A, and S have to each other?
   b) What is circle P called with respect to the triangle?
   c) What is point P called with respect to the triangle?
   d) What is the relationship of the angles of the triangle to the circle?
   e) What is the relationship of the triangle to the circle?

2. **Construction 11**
   To circumscribe a circle about a triangle.

   Since we have proved that every triangle is cyclic, it is always possible to circumscribe a circle about a triangle, regardless of its shape. A method is illustrated below.
   Trace the right and obtuse triangles shown at the top of the facing page and circumscribe a circle about each.

1

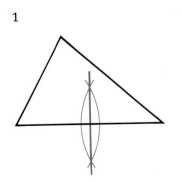

2

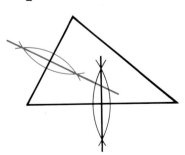

3

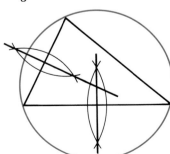

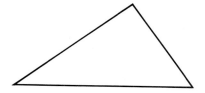 

3. The position of the circumcenter of a triangle is determined by the measures of its angles. For example, the circumcenter of an acute triangle lies in its interior.
   What is the relationship of the circumcenter of
   a) a right triangle to the triangle?
   b) an obtuse triangle to the triangle?

4. Two points do not determine a circle because there are an infinite number of circles that contain both of them.
   a) Mark two points on your paper that are about an inch apart and then construct at least five circles that contain both of them.
   b) What relationship do the centers of these circles have to the line segment that joins the two points?

SET II

1. If three points are collinear, there is no point equidistant from them. This can easily be proved by the indirect method. Answer the following questions about this proof.
   Suppose that L, U, and E are collinear and that point B is equidistant from them. Then BL = BU = BE.
   a) It follows that ∠1 = ∠2, ∠3 = ∠4, and ∠1 = ∠4. Why?
   b) Therefore, ∠2 = ∠3. Why?
   c) Therefore, ∢2 and ∢3 are right angles. Why?

   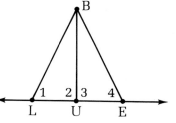

   Since ∠1 = ∠2 and ∠3 = ∠4, ∢1 and ∢4 are also right angles.
   d) Therefore, $\overline{BL} \perp \overleftrightarrow{LE}$, $\overline{BU} \perp \overleftrightarrow{LE}$, and $\overline{BE} \perp \overleftrightarrow{LE}$. Why?
   e) What theorem does this conclusion contradict?

2. We assumed in the proof of Corollary 2 that the center of the circumcircle of △ABC and the midpoint of side $\overline{AC}$ are two distinct points. What would happen if they were the same point? Suppose P is both the center of the circumcircle and the midpoint of side $\overline{AC}$.
   a) What would be the relationship of chord $\overline{AC}$ and circle P?
   b) Must a line through P that bisects $\overline{AC}$ be perpendicular to it? Explain.
   c) Must the perpendicular bisector of $\overline{AC}$, on the other hand, pass through P? Explain.

   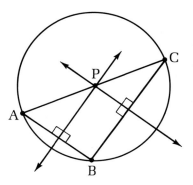

3. Given: △STA with circumcircle R.
   Prove: △WTA is obtuse.
   Par 9.

SET III

Obtuse Ollie put a tall ladder against a wall so that it made 45° angles with the wall and floor. After he had climbed halfway up the ladder, the top end started slipping and slid all the way down the wall. Ollie was too startled to do anything but hold on.

Draw two perpendicular line segments to represent the wall and floor. Make an accurate drawing of Ollie's path as the ladder slid down the wall by using one of the shorter edges of a file card to represent the ladder as shown below.

What kind of a path do you think Ollie traveled? The result of one of the exercises in this lesson implied that the midpoint of the hypotenuse of a right triangle is its circumcenter. On the assumption that this is true, can you explain why the path has the shape that it does?

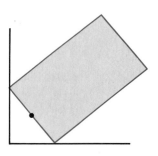

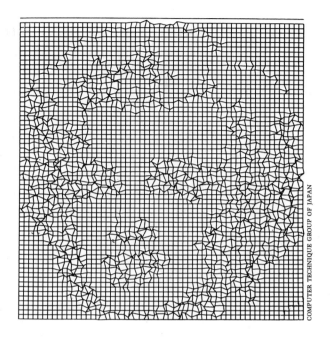

Lesson 2

Cyclic Quadrilaterals

This picture of Marilyn Monroe was produced by a computer programmed to transform a photograph into a net pattern. The net originally consisted entirely of squares, many of which have been deformed into triangles and quadrilaterals of various shapes.

We have proved that every triangle is cyclic—that a circle can be circumscribed about a triangle regardless of its shape. The adjoining diagram suggests that squares are also cyclic. What about quadrilaterals of other shapes?

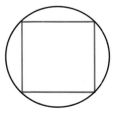

Whether or not a quadrilateral is cyclic depends upon the relationship of its opposite angles.

507

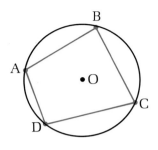

► Theorem 79

A quadrilateral is cyclic iff a pair of its opposite angles are supplementary.

Notice that this theorem consists of two statements. First we will prove that if a quadrilateral is cyclic, its opposite angles are supplementary.

Hypothesis: Quadrilateral ABCD is cyclic.
Conclusion: ∡A and ∡C are supplementary.

Since quadrilateral ABCD is cyclic, let circle O contain its vertices (if a quadrilateral is cyclic, there exists a circle that contains all of its vertices).

Now $\angle A = \frac{1}{2}m\overset{\frown}{BCD}$ and $\angle C = \frac{1}{2}m\overset{\frown}{BAD}$ (an inscribed angle is equal in measure to half its intercepted arc). Since $\angle A + \angle C = \frac{1}{2}m\overset{\frown}{BCD} + \frac{1}{2}m\overset{\frown}{BAD}$ (addition), $\angle A + \angle C = \frac{1}{2}(m\overset{\frown}{BCD} + m\overset{\frown}{BAD}) = \frac{1}{2}(360°) = 180°$ (a circle has a degree measure of 360°). Therefore, ∡A and ∡C are supplementary.

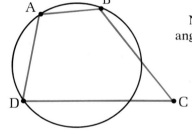

Now we will prove the converse; that if a pair of opposite angles of a quadrilateral are supplementary, then it is cyclic.

Hypothesis: Quadrilateral ABCD in which ∡A and ∡C are supplementary.
Conclusion: Quadrilateral ABCD is cyclic.

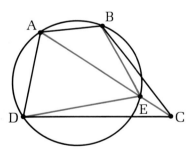

Draw a circle through points A, B, and D (three noncollinear points determine a circle). Point C lies either *outside* this circle, *on* it, or *inside* it. Suppose C lies outside the circle. Draw $\overline{AC}$ and let its second point of intersection with the circle be called E. Also draw $\overline{EB}$ and $\overline{ED}$.

Now ∡DAB and ∡DEB are supplementary (if a quadrilateral is cyclic, its opposite angles are supplementary). Since ∡DAB and ∡DCB are also supplementary (by hypothesis), $\angle DEB = \angle DCB$ (supplements of the same angle are equal).

But $\angle DEA > \angle DCA$ and $\angle AEB > \angle ACB$ (an exterior angle of a triangle is greater than either remote interior angle). Hence, $\angle DEA + \angle AEB > \angle DCA + \angle ACB$, so $\angle DEB > \angle DCB$.

Since we have just proved, however, that $\angle DEB = \angle DCB$, our original assumption must be false. In other words, C cannot lie outside the circle.

The rest of the proof, showing that C cannot lie inside the circle either, is left to you as the first exercise.

1. Answer the following questions about the rest of the proof of Theorem 79.

Exercises

     Hypothesis: Quadrilateral ABCD;
                  ∡A and ∡C are supplementary.
     Conclusion: Quadrilateral ABCD is cyclic.

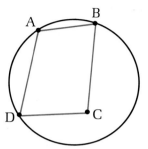

    Draw $\overleftrightarrow{AC}$ and let its second point of intersection with the circle be called E. Draw $\overline{EB}$ and $\overline{ED}$.
a) ∡DAB and ∡DEB are supplementary. Why?
b) Since ∡DAB and ∡DCB are also supplementary, ∠DEB = ∠DCB. Why?
c) ∠DCA > ∠DEA and ∠ACB > ∠AEB. Why?
d) ∠DCA + ∠ACB > ∠DEA + ∠AEB. Why?
e) ∠DCB = ∠DCA + ∠ACB and
   ∠DEB = ∠DEA + ∠AEB. Why?
f) ∠DCB > ∠DEB. Why?
g) What does this result contradict?

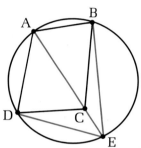

    So C cannot lie inside the circle.

2. Quadrilateral HUSN is inscribed in circle O, $\overline{SD}$ is tangent to the circle at S, and H-U-D. If ∠H = 130° and ∠HUS = 120°, find the measure of each of the following:
a) ∠NSU.
b) ∠N.
c) ∠NSD.
d) ∠D.

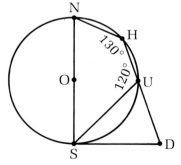

3. Explain why every rectangle is cyclic.

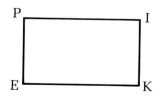

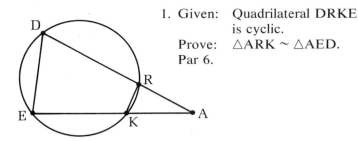

1. Given: Quadrilateral DRKE
   is cyclic.
   Prove: △ARK ~ △AED.
   Par 6.

2. Given: ▱MTEF and circle O.
   Prove: Quadrilateral NTER
   is cyclic.
   Par 10.

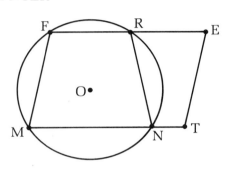

SET III

The Hindu mathematician Brahmagupta discovered several theorems about cyclic quadrilaterals. You have already seen one of them, an extension of Heron's theorem that can be used to find the area of a cyclic quadrilateral.*

Another of Brahmagupta's theorems concerns a cyclic quadrilateral whose diagonals are perpendicular to each other. Draw a large circle and construct two perpendicular chords in it. Label them $\overline{AB}$ and $\overline{CD}$ and their point of intersection point E. Join the four points on the circle to form a cyclic quadrilateral ACBD. Find the midpoint of each side of quadrilateral ACBD and draw the four lines determined by point E and these points.

These lines are related to the opposite sides of the quadrilateral in an interesting way. What is it?

*See page 360.

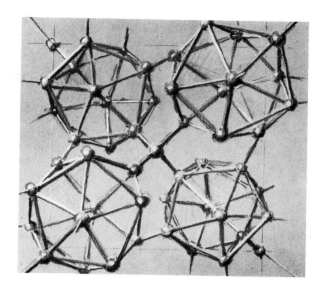

Lesson 3

Incircles

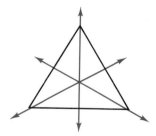

Crystals of the element boron are almost as hard as diamond crystals. The diagram shown here illustrates part of a boron crystal: the spheres represent boron atoms and the sticks represent the bonds that hold them together.* The atoms are arranged in three-dimensional arrays called icosahedra and each icosahedron consists of a set of equilateral triangles. The equilateral triangle appears in the structure of many crystals because it is a highly symmetric figure.

Every equilateral triangle has three lines of symmetry, which are the perpendicular bisectors of its sides. We have proved that the perpendicular bisectors of the sides of a triangle are concurrent in a point that, since it is equidistant from its *vertices*, is the center of the triangle's circumcircle.

This point is also equidistant from the three *sides* of the triangle. As a result, a second circle can be drawn with it as center that is tangent to all three sides of the triangle. This circle is shown in the figure at the right and is said to be *inscribed* in the triangle. The triangle, in turn, is said to be *circumscribed* about the circle.

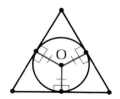

*From *The Architecture of Molecules* by Linus Pauling and Roger Hayward. W. H. Freeman and Company. Copyright © 1964.

511

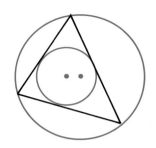

► Definition

A circle is **inscribed in a triangle** iff each side of the triangle is tangent to the circle.

The circle is called the *incircle* of the triangle and its center is called the *incenter* of the triangle.

Does every triangle have an incircle? If so, how can its center be found? The adjoining figure illustrates a scalene triangle for which both a circumcircle and an incircle have been drawn. Their centers, however, are obviously not the same point. We already know how to find the circumcenter; it is the point in which the *perpendicular bisectors of the sides* of the triangle are concurrent. We will prove that the *angle bisectors* of the triangle are also concurrent and that the point of concurrency is the incenter of the triangle.

► **Theorem 80**

The angle bisectors of a triangle are concurrent.

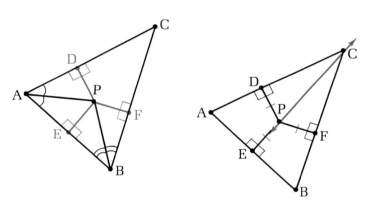

Let △ABC represent any triangle. Let two lines bisect ∡A and ∡B and let their point of intersection be called P.

Through P, draw $\overline{PD} \perp \overline{AC}$, $\overline{PE} \perp \overline{AB}$, and $\overline{PF} \perp \overline{BC}$ (through a point not on a line, there is exactly one perpendicular to the line).

Since △PAD ≅ △PAE (A.A.S.), PD = PE; also, since △PBE ≅ △PBF (A.A.S.), PE = PF. Hence, PD = PE = PF.

Draw $\overleftrightarrow{CP}$ (two points determine a line). Since △PCD ≅ △PCF (H.L.), ∠PCD = ∠PCF. So $\overleftrightarrow{CP}$ is the bisector of ∡ACB. Therefore, since all three angle bisectors pass through point P, they are concurrent.

► **Corollary**

Every triangle has an incircle.

Since PD = PE = PF, a circle can be drawn with P as center and PD as radius that contains all three points, D, E, and F. Since $\overline{AC} \perp \overline{PD}$, $\overline{AB} \perp \overline{PE}$, and $\overline{BC} \perp \overline{PF}$, the sides of the triangle are tangent to the circle (if a line is perpendicular to a radius at its outer endpoint, it is tangent to the circle).

Therefore, circle P is inscribed in △ABC and is its incircle by definition.

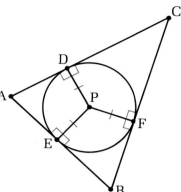

SET I

Exercises

1. In the adjoining figure, A is the incenter of △HOG and N is its circumcenter.

   Which point satisfies each of these conditions?
   a) It is equidistant from H and G.
   b) It lies on the bisector of ⊀O.
   c) It lies on the perpendicular bisector of $\overline{HG}$.
   d) It is equidistant from $\overline{HO}$ and $\overline{HG}$.

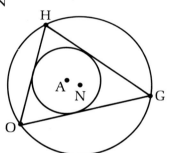

2. **Construction 12**
   To inscribe a circle in a triangle.

   A method is illustrated below.
   a) Why was the perpendicular line constructed in the third step?

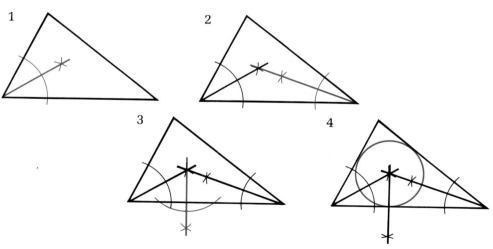

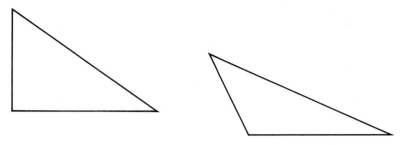

b) Trace the right and obtuse triangles shown here and inscribe a circle in each.

c) Is the position of the incenter of a triangle determined by the measures of its angles in the same way that the position of the circumcenter is?

3. Not every quadrilateral has an incircle. There is no way to draw a circle inside the rectangle shown here, for instance, so that it is tangent to each side of the rectangle.

Make a neat sketch of an example of each of the following quadrilaterals to decide whether it seems to have an incircle.

a) A square

b) An isosceles trapezoid.

c) A rhombus.

d) A parallelogram.

e) A kite.

SET II

Using an argument similar to the one in the proof of Theorem 80, it is easily shown that a quadrilateral has an incircle iff its angle bisectors are concurrent.

1. To show that the angle bisectors of a rhombus are concurrent, we might prove that the lines that contain its diagonals bisects its angles. Use the figure below to explain why, if SNEA is a rhombus, $\overleftrightarrow{AN}$ bisects ∡SAE.

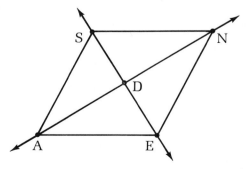

2. To show that the angle bisectors of a kite are concurrent,
   complete the following argument based upon the adjoining
   figure.
      Suppose JONE is a kite in which JO = JE and ON = EN.
   Draw $\overleftrightarrow{JN}$.
   a) Why does $\overrightarrow{JN}$ bisect ∡OJE and ∡ONE?

      Let $\overrightarrow{OS}$ bisect ∡JON; draw $\overleftrightarrow{SE}$.
   b) Use congruent triangles to explain why $\overleftrightarrow{SE}$ bisects ∡JEN.

3. Given: The circle is inscribed
                in quadrilateral NCLU.
   Prove: NC + UL = NU + CL.
   Par 6.

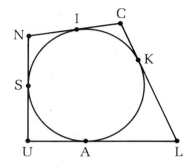

SET III

This problem from a Hebrew geometry book is about a relationship between the radius of the incircle of a right triangle and the
lengths of the triangle's sides. Can you explain why it is true?

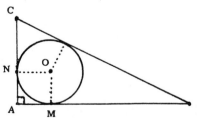

48) במשולש ישר הזווית ABC
חסום מעגל, שמחוגיו ON ו-OM.
הוכח, כי:

$$2OM = AB + AC - BC$$

CLEVER IDEA. WONDER WHY I DIDN'T THINK OF THAT?

Lesson 4

Ceva's Theorem

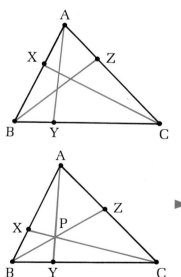

Most of the theorems that we have proved in our study of geometry were known to Euclid, who included them in the *Elements*. In fact, after the development of geometry in Greece between 600 and 200 B.C., very few significant additions to the subject were made until the seventeenth century. One of the new theorems that appeared at that time was discovered by an Italian mathematician and engineer named Giovanni Ceva.

We have proved that the perpendicular bisectors of the sides of every triangle are concurrent; also that the angle bisectors of every triangle are concurrent. Ceva's theorem concerns sets of concurrent line segments in a triangle, which we will refer to as *cevians*.

▶ Definition
A *cevian* of a triangle is a line segment that joins a vertex of the triangle to a point on the opposite side.

In each of the triangles shown here, three cevians have been drawn, one from each vertex. They are $\overline{AY}$, $\overline{BZ}$, and $\overline{CX}$.

In the second triangle on the preceding page, these three line segments are concurrent at point P. Ceva's theorem concerns the lengths of the six segments into which concurrent cevians such as these divide the sides of the triangle.

▶ **Theorem 81** (Ceva's Theorem)
Three cevians $\overline{AY}$, $\overline{BZ}$, and $\overline{CX}$ of $\triangle ABC$ are concurrent iff

$$\frac{AX}{XB} \cdot \frac{BY}{YC} \cdot \frac{CZ}{ZA} = 1.$$

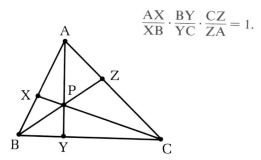

Proof that if the three cevians are concurrent, then

$$\frac{AX}{XB} \cdot \frac{BY}{YC} \cdot \frac{CZ}{ZA} = 1.$$

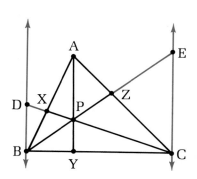

Through B and C, draw lines parallel to $\overline{AY}$. Let the points in which they intersect $\overleftrightarrow{CX}$ and $\overleftrightarrow{BZ}$ be called D and E, respectively.

Because of equal vertical angles and equal alternate interior angles, $\triangle AXP \sim \triangle BXD$ and $\triangle CZE \sim \triangle AZP$. Therefore,

$$\frac{AX}{BX} = \frac{AP}{BD} \quad \text{and} \quad \frac{CZ}{AZ} = \frac{CE}{AP}.$$

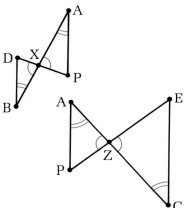

Because of the common angle and equal corresponding angles, $\triangle BYP \sim \triangle BCE$ and $\triangle BCD \sim \triangle YCP$. Therefore,

$$\frac{BY}{BC} = \frac{YP}{CE} \quad \text{and} \quad \frac{BC}{YC} = \frac{BD}{YP}.$$

We can now build the conclusion of Ceva's theorem by multiplying the left and right sides of these equations.

$$\frac{AX}{BX} \cdot \frac{BY}{BC} \cdot \frac{BC}{YC} \cdot \frac{CZ}{AZ} = \frac{AP}{BD} \cdot \frac{YP}{CE} \cdot \frac{BD}{YP} \cdot \frac{CE}{AP}.$$

Simplifying,

$$\frac{AX}{BX} \cdot \frac{BY}{\cancel{BC}} \cdot \frac{\cancel{BC}}{YC} \cdot \frac{CZ}{AZ} = \frac{\cancel{AP}}{\cancel{BD}} \cdot \frac{\cancel{YP}}{\cancel{CE}} \cdot \frac{\cancel{BD}}{\cancel{YP}} \cdot \frac{\cancel{CE}}{\cancel{AP}},$$

so

$$\frac{AX}{BX} \cdot \frac{BY}{YC} \cdot \frac{CZ}{AZ} = 1.$$

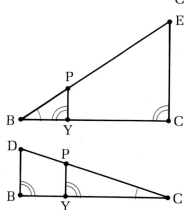

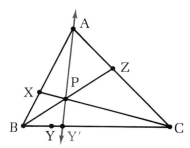

Proof of the converse:

If $\dfrac{AX}{XB} \cdot \dfrac{BY}{YC} \cdot \dfrac{CZ}{ZA} = 1$, then the three cevians $\overline{AY}$, $\overline{BZ}$, and $\overline{CX}$ of $\triangle ABC$ are concurrent.

Let $\overline{CX}$ and $\overline{BZ}$ intersect in point P. Draw $\overleftrightarrow{AP}$, and let the point in which it intersects $\overline{BC}$ be called Y′.

Now

$$\frac{AX}{XB} \cdot \frac{BY'}{Y'C} \cdot \frac{CZ}{ZA} = 1 \text{ (proved in the first part)},$$

and

$$\frac{AX}{XB} \cdot \frac{BY}{YC} \cdot \frac{CZ}{ZA} = 1 \text{ (by hypothesis)}.$$

So

$$\frac{AX}{XB} \cdot \frac{BY'}{Y'C} \cdot \frac{CZ}{ZA} = \frac{AX}{XB} \cdot \frac{BY}{YC} \cdot \frac{CZ}{ZA},$$

and

$$\frac{BY'}{Y'C} = \frac{BY}{YC} \text{ (division)}.$$

By denominator addition,

$$\frac{BY' + Y'C}{Y'C} = \frac{BY + YC}{YC}.$$

Since $BY' + Y'C = BC = BY + YC$, $\dfrac{BC}{Y'C} = \dfrac{BC}{YC}$. By the means-extremes theorem, $BC \cdot YC = BC \cdot Y'C$ and division, $YC = Y'C$. Since $CY = CY'$, Y and Y′ must be the same point. This follows from the unique point corollary, which says that on a given ray, $\overrightarrow{CB}$, there is exactly *one* point at any given distance from point C. Since Y and Y′ are the same point, $\overleftrightarrow{AY}$ and $\overleftrightarrow{AY'}$ are the same line. Therefore, $\overline{AY}$, $\overline{BZ}$, and $\overline{CX}$ all contain point P and are concurrent.

Although Ceva's Theorem is not at all obvious and its proof quite complex, it is a theorem well worth knowing as you will see from using it.

1. In △MCE, three cevians are concurrent at point H.
   a) Name them.

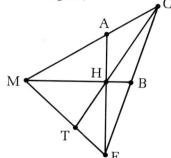

   Copy and complete the following equations for this figure, which follow from Ceva's Theorem:

   b) $\dfrac{CB}{BE} \cdot \dfrac{ET}{TM} \cdot \underline{\hphantom{mmm}} = 1$.

   c) $\dfrac{MT}{TE} \cdot \underline{\hphantom{mmm}} \cdot \underline{\hphantom{mmm}} = 1$.

2. Segments $\overline{JI}$, $\overline{LT}$, and $\overline{EU}$ are cevians in △JLE. Would they be concurrent if

   a) $\dfrac{JU}{UL} = \dfrac{4}{5}$, $\dfrac{LI}{IE} = \dfrac{3}{2}$, and $\dfrac{ET}{TJ} = \dfrac{5}{6}$?

   b) $JT = 5$, $TE = 8$, $EI = 4$, $IL = 3$, $LU = 6$, and $UJ = 5$?

   Suppose $\overline{JI}$, $\overline{LT}$, and $\overline{EU}$ are concurrent.

   c) If $\dfrac{LI}{IE} = \dfrac{7}{2}$ and $\dfrac{ET}{TJ} = \dfrac{1}{3}$, find $\dfrac{JU}{UL}$.

   d) If $LU = 2$, $UJ = 6$, $JT = TE$, and $EI = 9$, find IL.

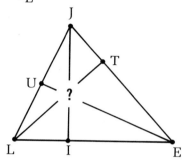

1. In proving Theorem 80, we verified that the angle bisectors of a triangle are concurrent. Ceva's Theorem can be used to prove this in a different way.

   Hypothesis: $\overrightarrow{HR}$, $\overrightarrow{NV}$, and $\overrightarrow{YE}$ bisect the angles of △HNY.

   Conclusion: $\overline{HR}$, $\overline{NV}$, and $\overline{YE}$ are concurrent.

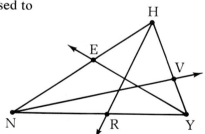

   a) Since $\overrightarrow{HR}$ bisects $\angle NHY$, $\dfrac{NR}{RY} = \dfrac{NH}{HY}$. Why?

   b) What proportion follows from the fact that $\overrightarrow{NV}$ bisects $\angle HNY$?

   c) What proportion follows from the fact that $\overrightarrow{YE}$ bisects $\angle HYN$?

d) Use the equations in parts a, b, and c to complete the following equation:

$$\frac{NR}{RY} \cdot \frac{YV}{VH} \cdot \frac{HE}{EN} = \text{▨}.$$

e) How does this prove that $\overline{HR}$, $\overline{NV}$ and $\overline{YE}$ are concurrent?

2. In $\triangle WNS$, $\overline{WD}$, $\overline{NO}$, and $\overline{SI}$ are concurrent in point R. If I and D are the midpoints of $\overline{WN}$ and $\overline{NS}$, respectively, is O necessarily the midpoint of $\overline{SW}$? Explain.

3. Given: Cevians $\overline{EA}$, $\overline{MT}$ and $\overline{HL}$ in $\triangle HEM$ are concurrent; TE = 2HT and LM = 3EL.
   Prove: MA = 6AH.
   Par 5.

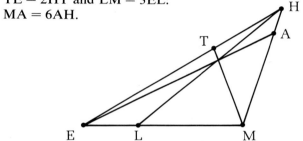

SET III

A nineteenth-century French mathematician named Joseph Gergonne proved that if a circle is inscribed in a triangle, then the cevians joining the vertices of the triangle to the points in which the circle is tangent to the opposite sides of the triangle are concurrent.

Can you explain how he did it?

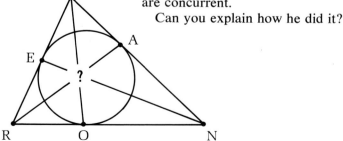

DRAWING BY O. SOGLOW; © 1959 THE NEW YORKER MAGAZINE, INC.

Lesson 5

The Centroid of a Triangle

At what point should a wooden stick be supported so that it is perfectly balanced? If the stick has a uniform thickness and density, it is obvious that this point, called its center of gravity, is its midpoint. Because of this, we might say that the center of gravity of a line segment is its midpoint, even though a line segment has no mass.

Where is the center of gravity of a triangle? Let us assume that by this we mean, At what point should a triangular board having uniform thickness and density be supported so that it will balance?

If we try to find this point by trial and error with a large board in the shape of a scalene triangle, we find that it is not the circumcenter, the point in which the perpendicular bisectors of the sides are concurrent. Neither is it the incenter, the point in which the angle bisectors are concurrent. Instead, it is the point in which the *medians* of the triangle intersect.

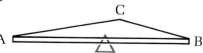

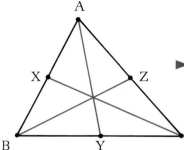

► Definition

A *median* of a triangle is a cevian joining a vertex to the midpoint of the opposite side.

It is very easy to prove that the three medians of every triangle are concurrent.

► **Theorem 82**

The medians of a triangle are concurrent.

Hypothesis: $\overline{AY}$, $\overline{BZ}$, and $\overline{CX}$ are medians of $\triangle ABC$.

Conclusion: $\overline{AY}$, $\overline{BZ}$, and $\overline{CX}$ are concurrent.

Proof.

Since $\overline{AY}$, $\overline{BZ}$, and $\overline{CX}$ are medians of $\triangle ABC$, it follows that X, Y, and Z are the midpoints of $\overline{AB}$, $\overline{BC}$, and $\overline{CA}$, respectively. Therefore, AX = XB, BY = YC, and CZ = ZA. Hence,

$$\frac{AX}{XB} = 1, \quad \frac{BY}{YC} = 1, \quad \text{and} \quad \frac{CZ}{ZA} = 1$$

so

$$\frac{AX}{XB} \cdot \frac{BY}{YC} \cdot \frac{CZ}{ZA} = 1 \cdot 1 \cdot 1 = 1.$$

It follows from Ceva's Theorem that $\overline{AY}$, $\overline{BZ}$, and $\overline{CX}$ are concurrent.

Archimedes showed that the point in which its medians are concurrent is the center of gravity of a triangular board of uniform thickness and density. Mathematicians refer to it as the *centroid* of the triangle.

► Definition

The *centroid* of a triangle is the point in which its medians are concurrent.

At this point it will probably come as no surprise to you that the lines containing the *altitudes* of a triangle are also concurrent. A couple of ways of proving this are included in the exercises.

► **Theorem 83**

The lines containing the altitudes of a triangle are concurrent.

► Definition

The *orthocenter* of a triangle is the point in which the lines containing its altitudes are concurrent.

In the adjoining figure, the three altitudes of △ABC are concurrent in point O, so O is the orthocenter of the triangle.

You may recall from its use in the term "orthogonal projection" that the prefix "ortho" means "right." The orthocenter of a triangle is determined by lines that form right angles with its sides.

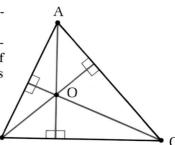

SET I

Exercises

1. Trace the triangles shown here.
   a) Use your straightedge and compass to find the centroid of each.
   b) Can the centroid of a triangle be outside the triangle?
   c) Find the orthocenter of each of the three triangles.
   d) What do you notice?

2. Trace the isosceles triangle shown here.
   a) Find, by construction, its circumcenter, incenter, centroid, and orthocenter.
   b) What do you notice?
   c) What do you think would have happened if the triangle were equilateral?

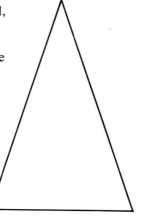

3. One way to prove that the three altitudes of a triangle are concurrent is shown below.

Hypothesis: △PIG in which $\overline{IE}$, $\overline{PR}$, and $\overline{GO}$ are altitudes.
Conclusion: $\overleftrightarrow{IE}$, $\overleftrightarrow{PR}$, and $\overleftrightarrow{GO}$ are concurrent.

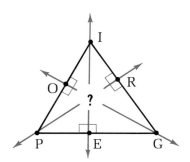

Proof.
Through each vertex of △PIG, draw a line parallel to the opposite side, forming a larger triangle, △SAN.
a) SIGP and INGP are parallelograms. Why?
b) SI = PG and IN = PG. Why?
c) SI = IN. Why?
d) $\overleftrightarrow{IE} \perp \overline{PG}$. Why?
e) $\overleftrightarrow{IE} \perp \overline{SN}$. Why?

So $\overleftrightarrow{IE}$ is the perpendicular bisector of side $\overline{SN}$ of △SAN. In the same way, it can be shown that $\overleftrightarrow{PR}$ is the perpendicular bisector of $\overline{SA}$ and that $\overleftrightarrow{GO}$ is the perpendicular bisector of $\overline{AN}$.
f) How does it follow that $\overleftrightarrow{IE}$, $\overleftrightarrow{PR}$, and $\overleftrightarrow{GO}$ are concurrent?

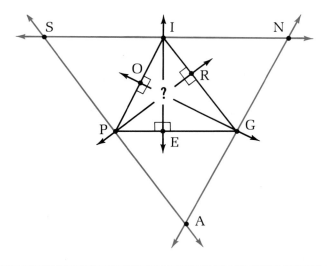

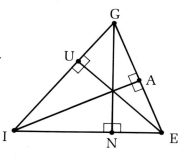

1. If all three altitudes of a triangle are cevians, Ceva's Theorem can be used to prove them concurrent.

      Hypothesis: $\triangle$IGE in which altitudes $\overline{GN}$, $\overline{IA}$, and $\overline{EU}$ are cevians.

      Conclusion: $\overleftrightarrow{GN}$, $\overleftrightarrow{IA}$ and $\overleftrightarrow{EU}$ are concurrent.

Proof.
Since $\overline{GN}$, $\overline{IA}$, and $\overline{EU}$ are altitudes, $\overline{GN} \perp \overline{IE}$, $\overline{IA} \perp \overline{GE}$, and $\overline{EU} \perp \overline{IG}$.

a) Explain why $\triangle GUE \sim \triangle GAI$.

b) $\dfrac{GU}{GA} = \dfrac{GE}{GI}$. Why?

   In the same way, $\triangle ING \sim \triangle IUE$ and $\triangle EAI \sim \triangle ENG$, so that $\dfrac{IN}{IU} = \dfrac{IG}{IE}$ and $\dfrac{EA}{EN} = \dfrac{EI}{EG}$.

c) Finish the proof.

2. The medians of a triangle are not only concurrent: they meet in a point that is a point of trisection of each one.

   In the figure below, $\overline{RU}$ and $\overline{TO}$ are medians of $\triangle RPT$ and points A and L are the midpoints of $\overline{RG}$ and $\overline{TG}$, respectively. Point G is a point of trisection of both $\overline{RU}$ and $\overline{TO}$ because RA = AG = GU and TL = LG = GO.

   To explain why this is so, we have added four segments to the figure. Since A is the midpoint of $\overline{RG}$, it is obvious that RA = AG. Give a convincing argument that AG = GU, and hence, RA = AG = GU.

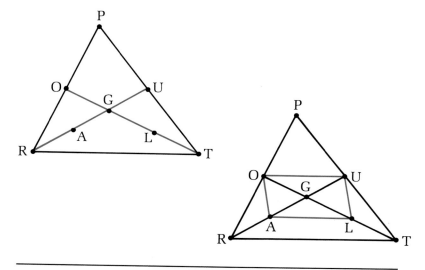

A board having uniform thickness and density that is triangular in shape can be balanced on its centroid, the point in which the medians of the triangle are concurrent.

If the board were in the shape of a quadrilateral instead, where would its balancing point be? The answer to this question was not discovered until as recently as the last century.

The first step in finding the balancing point is to trisect the sides of the quadrilateral. Trace the figure below, in which this has been done.

Now draw $\overline{EF}$, $\overline{GH}$, $\overline{IJ}$, and $\overline{KL}$ and extend them until they meet in four points outside the figure. Label these points M, N, O, and P.

1. What special type of quadrilateral do you suppose MNOP is?

The balancing point of the original quadrilateral ABCD is the same as the balancing point of MNOP.

2. Where do you suppose it is?

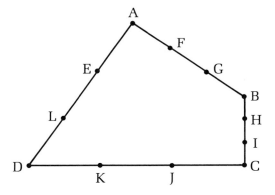

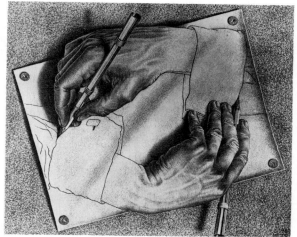

DRAWING HANDS BY M. C. ESCHER. ESCHER FOUNDATION,
HAAGS GEMEENTEMUSEUM, THE HAGUE

Lesson **6**

Some Triangle Constructions

The very first problem presented by Euclid in the *Elements* concerns a method for constructing an equilateral triangle with a straightedge and compass. Mathematicians ever since have been intrigued with determining exactly what constructions are possible: that is, what drawings can be made with these two tools alone. Construction problems can be very challenging and are, for anyone who likes to solve puzzles, a lot of fun. In this lesson we will consider a variety of construction problems concerning triangles.

Perhaps the best way to learn how to do original constructions is to study some examples. Our first one will be the construction of a right triangle having c as its hypotenuse and a as one of its legs.

Before attempting to do the construction itself, it is a good idea to make a rough sketch of the finished figure in order to get an idea of how to proceed. The completed triangle will look something like the one at the right.

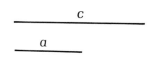

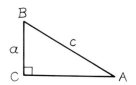

527

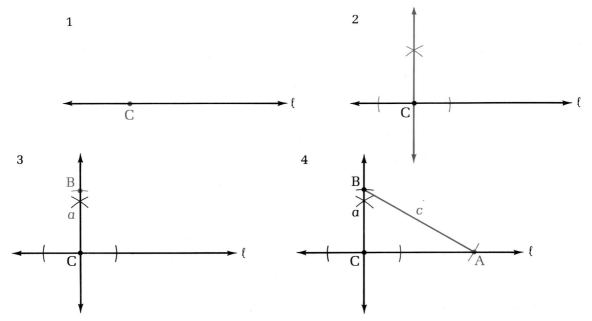

1

2

3

4

First, we can draw a line ℓ to contain the base and mark a point C on it for the vertex of the right angle. Next we can construct the line through point C that is perpendicular to line ℓ. This line will contain leg a, and, since we know its length, we can locate point B with a compass. Finally, with B as center, we can draw an arc with radius equal to c intersecting line ℓ at A. Drawing $\overline{BA}$ completes the triangle.

Our second example will be the construction of an isosceles triangle having a as one of its legs and r as the radius of its circumcircle.

First, we might draw the circumcircle with any point O as center and r as radius. Next we can choose a point A on the circle for the top vertex of the triangle. With A as center and a as radius, we can draw equal arcs intersecting the circle in points B and C. Finally, we can join A, B, and C with segments to form the triangle.

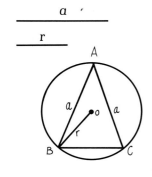

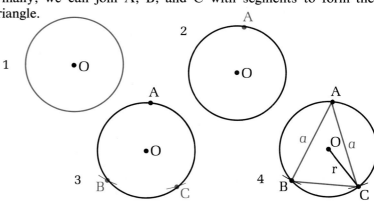

Draw and label segments having the following lengths for use in the exercises of this lesson: a, 2 cm; b, 3 cm; c, 4 cm; d, 5 cm; and e, 6 cm. Before doing each construction, make a sketch of the completed figure and then try to develop a plan based upon relationships that you notice among its parts.

Exercises

SET I

1. Construct a triangle having c, d, and e as its sides.

2. Construct a right triangle having b and d as its legs.

3. There is often more than one way to do a construction. The right triangle in the first example in this lesson, for instance, could also have been constructed as illustrated here.
   a) What was done in step 2?
   b) What was done in step 3?
   c) When $\overline{BC}$ and $\overline{CA}$ are drawn in step 5, how do we know that $\angle C$ is a right angle?

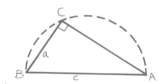

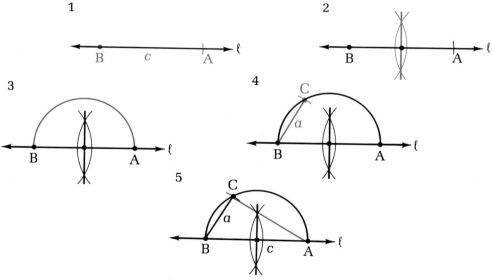

4. Use any method you wish to construct a right triangle having a as one leg and e as its hypotenuse.

5. Construct an isosceles right triangle having d as its hypotenuse.

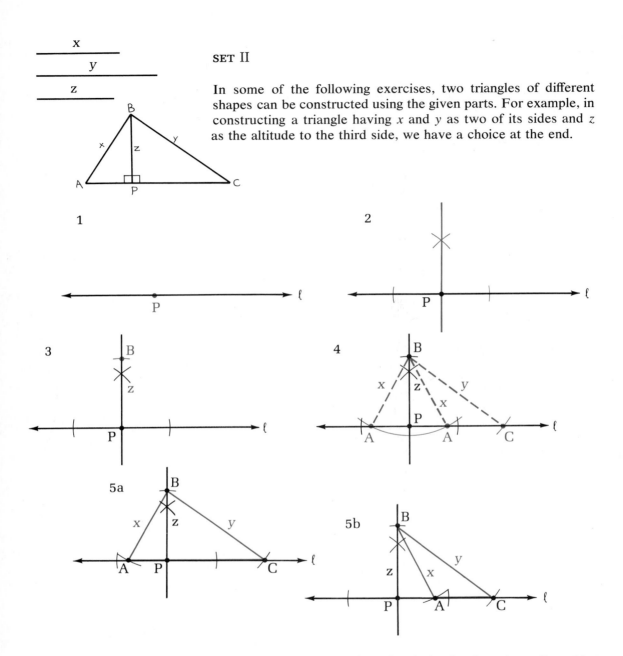

In some of the following exercises, two triangles of different shapes can be constructed using the given parts. For example, in constructing a triangle having x and y as two of its sides and z as the altitude to the third side, we have a choice at the end.

1

2

3

4

5a

5b

1. Construct an isosceles triangle having b as the radius of its circumcircle and d as its base. (There are two possibilities.)

2. Construct a 30°-60° right triangle having b as its shorter leg. (Hint: construct an equilateral triangle first.)

3. Construct a triangle having a and e as two of its sides and c as the median to side e.

4. Construct a triangle having *c* and *e* as two of its sides and *b* as the altitude to side *e*. (There are two possibilities.)

SET III

The French general and emperor Napoleon was also an amateur mathematician. He was especially interested in geometry and may have been the first to discover the following rather remarkable construction.

Draw a large scalene triangle of any shape. Construct three equilateral triangles, each sharing one side with the scalene triangle and each facing outward from it. Find the centers of the equilateral triangles and join them to form a fifth triangle.

What do you notice?

Chapter 13/Summary and Review

Basic Ideas

Centroid 522
Cevian 516
Circumcenter 503
Circumscribed circle 503
Concyclic points 502
Cyclic triangle or quadrilateral 502
Incenter 512
Inscribed circle 512
Median 522
Orthocenter 522

Constructions

11. To circumscribe a circle about a triangle. 504
12. To inscribe a circle in a triangle. 513

Theorems

78. Every triangle is cyclic. 503
    Corollary 1. Three noncollinear points determine a circle.
    503
    Corollary 2. The perpendicular bisectors of the sides of a
    triangle are concurrent. 504
79. A quadrilateral is cyclic iff a pair of its opposite angles are
    supplementary. 508
80. The angle bisectors of a triangle are concurrent. 512
    Corollary. Every triangle has an incircle. 512
81. *Ceva's Theorem*. Three cevians $\overline{AY}$, $\overline{BZ}$, and $\overline{CX}$ of $\triangle ABC$
    are concurrent iff $\dfrac{AX}{XB} \cdot \dfrac{BY}{YC} \cdot \dfrac{CZ}{ZA} = 1$. 517
82. The medians of a triangle are concurrent. 522
83. The lines containing the altitudes of a triangle are concurrent.
    522

Exercises

1. Lines $\overleftrightarrow{RE}$, $\overleftrightarrow{HK}$, and $\overleftrightarrow{IO}$ are the perpendicular bisectors of the sides of $\triangle ATC$.
   a) What is point S called with respect to $\triangle ATC$?
   b) Why is each side of $\triangle RHI$ parallel to one of the sides of $\triangle ATC$?
   c) Why is $\overline{RE} \perp \overline{IH}$, $\overline{HK} \perp \overline{RI}$, and $\overline{IO} \perp \overline{RH}$?
   d) What are $\overline{RE}$, $\overline{HK}$, and $\overline{IO}$ called with respect to $\triangle RHI$?
   e) What is point S called with respect to $\triangle RHI$?

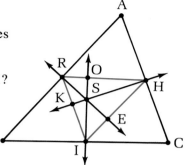

2. Draw a circle with your compass and label its center O. Choose any point on the circle and label it P.
   a) Draw at least ten chords of the circle that have P as one endpoint.
   b) Locate, as accurately as you can, the midpoint of each of these chords.
   c) What relationship do these midpoints seem to have to each other?

3. Copy $\triangle YAM$ and make each of the following constructions.
   a) Circumscribe a circle about it.
   b) Inscribe a circle in it.
   c) Find its centroid.

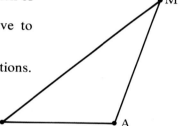

4. Copy $\triangle SIA$.
   a) Construct its altitudes; label them $\overline{SN}$, $\overline{IC}$, and $\overline{AP}$ and the point in which they intersect H. Draw $\triangle PCN$.
   b) It is possible to prove that point H is the incenter of $\triangle PCN$. What does this imply about the relationship of $\overleftrightarrow{SN}$, $\overleftrightarrow{IC}$, and $\overleftrightarrow{AP}$ to $\triangle PCN$?
   c) Which lines in the figure are equidistant from point H?

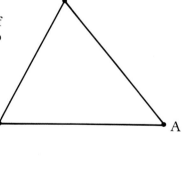

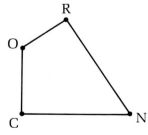

5. In quadrilateral CORN, $\overline{OC} \perp \overline{CN}$ and $\overline{OR} \perp \overline{RN}$. Explain why CORN is cyclic.

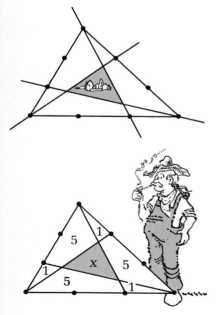

SET II

Draw and label segments having these lengths: a, 3 cm; b, 4 cm; c, 5 cm. Use them to make the following constructions:

1. Construct an isosceles triangle having a as its base and b as the corresponding altitude.

2. Construct a 30°-60° right triangle having c as its hypotenuse.

3. Construct a triangle having a and c as two of its sides and b as the radius of its circumcircle.

4. The three cevians in △TRI are concurrent. Find x.

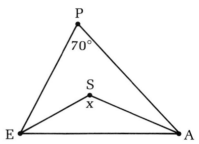

5. Point S is the incenter of △PEA. If ∠P = 70°, find ∠ESA. (Hint: Let ∠SEA = a° and ∠SAE = b°.)

SET III

Grandpa Dilcue's farm was in the shape of the triangular region shown above before the highway commission decided to build some roads across it. After the roads were built, Grandpa decided to sell all of the land except for the region in the middle. He is curious to know how the area of this region compares with the area of his original farm.

Can you figure out what fraction of the original farm it is from the following clues?

Each road passes through a point of trisection of one of the sides of the original farm.

Each of the three quadrilateral regions formed has an area of five acres.

Each of the remaining three triangular regions has an area of one acre.

Chapter **14**

REGULAR POLYGONS
AND THE CIRCLE

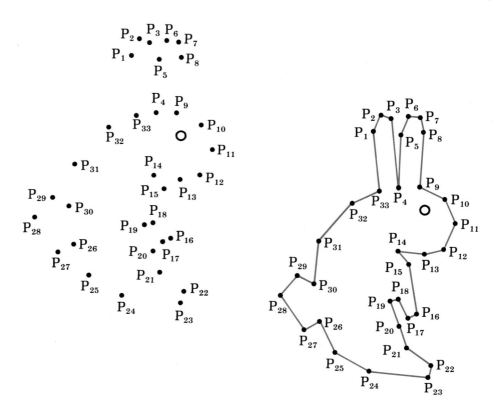

Lesson 1

Polygons

A popular children's puzzle consists of a set of numbered dots that are to be joined in sequence to make a picture. An example of such a puzzle, together with its solution, is shown here. This puzzle is somewhat unusual in that it looks like two different things, depending upon the way in which you view it. Turn the page 90° counterclockwise and you may see something entirely different!

In solving such a puzzle, a set of points is joined by a series of line segments in a certain order. In the example shown above, the last point has also been joined to the first to make a closed figure. The result, not including its "eye," is an example of a *polygon*. The term "polygon" is derived from a Greek word meaning "many-angled." We will base our definition upon the way in which the figure is drawn.

► Definition

Let P_1, P_2, . . . , P_n be a set of n distinct points in a plane where $n > 2$. Then the union of the segments $\overline{P_1P_2}$, $\overline{P_2P_3}$, . . . , $\overline{P_{n-1}P_n}$, $\overline{P_nP_1}$ is a *polygon* if the segments intersect only at their endpoints and no two segments with a common endpoint are collinear.

Although this definition is rather long, it has been carefully worded to include only those figures that we want to consider to be polygons. Notice that if n is 3, the polygon is a triangle and if n is 4, it is a quadrilateral.

The terms that we have already defined for these figures are also used in describing polygons in general. For example, polygon ABCDE shown here has five *vertices*, five *sides*, and five *angles*. A line segment joining any two nonconsecutive vertices, such as $\overline{AC}$, is one of its *diagonals*.

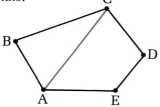

Number of Sides	Name of Polygon
3	Triangle
4	Quadrilateral
5	Pentagon
6	Hexagon
7	Heptagon
8	Octagon
9	Nonagon
10	Decagon
12	Dodecagon

Polygons are named according to the number of sides they have. Those to which we will refer the most frequently are listed in the adjoining table.

In general, a polygon having n sides is called an *n-gon* so that, for example, a decagon could also be called a 10-gon.

The "rabbit-duck" polygon is a 33-gon. Furthermore, it is *concave*. We can distinguish between convex and concave polygons in the same way that we have between convex and concave quadrilaterals.

► Definition

A polygon is *convex* iff for each line that contains a side of the polygon, the rest of the polygon lies in one of the half-planes that has the line for its edge. All other polygons are *concave*.

A convex polygon   A concave polygon

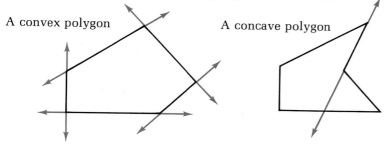

Exercises

1. Although each of the following figures consists of five line segments, only two of them are pentagons. Identify the two that are not and, in each case, explain why not.

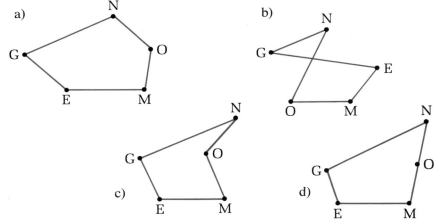

a)

b)

c)

d)

2. Every quadrilateral, whether it is convex or concave, has two diagonals. An example of each is shown at the left.

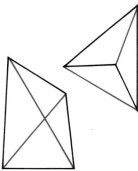

a) Copy the convex and concave pentagons shown here and draw all of the diagonals of each.

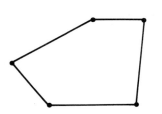
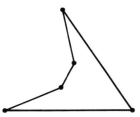

b) Do the same with the hexagons shown here.

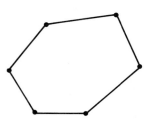
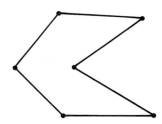

c) Does the number of diagonals that a polygon has seem to depend upon whether it is convex or concave?

3. The first polygon at the right is a convex quadrilateral; more specifically, a square.
   a) Give a specific name for the second polygon at the right.
   b) Does our definition of *perimeter* for a triangle or quadrilateral (see page 337) seem appropriate for polygons in general?
   c) Which of the two polygons at the right has the greater perimeter?
   d) Which has the greater area?
   e) Is it correct to say of two polygons that the polygon with the greater perimeter has the greater area?

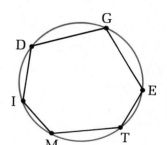

4. The hexagon shown here is an example of a *cyclic polygon*. It is said to be *inscribed* in the circle.
   a) Write a definition for the term *cyclic polygon*.
   b) Do you think a concave polygon can be cyclic? If so, draw an example of one.
   c) Is every convex polygon cyclic?

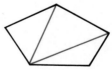

SET II

1. We have proved that the sum of the measures of the angles of a triangle is 180° and that the sum of the measures of the angles of a convex quadrilateral is 360°. The proof for the quadrilateral was based upon dividing it into two triangles by a diagonal from one vertex.
   a) Into how many triangles are a convex pentagon and a convex hexagon divided by the diagonals drawn from one vertex?

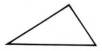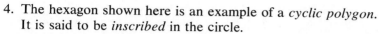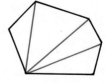

   b) What do you think is the sum of the measures of the angles of each figure?
   c) Copy and complete the following table.

Number of sides of polygon	3	4	5	6
Number of diagonals from one vertex	0	1		
Number of triangles formed	1	2		
Sum of the measures of the polygon's angles	180°	360°		

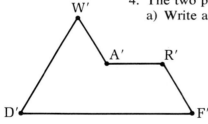

d) In general, into how many triangles is a convex *n*-gon divided by the diagonals from one vertex?

e) What do you think is a general formula for the sum of the measures of the angles of a convex *n*-gon?

2. We defined a *polygonal region* as the union of a finite number of triangular regions in a plane that have no interior points in common. Examples of polygonal regions are shown here.

a) Is it correct to say that every polygonal region is bounded by a polygon?

b) Do you think every polygon is the boundary of some polygonal region?

3. The two polygons in the figure below are congruent.

a) Write a definition for the term *congruent polygons*.

b) If two polygons are congruent, do they necessarily have equal perimeters?

c) Do they necessarily have equal areas?

4. The two polygons in the figure below are similar.

a) Write a definition of *similar polygons*.

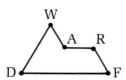

We have proved that the ratio of the perimeters of two similar *triangles* is equal to the ratio of the corresponding sides. It is easy to prove that this relationship holds true for similar polygons in general.

Suppose DWARF ~ D'W'A'R'F'.

b) Why is $\dfrac{DW}{D'W'} = \dfrac{WA}{W'A'} = \dfrac{AR}{A'R'} = \dfrac{RF}{R'F'} = \dfrac{DF}{D'F'}$?

c) Why is $\dfrac{DW + WA + AR + RF + DF}{D'W' + W'A' + A'R' + R'F' + D'F'} = \dfrac{DW}{D'W'}$?

d) Why is $\dfrac{\rho DWARF}{\rho D'W'A'R'F'} = \dfrac{DW}{D'W'}$?

We have also proved that the ratio of the *areas* of two similar triangles is equal to the *square* of the ratio of the corresponding sides. This also is true of similar polygons in general.

   e) Write an equation in terms of the similar polygons on the facing page to illustrate this fact.

SET III

This figure was drawn by George Washington at the age of seventeen in surveying a piece of land. He walked from the point marked A at the lower right corner to the point marked B and so forth around the hexagon until he returned to A, measuring the direction of each side with a magnetic compass and the length of each side in terms of his strides.

   A protractor reveals that the six angles of the hexagon have the following measures: $\angle A = 82°$, $\angle B = 100°$, $\angle C = 44°$, $\angle D = 107°$, $\angle E = 107°$, and $\angle F = 120°$.

   On the basis of a previous exercise in this lesson, it might seem that the sum of the measures of the six angles should be 720°, yet their actual sum is only 560°. Can you explain why there is a discrepancy and also why it is 160°?

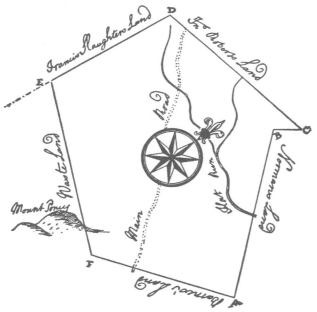

THE GEORGE WASHINGTON ATLAS, WASHINGTON, D. C., 1932

Lesson 2

Regular Polygons

Snowflakes are a beautiful example of geometry in nature. A farmer-meteorologist in Vermont spent many winters taking photographs of thousands of them through a microscope; some of his pictures are shown here.

Although it has been said that no two snowflakes are alike, those illustrated here have several basic properties in common. They are all six-sided and hence hexagonal in shape. Furthermore, each is convex and has equal sides and equal angles; such polygons are called *regular*.

▶Definition

A ***regular polygon*** is a convex polygon that is both equilateral and equiangular.

You know that one of these properties cannot exist without the other in polygons having three sides. We have proved that all equilateral triangles are equiangular and conversely.

If a polygon has more than three sides, however, having one property does not imply the other. Consider quadrilaterals, for example. A rhombus is equilateral but not necessarily equiangular, whereas a rectangle is equiangular but not necessarily equilateral. Only squares are both.

AFTER YEARS OF RELENTLESS STUDY,

LABORIOUS RESEARCH,

AND INNUMERABLE CALCULATIONS,......

I HAVE HEREBY ASCERTAINED THAT,....

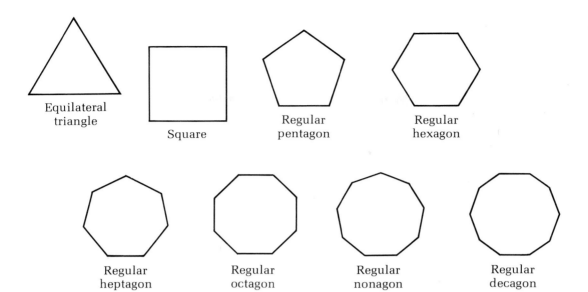

Equilateral
triangle

Square

Regular
pentagon

Regular
hexagon

Regular
heptagon

Regular
octagon

Regular
nonagon

Regular
decagon

Figures representing regular polygons having from three through ten sides are shown above. It is evident that, as the number of sides of a regular polygon increases, it looks more and more like a circle. This is a consequence of the fact that every regular polygon is cyclic.

▶ **Theorem 84**
Every regular polygon is cyclic.

Hypothesis: ABCDE is a regular polygon.
Conclusion: ABCDE is cyclic.

To illustrate our proof, we will use a regular pentagon. The proof, however, applies to all regular polygons because it does not depend upon the number of sides of the pentagon.

The idea is to draw a circle through three vertices of the polygon and then prove, by means of congruent triangles, that the distance from each of the remaining vertices to the center of this circle is equal to its radius. From this it follows that they also lie on the circle.

NO TWO SNOWFLAKES......

...ARE ALIKE....

CRUNCH

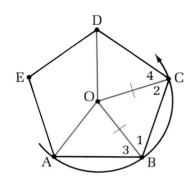

Proof.
Draw the circle determined by points A, B, and C and let its center be called O. (It follows from the definition of a polygon that A, B, and C are noncollinear; three noncollinear points determine a circle.) Draw $\overline{OA}$, $\overline{OB}$, $\overline{OC}$, and $\overline{OD}$.

We know that OB = OC (all radii of a circle are equal), so $\angle 1 = \angle 2$ (if two sides of a triangle are equal, the angles opposite them are equal.) Now $\angle ABC = \angle BCD$ (a regular polygon is equiangular), so $\angle 3 = \angle 4$ (subtraction). Also AB = CD (a regular polygon is equilateral), so $\triangle OBA \cong \triangle OCD$ (S.A.S.). Therefore, OD = OA. Since OA is the radius of the circle, it follows that OD is also. Hence D lies on circle O, because a circle is the set of all points in a plane at a distance of one radius from its center.

To prove that E also lies on circle O, $\overline{OE}$ can be drawn and $\triangle ODE$ proved congruent to $\triangle OCB$ in the same fashion.

A polygon whose vertices lie on a circle is said to be *inscribed* in the circle and the circle is said to be *circumscribed* about the polygon. Other terms that we will use in referring to regular polygons are illustrated and defined below.

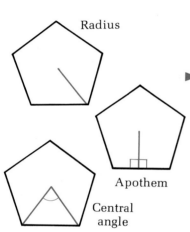

▶ Definitions
The *center* of a regular polygon is the center of its circumscribed circle.

A *radius* of a regular polygon is a line segment that joins its center to a vertex.

An *apothem* of a regular polygon is a perpendicular line segment from its center to one of its sides.

A *central angle* of a regular polygon is an angle formed by radii drawn to two consecutive vertices.

Exercises Set I

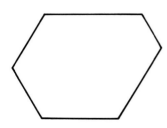

1. The hexagon shown here is equiangular but not equilateral.
   a) Does it seem to be cyclic?
   b) Draw a hexagon that is equilateral but not equiangular.
   c) Does it seem to be cyclic?
   d) Draw a regular hexagon.
   e) Is it necessarily cyclic?

2. Circle E is circumscribed about equilateral △FUD; $\overline{EG}$ ⊥ $\overline{FD}$.

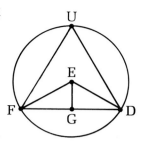

   a) What is point E called with respect to △FUD?
   b) What is $\overline{EG}$ called?
   c) How do we know that $\overline{EG}$ bisects $\overline{FD}$?
   d) What is $\overline{ED}$ called with respect to △FUD?
   e) What is ⊿FED called with respect to △FUD?

3. Every regular polygon is symmetric with respect to the perpendicular bisectors of its sides and to the bisectors of its angles. Square MINT, for example, has four symmetry lines. Lines ℓ_1 and ℓ_3 are the perpendicular bisectors of its sides and ℓ_2 and ℓ_4 bisect its angles.
   a) At what point do these four lines seem to be concurrent?
   b) Does MINT seem to have rotational symmetry? Explain.

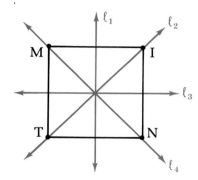

4. Trace the regular polygons below and draw all of the symmetry lines of each.
   a) If a regular polygon has n sides, how many lines of symmetry do you think it has?
   b) In what way are the symmetry lines of regular polygons that have an even number of sides different from the symmetry lines of regular polygons that have an odd number of sides?
   c) Not all regular polygons have point symmetry. Upon what does this property seem to depend?

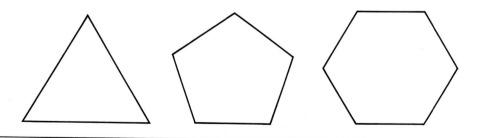

The fact that all regular polygons are cyclic provides a convenient way to construct some of these figures. A circle is drawn first and then the vertices of the polygon are located on it.

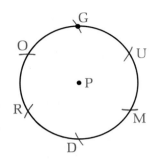

1. A polygon that is especially easy to construct in this way is the regular hexagon.
   a) Draw a circle P and choose a point G on it. Starting at G, mark off equal arcs around the circle, each with radius equal to PG, as shown in the first figure. If you do this accurately, the last arc should intersect the circle in point G. The points in which the arcs intersect the circle are the vertices of a regular hexagon. Join them in order to complete the construction.

   To explain why the construction works, draw the six radii of the hexagon.
   b) What kinds of triangles are △GPU, △UPM, △MPD, △DPR, and △RPO?
   c) In each of these triangles, what is the measure of the angle with its vertex at P?
   d) Why does ∡OPG have the same measure?
   e) Why is △OPG equilateral?
   f) Why is OG equal in length to the other sides of the hexagon?
   g) Why is the hexagon equiangular?

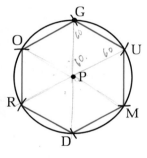

2. Construct each of the following regular polygons by first locating its vertices on a circle.
   a) A square.
   b) An octagon.
   c) A dodecagon.

3. Use the figure below to prove that any two central angles of a regular polygon are equal.

   Hypothesis: NOUGA is a regular polygon with central ∡s 1 and 2.
   Conclusion: ∠1 = ∠2.
   Par 7. (Hint: Circumscribe circle T about NOUGA.)

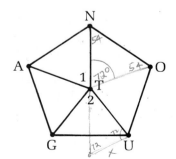

4. Use the figures below to answer the following questions.
   a) As the number of sides of a regular polygon increases, how does the measure of one of its central angles change?
   b) Find the measure of a central angle of each figure shown.
   c) What do you think is a general formula for the measure of a central angle of a regular *n*-gon?

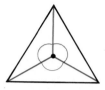

SET III

*Three Dissection Puzzles.** Carefully trace the figures shown here and cut each one apart.

Can you rearrange the pieces of

1. the hexagon to form a square?

2. the star to form an equilateral triangle?

3. the cross to form a regular dodecagon?

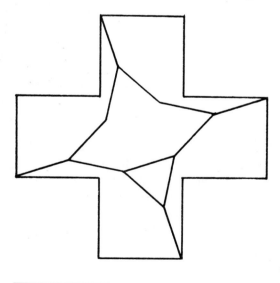

_____

*From "Geometric Dissections," in *The Unexpected Hanging* by Martin Gardner (Simon & Schuster, 1969), Chap. 4. Originally published in "Mathematical Games" by Martin Gardner. Copyright © 1961 by Scientific American, Inc.

Lesson 3

The Perimeter of a
Regular Polygon

General Turtle, Inc. is a company that makes a small device known as a "turtle." If one of these turtles is linked to a computer, it can be ordered to do such things as move in a given direction, turn through a given angle, and trace the path it follows with a pen. Its purpose is to help elementary school children learn mathematics.

Among the tricks that the turtle can perform when given the correct directions is the drawing of regular polygons. For example, if it is told to move 10 inches and turn right 90°, and then to repeat these two actions until it returns to its starting point, it will produce a square. In the photograph, it has drawn a regular pentagon, after having been given slightly different directions.

How far does the turtle travel in drawing a regular polygon? The answer to this question depends upon two things: the number and length of the polygon's sides. The total distance covered is the perimeter of the polygon and can be expressed by the equation

$$\rho = ns,$$

where n is the number of sides of the polygon and s is the length of one side. For example, the turtle would move 50 inches in drawing a regular pentagon that has 10-inch sides.

548

The perimeter of a regular polygon can also be expressed in terms of the length of its radius. In the figure shown here, a regular pentagon with sides of length s has been inscribed in a circle with radius r. Segment $\overline{OM}$ is an apothem of the pentagon, so $\overline{OM}$ is perpendicular to $\overline{AB}$ and hence bisects it. (If a line through the center of a circle is perpendicular to a chord, it also bisects it.) So $AM = \frac{1}{2}AB = \frac{1}{2}s$.

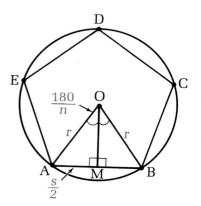

It is easy to see that $\triangle AOM \cong \triangle BOM$, so $\angle AOM = \angle BOM$. This means that $\overrightarrow{OM}$ bisects central $\angle AOB$. Since a central angle of a regular n-gon has a measure of $\dfrac{360°}{n}$, $\angle AOM = \frac{1}{2}\left(\dfrac{360}{n}\right) = \dfrac{180°}{n}$.

Now to find a relationship between r and s we can apply the sine ratio to right $\triangle AOM$.

$$\sin \frac{180}{n} = \frac{\frac{s}{2}}{r}.$$

Multiplying both sides of the equation by r,

$$r \sin \frac{180}{n} = \frac{s}{2},$$

and by 2,

$$s = 2r \sin \frac{180}{n}.$$

Since $\rho = ns$, we get, by substitution,

$$\rho = n\left(2r \sin \frac{180}{n}\right) = 2\left(n \sin \frac{180}{n}\right)r.$$

The product $n \sin \dfrac{180}{n}$ depends only upon n, the number of sides of the polygon. We will borrow another letter from the Greek alphabet, ν^*, to represent this product so that

$$\nu = n \sin \frac{180}{n}.$$

Substituting into the equation above, we get

$$\rho = 2\nu r.$$

We have developed a formula for the perimeter of a regular polygon having a given number of sides in terms of its radius. Since this formula does not depend upon the fact that we used a polygon with five sides to derive it, we will state it as a general theorem.

---

*The name of this letter is "nu"; it is the equivalent of the letter "n" in the English alphabet.

▶ **Theorem 85**

The perimeter of a regular n-gon is $2\nu r$, where r is the length of its radius and $\nu = n \sin \dfrac{180}{n}$.

As an example of how to use this theorem, consider the following problem. Suppose the turtle has been told to draw a regular pentagon with a 10-inch radius. How far would it travel?

$$\rho = 2\nu r = 2\left(n \sin \frac{180}{n}\right)r$$

$$= 2\left(5 \sin \frac{180}{5}\right)(10)$$

$$= 100 \sin 36°$$

$$\approx 100(.588)*$$

$$\approx 58.8$$

The turtle would move approximately 58.8 inches.

Exercises

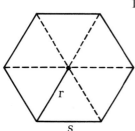

1. The regular hexagon is the only regular polygon whose radius is equal in length to its side.
   a) If one side of a regular hexagon is 7 units long, what is its perimeter?
   b) The perimeter can also be found by means of the equation $\rho = 2\nu r$, where $\nu = n \sin \dfrac{180}{n}$. Show that this equation gives the same result that you got in part a.
   c) If one of the longest diagonals of a regular hexagon is 11 units long, what is its perimeter?
   d) What is the ratio of the perimeter of a regular hexagon to the length of one of its longest diagonals?

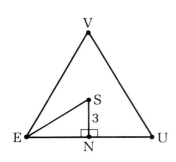

2. Triangle EVU is equilateral with radius $\overline{SE}$ and apothem $\overline{SN}$; SN = 3.
   a) What kind of triangle is $\triangle ESN$?
   b) Find the length of radius $\overline{SE}$.
   c) Find EN.
   d) Find EU.

---

*From the table of trigonometric ratios on page 439.

e) Find $\rho\triangle EVU$.

f) Change your answer to part e to decimal form. (Use $\sqrt{3}\approx 1.732$.)

g) Use the formula $\rho = 2vr$ to find $\rho\triangle EVU$.

3. Polygon APHRODITE is a regular nonagon with radius $\overline{XA}$ and apothem $\overline{XY}$; $XA = 10$.

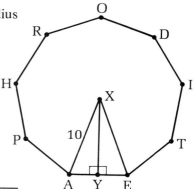

a) Find $\angle AXE$.

b) Find $\angle AXY$.

c) Use the equation $\sin\angle AXY = \dfrac{AY}{AX}$ to find AY.

d) Find AE.

e) Find ρAPHRODITE.

f) Use the formula $\rho = 2vr$ to find ρAPHRODITE.

SET II

1. Quadrilateral ADIS is a square with radius $\overline{OD}$ and apothem $\overline{ON}$; $ON = 1$.

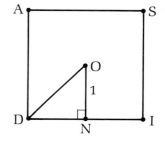

a) Explain why $OD = \sqrt{2}$.

b) Use right $\triangle DON$ to show that $\sin 45° = \dfrac{1}{\sqrt{2}}$.

c) Use the fact that $\sin 45° = \dfrac{1}{\sqrt{2}}$ and the formula $\rho = 2vr$ to find the *exact* value of ρADIS.

d) Verify that your answer to part c is correct by finding ρADIS in a different way.

2. The triangle, hexagon, and dodecagon inscribed in this circle are all regular and the radius of the circle is 10 units.

a) Find the perimeter of the triangle.

b) Find the perimeter of the hexagon.

c) Find the perimeter of the dodecagon.

d) If the number of sides of a regular polygon inscribed in a circle is doubled, is its perimeter doubled?

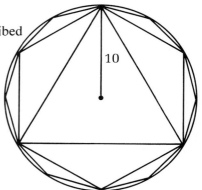

3. This figure represents a regular 180-gon. If the radius of the polygon is $\frac{1}{2}$, find its perimeter. Use the value $\sin 1° = .0175$.

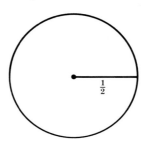

SET III

Suppose the turtle described in this lesson was programmed to move one foot and then turn right one second, repeating these actions until it returns to its starting point.

1. What regular polygon would it draw?

2. How many miles would it travel before returning to the point from which it began?

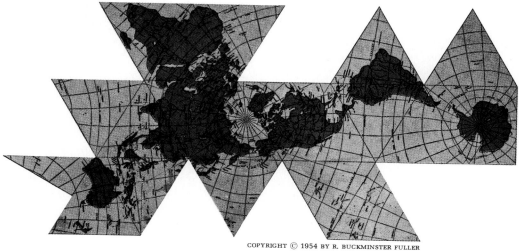

Lesson 4

The Area of a Regular Polygon

Perhaps the cleverest map of the earth ever devised is one created by Buckminster Fuller, the inventor of the geodesic dome. It consists of twenty equilateral triangles, each of which contains an equal amount of the earth's surface. The map can be folded along the sides of these triangles into a three-dimensional "globe" called an *icosahedron.**

The orientation of the earth's land masses on the map's surface has been arranged so that the corners of the icosahedron, which are the regions of greatest distortion, fall on either the ocean or a shore line. This is evident in the flat version of the map.

If a model of Mr. Fuller's icosahedron map were made in which each edge was 1 foot long, what would its surface area be? We can easily answer this question by using the formula for the area of an equilateral triangle,

$$\alpha_{\text{eq. triangle}} = \frac{a^2}{4}\sqrt{3},$$

where a is the length of a side. The map consists of 20 equilateral triangles, so its area is

$$20\left(\frac{1^2}{4}\sqrt{3}\right) = 5\sqrt{3} \approx 5(1.732) = 8.66 \text{ sq ft.}$$

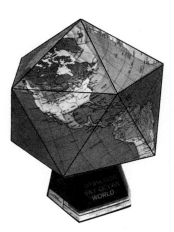

---

*You may recall that atoms in boron crystals are also arranged in icosahedra. Look again at the figure on page 511.

The solution of this problem required a formula for the area of a regular polygon: the equilateral triangle. The only other regular polygon for which we have such a formula is the square. If a is the length of a side,

$$\alpha_{\text{eq. triangle}} = \frac{a^2}{4}\sqrt{3} \quad \text{and} \quad \alpha_{\text{square}} = a^2.$$

To find the areas of other regular polygons, it would be useful to have a more general formula. We will derive such a formula, not in terms of the length of a *side*, but rather in terms of the length of the *radius* of the polygon.

We will use a regular pentagon to develop the formula, but our argument will not depend upon the fact that it has five sides.

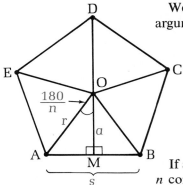

If all of the radii of a regular n-gon are drawn, they divide it into n congruent isosceles triangles. (In this pentagon, there are five of these triangles.)

The area of one of these triangles, such as $\triangle AOB$, is $\frac{1}{2}sa$ because apothem $\overline{OM}$ is the altitude to base $\overline{AB}$. It follows that the total area of the n triangles in the figure is $n(\frac{1}{2}sa)$, so that

$$\alpha_{\text{regular } n\text{-gon}} = \tfrac{1}{2}nsa.$$

Since the perimeter of a regular n-gon is ns, we can also write

$$\alpha_{\text{regular } n\text{-gon}} = \tfrac{1}{2}\rho a.$$

We now have a formula for the area of a regular n-gon in terms of its perimeter and apothem. We will now restate it in terms of the polygon's radius.

By Theorem 85, $\rho = 2\nu r$. By applying the cosine ratio to right $\triangle AOM$, $\cos \dfrac{180}{n} = \dfrac{a}{r}$. Multiplying both sides of this equation by r, $a = r \cos \dfrac{180}{n}$. Finally, substituting these results into the equation, $\alpha_{\text{regular } n\text{-gon}} = \frac{1}{2}\rho a$, we get

$$\alpha_{\text{regular } n\text{-gon}} = \tfrac{1}{2}(2\nu r)\left(r \cos \frac{180}{n}\right) = \nu \cos \frac{180}{n} r^2.$$

▶ **Theorem 86**

The area of a regular n-gon is $\nu \cos \dfrac{180}{n} r^2$, where r is the length of its radius and $\nu = n \sin \dfrac{180}{n}$.

To illustrate how this theorem is used, we will find the area of the regular pentagon with a 10-inch radius drawn by the turtle that is described on page 550.

$$\alpha_{\text{pentagon}} = v \cos \frac{180}{n} r^2 = n \sin \frac{180}{n} \cos \frac{180}{n} r^2$$

$$= 5 \sin \frac{180}{5} \cos \frac{180}{5} (10)^2$$

$$= 500 \sin 36° \cos 36°$$

$$\approx 500(.588)(.809) \approx 238.$$

The area of the regular pentagon would be about 238 square inches.

SET I

Exercises

1. Quadrilateral WILO is a square with radius $\overline{\text{NO}}$ and apothem $\overline{\text{NS}}$; WI = 6.
   a) Use WI to find αWILO.
   b) Find NS.
   c) Find the exact length of $\overline{\text{NO}}$. (In other words, leave it in simple radical form.)
   d) Use right $\triangle$NOS to show that $\sin 45° = \dfrac{1}{\sqrt{2}} = \cos 45°$.
   e) Use the formula, $\alpha_{\text{regular } n\text{-gon}} = v \cos \dfrac{180}{n} r^2$ to find αWILO.

After comparing your answers to parts a and e, you will agree that the second method for finding the area of a square is somewhat like using a sledgehammer to crack a peanut shell. But in doing the problem both ways, we have verified that the sledgehammer works!

The table on the next page will save you some time in doing the exercises in this and the following lessons. It gives the product of the sine and cosine of the angles from 1° through 45°, each to four significant figures.

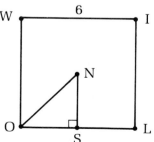

A	sin A cos A	A	sin A cos A	A	sin A cos A
1°	.01745	16°	.2650	31°	.4415
2°	.03488	17°	.2796	32°	.4494
3°	.05226	18°	.2939	33°	.4568
4°	.06959	19°	.3078	34°	.4636
5°	.08682	20°	.3214	35°	.4698
6°	.1040	21°	.3346	36°	.4755
7°	.1210	22°	.3473	37°	.4806
8°	.1378	23°	.3597	38°	.4851
9°	.1545	24°	.3716	39°	.4891
10°	.1710	25°	.3830	40°	.4924
11°	.1873	26°	.3940	41°	.4951
12°	.2034	27°	.4045	42°	.4973
13°	.2192	28°	.4145	43°	.4988
14°	.2347	29°	.4240	44°	.4997
15°	.2500	30°	.4330	45°	.5000

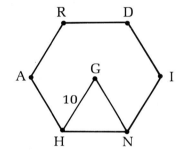

2. Polygon HARDIN is a regular hexagon with radius $\overline{GH}$ of length 10.
   a) Find the exact area of △HGN.
   b) Find the exact area of HARDIN.
   c) Change your answer to part b to decimal form. (Use $\sqrt{3} \approx 1.732$.)
   d) Use the formula, $\alpha_{\text{regular } n\text{-gon}} = \nu \cos \dfrac{180}{n} r^2$ to find αHARDIN. (You can find the value of $\sin \dfrac{180}{n} \cos \dfrac{180}{n}$ in the table above.)

3. The ratio of the areas of two regular polygons that have the same number of sides is equal to the square of the ratio of their radii.
   a) Show why this is true.
   b) What relationship exists between the ratio of the perimeters of two such polygons and the ratio of their radii?
   c) The radius of one regular octagon is twice that of another. How do they compare in area?
   d) How do their perimeters compare?

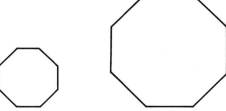

1. This figure is a regular decagon with radius of length 11.
   a) Find its perimeter.
   b) Find its area.

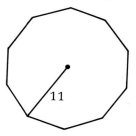

2. The figure at the right represents a regular 30-gon with radius r.
   a) Show that its approximate area is given by the following formula: $a_{\text{regular } n\text{-gon}} = 3.12r^2$.
   b) Find a comparable formula for the approximate area of a regular 90-gon with radius r.
   c) Even though a regular 90-gon has three times as many sides as a regular 30-gon, the formulas for their areas are almost the same. Explain why.

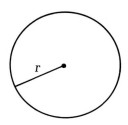

SET III

There is an old puzzle about a man who had a window that was a yard square.* He decided that it let in too much light, so he boarded up half of it and still had a square window a yard high and a yard wide. The puzzle is to figure out how he did it and the answer is shown here. As you can see, the man formed a smaller square by joining the midpoints of the sides of the original square.

Now suppose the original window was in the shape of a regular hexagon and that it was boarded up in the same way to form a smaller hexagon. Can you figure out how the amount of light let in by the smaller of the hexagonal windows would compare with that let in by the larger?

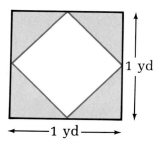

*Adapted from *536 Puzzles and Curious Problems* by Henry Ernest Dudeney (Scribner's, 1967).

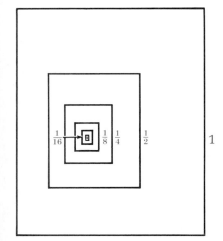

Lesson 5

Limits

This photograph of a boy holding a mirror was taken by pointing the camera toward another mirror facing him. As a result, his image is reflected back and forth between the two mirrors a seemingly infinite number of times.

In the picture, each image is half the height of the previous one so that if the first image is 1 unit tall, the successive heights of the other images are

$$\frac{1}{2}, \frac{1}{4}, \frac{1}{8}, \frac{1}{16}, \text{ and so on.}$$

The drawing represents the images as a sequence of similar rectangles. After a certain point, it is difficult to draw any more because the rectangles become so small. The further along we go, the closer and closer their heights approach zero. The tenth rectangle, for instance, has a height of 1/512 or about 0.002 unit. The hundredth rectangle has a height of about

0.000000000000000000000000002 unit.

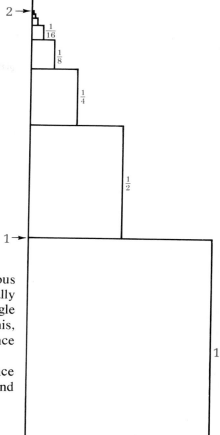

The height of the stack of
the first two rectangles $= 1 + \frac{1}{2} = 1\frac{1}{2}$;

the first three rectangles $= 1 + \frac{1}{2} + \frac{1}{4} = 1\frac{3}{4}$;

the first four rectangles $= 1 + \frac{1}{2} + \frac{1}{4} + \frac{1}{8} = 1\frac{7}{8}$;

the first five rectangles $= 1 + \frac{1}{2} + \frac{1}{4} + \frac{1}{8} + \frac{1}{16} = 1\frac{15}{16}$.

If each rectangle in the sequence is half as tall as the previous one, we will never come to a rectangle whose height is actually zero. However, if we go far enough, we can find a rectangle whose height is *as close to zero* as we please. Because of this, we say that the heights of the rectangles in this sequence *approach zero as a limit.*

If we represent the height of the nth rectangle in the sequence as h_n, then as n, the number of the rectangle, gets larger and larger, h_n approaches zero. In symbols, this is written

$$\lim_{n \to \infty} h_n = 0.$$

Another example of a limit consists of arranging the rectangles in a vertical stack as shown here. If "all" of the rectangles were included, how tall would the stack be? From the figure, the answer seems to be 2, and the list beside it seems to confirm this.

A stack of the first ten rectangles would be about 1.998 units high, whereas a stack of the first hundred rectangles would be about 1.99999999999999999999999999998 units high!

The height of a finite stack of rectangles will never actually be 2, but if we include enough of them in the stack, we can make its height *as close to 2* as we please. So the height of the stack *approaches 2 as a limit.* If the height of the stack of n rectangles is represented as H_n, we have

$$\lim_{n \to \infty} H_n = 2.$$

Since a mathematically precise definition of the word *limit* would be quite difficult to understand, we will give an informal explanation instead. The two examples you have just seen should help you understand it.

▶ An Informal Explanation of "Limit"

Let $a_1, a_2, a_3, \ldots, a_n, \ldots$ be a sequence of numbers. Then the number L is the *limit* of this sequence if, by letting n get sufficiently large, the successive differences between L and a_n can be made as small as we please.

In the example of the successive heights of the mirror images of the boy, the sequence is $\dfrac{1}{2}, \dfrac{1}{4}, \dfrac{1}{8}, \dfrac{1}{16}, \ldots$ The numbers in a sequence are called its *terms*. The limit of this sequence is 0 because, no matter how small a number you choose, the difference between 0 and all of the terms of the sequence beyond a certain point will be less than the number you chose.

$$\frac{1}{2}, \frac{1}{4}, \frac{1}{8}, \frac{1}{16}, \frac{1}{32}, \frac{1}{64}, \left[\frac{1}{128}, \frac{1}{256}, \frac{1}{512}, \left[\frac{1}{1028}, \ldots \right. \right.$$

If you choose $\dfrac{1}{100}$, for instance, the difference between every term beyond the sixth and zero is less than $\dfrac{1}{100}$. If you choose $\dfrac{1}{1000}$, the difference between every term beyond the ninth and zero is less than $\dfrac{1}{1000}$, and so forth.

You know that, as the number of sides of a regular polygon increases, it looks more and more like a circle. We will use limits in the next lesson to define and develop formulas for the circumference and area of a circle by relating these numbers to the perimeter and area of a regular polygon.

Exercises

Some of the sequences in this lesson will be illustrated by the first few terms. From these terms, guess a pattern for the sequence and base your answers on this pattern.

SET I

1. Consider the sequence 5.1, 5.01, 5.001, ...
   a) What do you think is the fourth term of this sequence?
   b) What is the difference between the fourth term of the sequence and the number 5?
   c) What is the first term in the sequence such that the difference between it and 5 is *less* than .000001?
   d) Is the difference between *every* term following the one you named in part c and the number 5 less than .000001?
   e) What does the limit of the sequence seem to be?

2. List the next three terms for each of the following sequences.
   a) 1.9, 1.99, 1.999, . . .
   b) 2, 4, 8, 16, . . .
   c) 7, 7, 7, 7, . . .
   d) 81, 27, 9, 3, . . .
   e) 0, 1, 0, 1, 0, 1, . . .

3. Many sequences do not have limits. For example, the terms of the sequence 5, 10, 15, 20, 25, . . . seem to be getting larger and larger. They do not seem to be getting closer and closer to a definite number.
   a) Which sequences in Exercise 2 do not seem to have limits?
   b) Name the numbers that you think are the limits of the rest.

4. The nth term of a certain sequence can be found by the formula $\dfrac{1}{n}.$
   a) What are the first five terms of this sequence?
   b) As n becomes larger and larger, what happens to $\dfrac{1}{n}$?
   c) What do you think is the value of $\lim\limits_{n \to \infty} \dfrac{1}{n}$?

5. The nth term of a certain sequence can be found by the formula $\dfrac{n}{n+1}.$
   a) What are the first five terms of this sequence?
   b) As n becomes larger and larger, what happens to $\dfrac{n}{n+1}$?
   c) What do you think is the value of $\lim\limits_{n \to \infty} \dfrac{n}{n+1}$?

SET II

The following exercises concern sequences that include trigonometric ratios. We will use the conclusions that you draw from them in the next lesson.

1. The s ne of an acute angle of a right triangle is the ratio of the length of the opposite leg to the length of the hypotenuse.

   In right $\triangle ABC$, $\sin A = \dfrac{a}{c}.$ If we let $c = 1$, then $\sin A = \dfrac{a}{1} = a.$

   In other words, the sine of $\angle A$ is the length a.

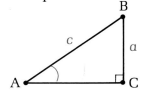

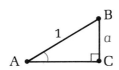 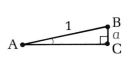

a) What happens to a as $\angle A$ gets smaller and smaller?

b) As the measure of $\angle A$ gets very close to 0°, what number do you think a, and hence $\sin A$, gets very close to?

c) What do you think is the limit of the sequence

$$\sin 4°, \ \sin 2°, \ \sin 1°, \ \sin \tfrac{1}{2}°, \ . \ . \ . \ ?$$

The terms of this sequence might also be written as

$$\sin \frac{180}{45}, \ \sin \frac{180}{90}, \ \sin \frac{180}{180}, \ \sin \frac{180}{360}, \ \cdot \ \cdot \ \cdot$$

d) What do you think is the value of $\displaystyle\lim_{n \to \infty} \sin \frac{180}{n}$?

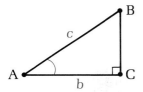

2. The cosine of an acute angle of a right triangle is the ratio of the length of the adjacent leg to the length of the hypotenuse.

In right $\triangle ABC$, $\cos A = \dfrac{b}{c}$. If we let $c = 1$, then $\cos A = \dfrac{b}{1} = b$. In other words, the cosine of $\angle A$ is the length b.

a) What happens to b as $\angle A$ gets smaller and smaller?

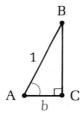 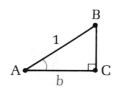 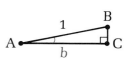

To find out what number b, and hence $\cos A$, gets very close to as the measure of $\angle A$ gets very close to 0°, consider the following sequence.

$$\cos 4°, \ \cos 2°, \ \cos 1°, \ \cos \tfrac{1}{2}°, \ . \ . \ .$$

To five significant figures, the decimal values of these numbers are

$$.99756, \ .99939, \ .99985, \ \text{and} \ .99996.$$

b) What do you think is the limit of this sequence?

c) What do you think is the value of $\displaystyle\lim_{n \to \infty} \cos \frac{180}{n}$?

3. In our formulas for the perimeter and area of a regular n-gon in terms of its radius, the Greek letter ν represents the product $n \sin \dfrac{180}{n}$. In this exercise, we will show that the sequence

$$45 \sin \frac{180}{45}, \ 90 \sin \frac{180}{90}, \ 180 \sin \frac{180}{180}, \ 360 \sin \frac{180}{360}, \ \dots$$

seems to have a limit.

a) Use the adjoining table, which is more precise than the one on page 439, to find the value of each of the given terms in this sequence to four decimal places.

b) Your results to part a should indicate that the sequence seems to have a limit. On the assumption that this is so, what do you think is the value of $\lim\limits_{n \to \infty} n \sin \dfrac{180}{n}$ to two decimal places?

c) Although the value of the above limit is not equal to $\dfrac{22}{7}$, this fraction is sometimes used to approximate it. Express $\dfrac{22}{7}$ as a decimal to the nearest hundredth to show why.

d) What do you think is the name of the number that is the *exact* value of $\lim\limits_{n \to \infty} n \sin \dfrac{180}{n}$?

A	$\sin A$
0.5°	.0087265
1°	.017452
2°	.034899
4°	.069756

SET III

This photograph seems to show an infinite sequence of elephants. Suppose the biggest elephant is 10 feet tall and that each succeeding elephant is half as tall as the one before. If, starting with the 10-foot elephant, there were an infinite number of such smaller elephants and each stood on the back of the next larger one, do you think the stack of elephants would be infinitely tall or would it have a finite height. Can you explain your reasoning?

Lesson 6

The Circumference and Area
of a Circle

Dissatisfied with the lack of any wheellike creatures in nature, Escher decided to invent the one illustrated in this print. It can move forward in three different ways. The figure at the upper left shows the creature walking on its six legs. The third figure shows it rolled up into a wheel, but still using its legs to push itself ahead. The last shows the creature with its legs tucked in for rolling down a slope.

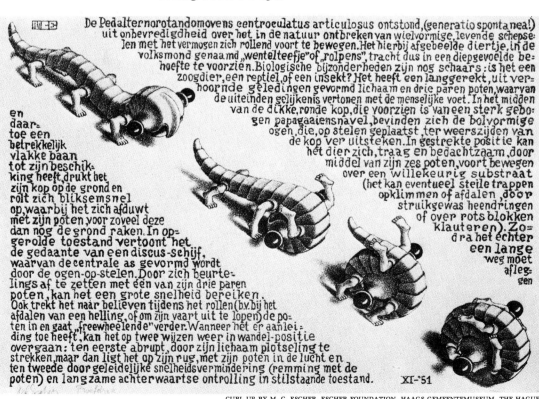

De Pedalternorotandomovens centroculatus articulosus ontstond,(generatio spontanea!) uit onbevredigdheid over het in de natuur ontbreken van wielvormige,levende schepse: len met het vermogen zich rollend voort te bewegen.Het hierbij afgebeelde diertje,in de volksmond genaamd „wentelteefje"of „rolpens", tracht dus in een diepgevoelde be- hoefte te voorzien.Biologische bijzonderheden zijn nog schaars:is het een zoogdier,een reptiel,of een insekt? Het heeft een langgerekt, uit ver- hoornde geledingen gevormd lichaam en drie paren poten,waarvan de uiteinden gelijkenis vertonen met de menselijke voet. In het midden van de dikke,ronde kop,die voorzien is van een sterk gebo: gen papagaaiensnavel,bevinden zich de bolvormige ogen,die,op stelen geplaatst,ter weerszijden van de kop ver uitsteken.In gestrekte positie kan het dier zich,traag en bedachtzaam,door middel van zijn zes poten,voort bewegen over een willekeurig substraat (het kan eventueel steile trappen opklimmen of afdalen,door struikgewas heendringen of over rotsblokken klauteren).Zo- dra het échter een lange weg moet afleg- gen

en daar: toe een betrekkelijk vlakke baan tot zijn beschik: king heeft,drukt het zijn kop op de grond en rolt zich bliksemsnel op,waarbij het zich afduwt met zijn poten,voor zoveel deze dan nog de grond raken.In op- gerolde toestand vertoont het de gedaante van een discus-schijf, waarvan de centrale as gevormd wordt door de ogen-op-stelen.Door zich beurte- lings af te zetten met één van zijn drie paren poten,kan het een grote snelheid bereiken. Ook trekt het naar believen tijdens het rollen(b.v.bij het afdalen van een helling,of om zijn vaart uit te lopen)de po- ten in en gaat „freewheelende"verder.Wanneer het er aanlei: ding toe heeft,kan het op twee wijzen weer in wandel-positie overgaan: ten eerste abrupt,door zijn lichaam plotseling te strekken,maar dan ligt het op zijn rug,met zijn poten in de lucht en ten tweede door geleidelijke snelheidsvermindering (remming met de poten) en langzame achterwaartse ontrolling in stilstaande toestand.

XI-'51

CURL UP BY M. C. ESCHER. ESCHER FOUNDATION, HAAGS GEMEENTEMUSEUM, THE HAGUE

In its rolled-up position, the number of revolutions that this wheel creature must make to travel any given distance depends upon its *circumference*, which in turn depends upon its radius. By its "circumference," we mean the "distance around it." Since we have defined distance in terms of straight lines rather than circles however, this won't do for a definition of circumference. The sequence of figures below suggests a way to get around this difficulty.

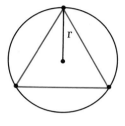 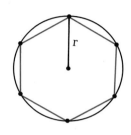 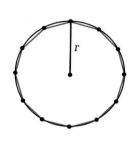 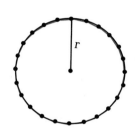

As the number of sides of a regular polygon inscribed in a circle is increased, the polygon looks more and more like the circle itself. The perimeter of the polygon apparently becomes a better and better approximation of the circumference of the circle.

▶ Definition

The *circumference* of a circle is the limit of the perimeters of the inscribed regular polygons.

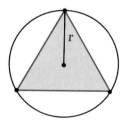  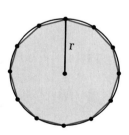 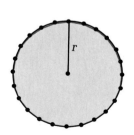

It also appears that as the number of sides of a regular polygon inscribed in a circle increases, the *area* of the polygon becomes a better and better approximation of the *area* of the circle.

▶ Definition

The *area* of a circle is the limit of the areas of the inscribed regular polygons.

Since the formulas for the perimeter and area of a regular
n-gon with radius of length r are

$$p_{n\text{-gon}} = 2vr \quad \text{and} \quad \alpha_{n\text{-gon}} = v \cos \frac{180}{n} r^2,$$

we can find the circumference and area of a circle by determining
the following limits:

$$c_{\text{circle}} = \lim_{n \to \infty} 2vr \quad \text{and} \quad \alpha_{\text{circle}} = \lim_{n \to \infty} v \cos \frac{180}{n} r^2.$$

In order to find out just what these limits are, we need the
following facts:

1. The limit of a constant is the constant itself. For example,

$$\lim_{n \to \infty} 2 = 2.$$

2. The limit of a product is equal to the product of the limits of
its factors. For example,

$$\lim_{n \to \infty} 2r = (\lim_{n \to \infty} 2)(\lim_{n \to \infty} r)$$

3. $\lim_{n \to \infty} v = \pi$, where $v = n \sin \dfrac{180}{n}$.

4. $\lim_{n \to \infty} \cos \dfrac{180}{n} = 1$.

The first two of these facts can be proved as theorems in higher
mathematics and the other two were implied by exercises in the
previous lesson.

We are now ready to derive formulas for the circumference and
area of a circle.

▶ **Theorem 87**
The circumference of a circle is $2\pi r$, where r is the length of its
radius.

By definition,

$$c_{\text{circle}} = \lim_{n \to \infty} 2vr.$$

By fact 2 about limits,

$$c_{\text{circle}} = (\lim_{n \to \infty} 2)(\lim_{n \to \infty} v)(\lim_{n \to \infty} r).$$

Since 2 and r are constants (r, the radius of the circle does not
change as n changes), it follows from fact 1 that

$$\lim_{n \to \infty} 2 = 2 \quad \text{and} \quad \lim_{n \to \infty} r = r.$$

By fact 3,

$$\lim_{n \to \infty} v = \pi.$$

So,

$$c_{\text{circle}} = 2\pi r.$$

► **Theorem 88**

The area of a circle is πr^2, where r is the length of its radius.

By definition,

$$\alpha_{\text{circle}} = \lim_{n \to \infty} \nu \cos \frac{180}{n} r^2$$

$$= \left(\lim_{n \to \infty} \nu \right) \left(\lim_{n \to \infty} \cos \frac{180}{n} \right) \left(\lim_{n \to \infty} r^2 \right)$$

$$= \pi \cdot 1 \cdot r^2 = \pi r^2.$$

The number π, which relates both the circumference and the area of a circle to its radius, is irrational and so its decimal form neither ends nor repeats. Its value has been calculated on modern computers to 500,000 decimal places! Rounded to 10 places, it is:

$$3.1415926536.$$

SET I

Exercises

1. We have assumed that $\lim_{n \to \infty} \nu$ exists and that it is the number π. The number π is usually defined as the ratio of two measurements in a circle. This definition can be derived from the equation $c = 2\pi r$, where c is the circumference of a circle and r is its radius.
   a) Solve this equation for π.
   b) What does the equation $d = 2r$ mean?
   c) Write another equation for π, using this fact.
   d) How can π be defined as the ratio of two measurements in a circle?

2. The following description from the Bible of a circular pool in Solomon's temple suggests a very simple approximation of π.

   "Also he made a molten sea of ten cubits from brim to brim . . . and a line of thirty cubits did compass it round about."*

   a) What was the circumference of this pool?
   b) What was its diameter?
   c) What approximation of π follows from this?

_____

*II Chronicles 4:2.

3. Two commonly used approximations of π are 3.14 and $3\frac{1}{7}$.
Expressed to five decimal places, 3.14 is 3.14000.
a) Find $3\frac{1}{7}$ to five decimal places.
b) To five decimal places, π is 3.14159. Which approximation of it is better: 3.14 or $3\frac{1}{7}$?

4. The Hindu mathematician Brahmagupta thought that π was equal to the square root of 10.
a) Find the square of 3.14.
b) Find the square of 3.15.
c) Is π *larger* or *smaller* than $\sqrt{10}$?

5. Since π is irrational, there is no fraction to which it is exactly equal. The fraction $\frac{355}{113}$, however, is an astonishingly good approximation of π. To how many decimal places does it give the correct value?

SET II

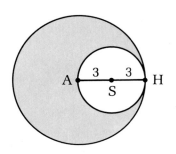

1. Circles F, I, and R have radii of 1, 2, and 3 units, respectively.
a) Find the exact circumference of each circle.
b) How does the ratio of the circumferences of two circles compare with the ratio of their radii?
c) Find the exact area of each circle.
d) How does the ratio of the areas of two circles compare with the ratio of their radii?

2. Find the exact area of the shaded region in each of the figures below.
a) The region bounded by two circles with centers at A and S.
b) SUMA is a square whose sides are tangent to circle C.
c) LARC is a rectangle inscribed in circle H.

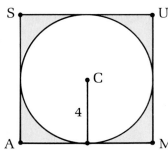

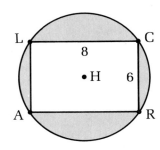

3. Obtuse Ollie measured his circumference and found that it was 47 inches. On the assumption that he is perfectly round in the middle, what is Ollie's approximate radius? (Use $\pi \approx 3.14$.)

4. Regular hexagon HEMLCK is inscribed in circle O with a radius of 5 centimeters.
   a) Find the perimeter of the hexagon.
   b) Find the approximate circumference of the circle.
   c) Find the approximate area of the hexagon. (Use $\sqrt{3} \approx 1.73$.)
   d) Find the approximate area of the circle.

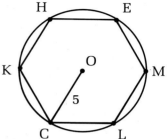

5. The earth is about 93 million miles from the sun. On the assumption that the earth's orbit is circular, find the approximate distance that the earth moves each day in traveling around the sun.

SET III

As a wheel rolls along a line, a point on its rim follows a path called a *cycloid*. The figure below shows part of this path.

The part of the curve beginning at point A and ending at point B is called an "arc" of the cycloid. Christopher Wren, a famous English architect, made the surprising discovery that the length of one arc of a cycloid is exactly four times the length of the diameter of the wheel that generates it.

Suppose Escher's wheel-creature curls itself up into a circular shape with a radius of 2 inches and rolls 100 *feet* along a line. Can you figure out the approximate length of the path traveled by a point on its rim?

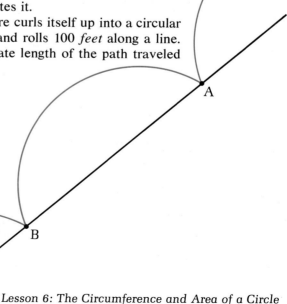

DRAWING BY TOM HENDERSON; PARADE MAGAZINE

Lesson 7

Sectors and Arcs

Suppose that a slice is cut from a large pizza as shown in the figure at the left. If the central angle of the slice is 60°, how does it compare in size with the pizza before it was cut?

Since 60° is one-sixth of 360°, it seems evident that the area of the slice would be one-sixth the area of the entire pizza. If the radius of the pizza is 12 inches, its area would be $\pi(12)^2$ or 144π square inches and the area of the slice would be $\frac{1}{6}(144\pi)$ or 24π square inches.

The slice is an example of a *sector* of a circle.

▶ Definition

A *sector* of a circle is a region bounded by an arc of the circle and the two radii to the endpoints of the arc.

The method we used to find the area of the slice illustrates the following theorem, which we will state without proof.

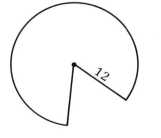

570

► **Theorem 89**

The area of a sector whose arc has a measure of $m°$ is $\frac{m}{360}\pi r^2$, where r is the radius of the circle.

As another example of how this theorem is used, consider the problem of finding the area of the rest of the pizza. It is also a sector, and its arc has a measure of 300°. Its area, then, is

$$\frac{300}{360}\pi (12)^2 = \frac{5}{6}(144\pi) = 120\pi \text{ sq in.}$$

Now that we have considered how the *area* of a sector can be determined, let's turn to the problem of finding the *length of its arc*.

For the pizza slice, we know that the *measure* of its arc is the same as the measure of its central angle: 60°. As we have already noted, this is one-sixth of 360°, the measure of the entire circle. Thus, it is reasonable that the *length* of the arc would be one-sixth of the length of the circle; in other words, one-sixth of its circumference. Since the circumference of the circle is $2\pi (12)$ or 24π inches, the length of the arc would be

$$\frac{1}{6}(24\pi) \quad \text{or} \quad 4\pi \text{ in.}$$

Again we will state our method as a theorem without proof.

► **Theorem 90**

The length of an arc whose measure is $m°$ is $\frac{m}{360}2\pi r$, where r is the radius of the circle.

We will use the letter ℓ to denote the length of an arc in the same way that we use the letter m to denote its measure.

SET I

Exercises

Express your answers to the following exercises in terms of π where appropriate.

1. In circle O, $OD = 4$ and $\angle O = 135°$.
   a) What is $m\widehat{RD}$?
   b) Find the circumference of the circle.
   c) Find $\ell\widehat{RD}$.
   d) Find the area of the circle.
   e) Find the area of the shaded sector.

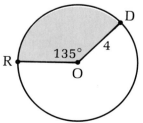

2. The radii of $\overset{\frown}{\text{MI}}$, $\overset{\frown}{\text{CR}}$, and $\overset{\frown}{\text{ON}}$ are 1, 6, and 3 centimeters, respectively; $m\overset{\frown}{\text{MI}} = 180°$, $m\overset{\frown}{\text{CR}} = 30°$, and $m\overset{\frown}{\text{ON}} = 30°$.
   a) Find $\ell\overset{\frown}{\text{MI}}$.
   b) Find $\ell\overset{\frown}{\text{CR}}$.
   c) Find $\ell\overset{\frown}{\text{ON}}$.
   d) If two arcs have equal measures, does it follow that they have equal lengths?
   e) If two arcs have equal lengths, does it follow that they have equal measures?

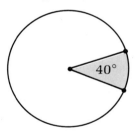

SET II

1. Obtuse Ollie baked Acute Alice a pie for her birthday. Alice ate the shaded sector shown in the diagram and Ollie ate the rest. The diameter of the pie was 10 inches.
   a) Find the area of the part Alice ate.
   b) Find the area of the part Ollie ate.

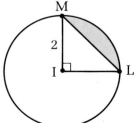

2. The shaded region in the figure below is called a *segment* of the circle. Its area can be found by subtracting the area of △MIL from the area of sector MIL.

   Find the area of the segment if the radius of the circle is 2 units and $\angle I = 90°$.

"All right, Joe, you can knock off."

DRAWING BY GEO. PRICE; © 1950 THE NEW YORKER MAGAZINE, INC.

3. Suppose the minute hand of a clock is 30 inches long. Through what distance would the tip of the minute hand move during one minute? Express your answer to the nearest inch.

4. Find the area of the shaded region in each of the following figures.
   a) Equilateral △INC is inscribed in circle H.
   b) The cross is equilateral.
   c) YARD is a square; A and D are the centers of the arcs.

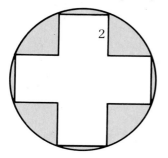

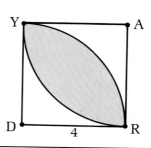

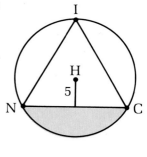

Dilcue tied his pet bulldog with a rope to one corner of a shed 12 feet long and 10 feet wide. If the rope is 15 feet long, can you figure out the exact area within biting distance?

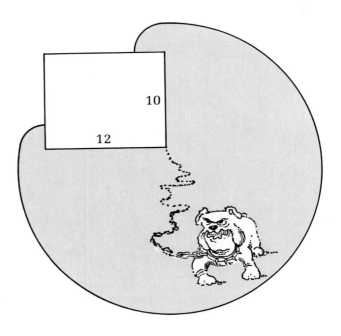

Chapter 14/Summary and Review

Basic Ideas

Apothem   544
Area of a circle   565
Central angle of a regular polygon   544
Circumference   565
Convex polygon   537
Limit   560
π 566, 567
Polygon   537
Polygons, names of   537
Radius of a regular polygon   544
Regular polygon   542
Sector   570

Theorems

84. Every regular polygon is cyclic.   543
85. The perimeter of a regular n-gon is $2vr$, where r is the length of its radius and $v = n \sin \dfrac{180}{n}$.   550
86. The area of a regular n-gon is $v \cos \dfrac{180}{n} r^2$, where r is the length of its radius and $v = n \sin \dfrac{180}{n}$.   554
87. The circumference of a circle is $2\pi r$, where r is the length of its radius.   566
88. The area of a circle is πr^2, where r is the length of its radius.   567
89. The area of a sector whose arc has a measure of $m°$ is $\dfrac{m}{360}\pi r^2$, where r is the radius of the circle.   571
90. The length of an arc whose measure is $m°$ is $\dfrac{m}{360}2\pi r$, where r is the radius of the circle.   571

Exercises

1. A boomerang can be thrown so that it moves in a circular path. Suppose a boomerang travels once around such a path so that its greatest distance from the thrower is 50 yards. How far does it travel?

"Come to bed, Ridgely. If your boomerang were going to return, it would have been back hours ago."

DRAWING BY SHIRVANIAN; © 1971 THE NEW YORKER MAGAZINE, INC.

2. The nth term of a sequence can be found by the formula $\frac{2n + 1}{n}$. Find the following terms of the sequence and express each in decimal form.
   a) The 10th term.
   b) The 100th term.
   c) The 1,000,000th term.
   d) What do you think is the value of $\lim_{n \to \infty} \frac{2n + 1}{n}$?

3. A regular hexagon has three pairs of parallel sides.
   a) How many pairs of parallel sides does a regular octagon have?
   b) How many pairs of parallel sides does a regular nonagon have?
   c) In general, how many pairs of parallel sides does a regular n-gon have?

4. A Japanese mathematician of the seventeenth century found the area of a circle by dividing it into rectangles as shown here. If the entire circle were divided into rectangles in this way, however, the rectangles would not fill the circle.
   a) If *n* represents the number of rectangles and all the rectangles have the same width, what happens to this width as *n* increases?
   b) How does the sum of the areas of the rectangles change as *n* increases?
   c) If S_n is the sum of the areas of the rectangles, what does $\lim_{n \to \infty} S_n$ seem to be?

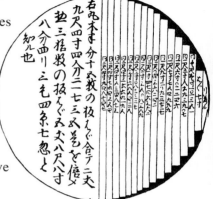

5. Suppose that a regular nonagon and a regular decagon have the same radius.
   a) Which polygon do you think has the greater area?
   b) Verify your answer to part a by computing the areas of a regular nonagon and a regular decagon, each with a radius of 10 units.

SET II

1. Given: DRAGON is a regular hexagon with diagonals $\overline{DO}$ and $\overline{RG}$.
   Prove that $\overline{DR} \parallel \overline{OG}$ without adding anything to the figure.
   Par 8.

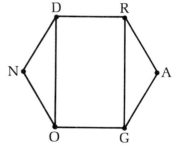

2. If a regular hexagon is inscribed in a circle, it is easy to distinguish the two figures because the sides of the hexagon are considerably shorter than the arcs of the circle that correspond to them.
   Suppose the radius of the hexagon is 3 units.
   a) How long is one of its sides?
   b) How long is the arc corresponding to a side? Express your answer to the nearest hundredth.
   c) How much longer than the side is the arc? Express your answer to the nearest hundredth.

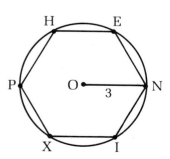

If a regular 60-gon is inscribed in a circle, on the other hand, it is hard to distinguish the two figures because the sides of the polygon are almost the same length as their corresponding arcs.

Suppose the radius of the 60-gon is 30 units.

d) How long is one of its sides? Express your answer to the nearest hundredth.

e) How long is the arc corresponding to a side? Express your answer to the nearest hundredth.

f) How much longer than the side is the arc? Express your answer to the nearest hundredth.

3. Find the area of the shaded region in each of these figures.

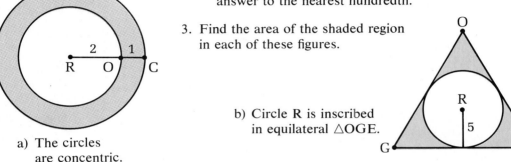

a) The circles are concentric.

b) Circle R is inscribed in equilateral △OGE.

A novel titled *Flatland** is about a two-dimensional world inhabited by creatures that are geometric figures. The following passage, somewhat abridged, describes some of them.

"Our middle class consists of equilateral triangles. Our professional men are squares and pentagons. Next above these come the nobility, of whom there are several degrees, beginning at hexagons and from thence rising in the number of their sides. When the number of the sides becomes so numerous, and the sides themselves so small that the figure cannot be distinguished from a circle, he is included in the circular or priestly order; and this is the highest class of all."

Suppose that every resident of Flatland has the *same perimeter.*

1. Who do you think would have the longest radius?

2. Who do you think would occupy the largest area?

3. To check the accuracy of your guesses, suppose that an equilateral triangle and a square each have a perimeter of 12 units. Draw figures to illustrate and determine the radius and area of each. (Use $\sqrt{2} \approx 1.41$ and $\sqrt{3} \approx 1.73$.)

*Edwin A. Abbott, *Flatland* (Dover, 1952).

Chapter 15

GEOMETRIC SOLIDS

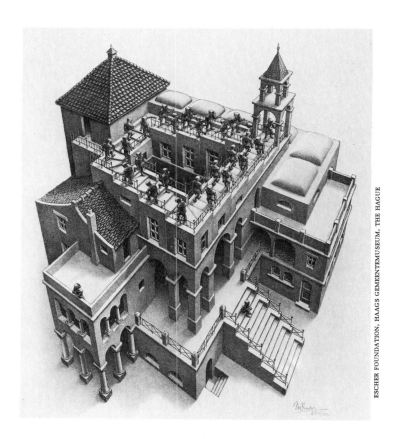

Lesson 1

Lines and Planes in Space

The staircase on the roof of the building in this picture is one that can exist only in the imagination. The picture, by Maurits Escher, is titled *Ascending and Descending* and the men walking the stairs are doing just that. But those on the outside are *always* ascending, whereas those on the inside are *always* descending! Two men, one watching from a lower level and the other sitting on the stairs at the bottom, see the futility of it all and refuse to join the rest.

Escher's picture is a complex perspective drawing that illustrates a number of special line and plane relationships in space. In this lesson, we will study some of these relationships.

First, look at the line of the railing that the fellow at the lower left is leaning against. It is *parallel* to the plane of the ground. The adjoining diagram shows this line-plane relationship. Remember that a plane, being infinite in extent, cannot be represented in its entirety in a picture. Instead, we draw a rectangle (usually a perspective view of one) that lies in the plane.

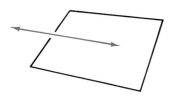

580

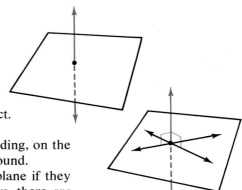

▶ Definition
A *line and a plane are parallel* iff they do not intersect.

The lines containing the vertical edges of the building, on the other hand, are *perpendicular* to the plane of the ground.

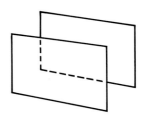

We cannot say that a line is perpendicular to a plane if they form right angles, because, as the figure above shows, there *are* no angles. However, the line *does* form right angles with lines in the plane that pass through the point of intersection and we can base our definition upon this fact.

▶ Definition
A *line and a plane are perpendicular* iff they intersect and the line is perpendicular to every line in the plane that passes through the point of intersection.

The walls of the building that face us lie in *parallel planes.* The adjoining figure illustrates two of these planes.

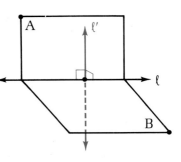

▶ Definition
Two *planes are parallel* iff they do not intersect.

We will assume that, if two planes do intersect, their intersection is a line. The floor and one of the walls of Escher's building lie in intersecting planes that are *perpendicular* to each other. In the adjoining figure, these planes are represented as A and B and the line in which they intersect as ℓ. If a line ℓ′ is drawn in plane A so that it is perpendicular to line ℓ, it will also be perpendicular to plane B. It is on this basis that we define *perpendicular planes.*

▶ Definition
Two *planes are perpendicular* iff one plane contains a line that is perpendicular to the other plane.

We will use the term "oblique" to refer to a line and a plane, or two planes, that intersect without being perpendicular to each other.

▶ Definition
A *line and a plane (or two planes) are oblique* iff they are neither parallel nor perpendicular.

SET I

1. The diagram below is a somewhat simplified version of Escher's impossible staircase. Some of the planes and lines containing parts of the figure have been named with capital and small letters, respectively. Refer to them in answering the following questions. Remember that lines and planes are infinite in extent.
   a) D ‖ G. Name three more pairs of parallel planes.
   b) A ⊥ F. Name three more pairs of perpendicular planes.
   c) x ‖ B. Name three more examples of a line parallel to a plane.
   d) y ⊥ D. Name three more examples of a line perpendicular to a plane.

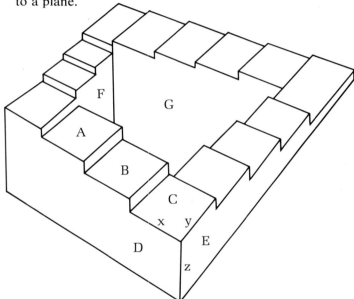

2. Answer the following questions on the basis of the same figure.
   a) If two lines are perpendicular to the same plane, do you think they must be parallel to each other?
   b) If a line is perpendicular to one of two parallel planes, must it be perpendicular to the other?
   c) If two planes are perpendicular to a third plane, must they be parallel to each other?
   d) If two planes are parallel to a third plane, must they be parallel to each other?
   e) If two lines lie in parallel planes, must the lines be parallel to each other?

3. In the figure shown here, $\overleftrightarrow{FD} \perp A$ and $\overrightarrow{OL}$, $\overrightarrow{OI}$, and $\overrightarrow{OR}$ lie in the plane.
   a) From the definition of a line perpendicular to a plane, what can you conclude about this figure?
   b) Name every angle in the figure that you know for certain to be a right angle.

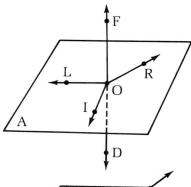

SET II

1. In this figure, planes V and M intersect in $\overleftrightarrow{ER}$ and planes O and M intersect in $\overleftrightarrow{NT}$. If $V \parallel O$, explain why $\overleftrightarrow{ER} \parallel \overleftrightarrow{NT}$.

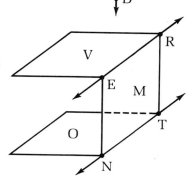

2. Given: $\overline{TX} \perp S$; $TE = TA$.
   Prove: $\angle E = \angle A$.
   Par 8.

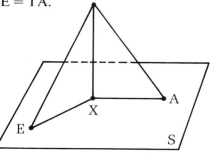

3. In the figure at the left below, planes M and I are perpendicular to $\overleftrightarrow{AN}$. We can prove that the planes are parallel to each other by using the indirect method.

   Suppose the planes are not parallel. Then they intersect in a line as shown in the figure at the right. Let E be a point on this line.
   a) Explain why this leads to a contradiction.
   b) What does the contradiction imply?

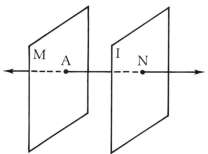

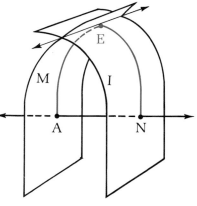

This picture by Escher, titled *Belvedere*, illustrates another impossible building. Can you explain why?

ESCHER FOUNDATION, HAAGS GEMEENTEMUSEUM, THE HAGUE

Lesson 2
Rectangular Solids

Years ago, students spent a full year learning plane geometry before taking a separate course in solid geometry. In this chapter, we will study in an informal way some of the basic topics included in that second course. Instead of continuing to define every term precisely as we did in the preceding lesson, we will take for granted the meanings of words in some of the definitions throughout the rest of this chapter. Furthermore, we will not attempt to prove every theorem but will merely present informal arguments to make them seem reasonable. This approach will enable us to explore a wider variety of topics than would otherwise be possible.

We begin our study with the geometric solid illustrated in this cartoon. A brick is a good model of a *polyhedron* because its faces are flat and polygonal in shape.

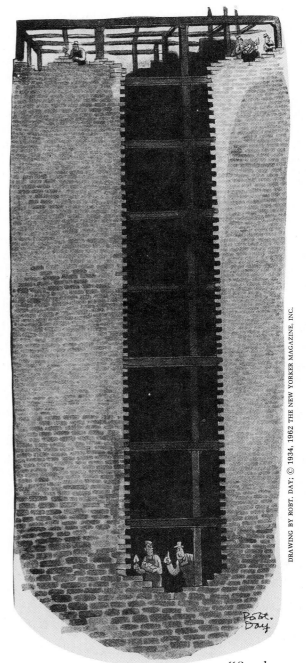

*"See here,
Pritchard, you're falling behind."*

▶ Definition
A *polyhedron* is a solid bounded by parts of intersecting planes.

The intersecting planes form polygonal regions that are called the *faces* of the polyhedron. Their sides are called the *edges* of the polyhedron and their vertices are called its *vertices*.

A brick illustrates a special type of polyhedron called a *rectangular solid*.

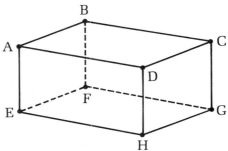

▶ Definition
A *rectangular solid* is a polyhedron that has six rectangular faces.

It is evident from the diagram shown above that the intersecting faces of a rectangular solid lie in perpendicular planes and that its opposite faces lie in parallel planes. It is also apparent that a rectangular solid has eight vertices. Two vertices of the solid that are not vertices of the same face are called *opposite vertices*. For example, in the diagram, one pair of opposite vertices is A and G and another pair is B and H.

A line segment that joins two opposite vertices of a rectangular solid is called a *diagonal* of the solid. Every rectangular solid has four diagonals and it is easy to prove that they have equal lengths.

The lengths of the three edges of a rectangular solid that meet at one of its vertices are the *dimensions* of the solid and are usually called its *length*, *width*, and *height*.

The dimensions of the solid shown here are l, w, and h. It is easy to find the length of one of the diagonals of the solid, such as $\overline{AC}$, in terms of the solid's dimensions by using the Pythagorean Theorem.

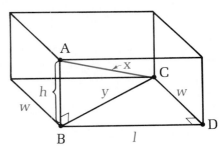

If $\overline{BC}$ is drawn, then $\triangle ABC$ and $\triangle BCD$ are right triangles ($\overline{AB}$ is perpendicular to the plane of the base of the solid, so it must be perpendicular to $\overline{BC}$). Now in right $\triangle ABC$, $x^2 = y^2 + h^2$, and in right $\triangle BCD$, $y^2 = l^2 + w^2$. Substituting, we get

$$x^2 = l^2 + w^2 + h^2,$$

and taking square roots,

$$x = \sqrt{l^2 + w^2 + h^2}.$$

► **Theorem 91**

The length of a diagonal of a rectangular solid is $\sqrt{l^2 + w^2 + h^2}$, where l, w, and h are its dimensions.

If all three dimensions of a rectangular solid are equal, it is a *cube*. If we let e represent the length of one edge of a cube, it follows that the length of one of its diagonals is

$$\sqrt{e^2 + e^2 + e^2} = \sqrt{3e^2} = e\sqrt{3}.$$

► **Corollary**

The length of a diagonal of a cube is $e\sqrt{3}$, where e is the length of one of its edges.

SET I

Exercises

1. The adjoining figure represents a cube.
   a) How many vertices and how many edges does a cube have?
   b) Name the vertex of the cube that is opposite I.
   c) Name the edges that are perpendicular to $\overline{AC}$.
   d) Name the edges that are parallel to $\overline{MA}$.
   e) Two edges of a cube that are neither perpendicular nor parallel are said to be *skew*. Name the edges that are skew to $\overline{AR}$.

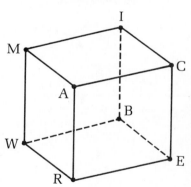

2. The adjoining figure represents a rectangular solid.
   a) Copy it and then add the three hidden edges to your drawing as dotted line segments. Also draw all four diagonals.
   b) The diagonals of a rectangular solid seem to have several relationships to each other. What are they?
   c) How many isosceles triangles do the diagonals seem to form with the edges of the solid?

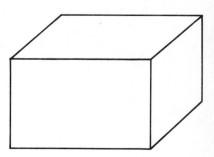

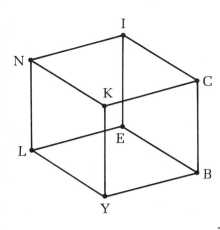

3. Copy the figure shown here, which represents a cube. Also draw $\overline{LI}$ and $\overline{LB}$.
   a) Find the measures of ∡ILE and ∡ELB and explain how you got your answers.
   b) Is ∠ILB = ∠ILE + ∠ELB? Explain.
   c) Find the measure of ∡ILB. (Hint: Add $\overline{IB}$ to the figure. What kind of triangle is △ILB?)

   For parts d and e, suppose LY = 8 and use $\sqrt{2} \approx 1.41$ and $\sqrt{3} \approx 1.73$.

   d) Find the exact and approximate length of $\overline{LI}$.
   e) Draw $\overline{LC}$ and find its exact and approximate length.

SET II

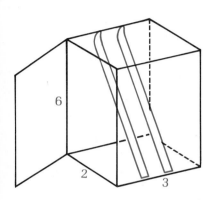

1. Dilcue bought a pair of skis that are 6 feet 10 inches long. He wanted to keep them in a broom closet that is 2 feet wide, 3 feet deep, and 6 feet high, but they are too long to fit as shown.
   a) Explain why.
   b) Are there any other ways he could put the skis in the closet so they might fit? ($\sqrt{5} \approx 2.24$ and $\sqrt{10} \approx 3.16$.)

2. Since the diagonals of a square are perpendicular to each other, it might seem that the diagonals of a cube would also be perpendicular to each other.

   It is easy to see, however, why this cannot be true. Even though the four diagonals of a cube are concurrent in the point that is the center of the cube, they are not perpendicular. Use indirect reasoning and the fact that space has only three dimensions to explain why not.

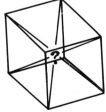

3. Obtuse Ollie drew the adjoining figure, which he intended to be a regular hexagon. Acute Alice claims that it is also a picture of a cube. Can this be correct?

Some people use the word "square" when they really mean "cube." This mistake is understandable because a cube seems to be the three-dimensional equivalent of the two-dimensional square. Since it is easy to picture figures in both two and three dimensions, it is natural to wonder whether there is such a thing as a four-dimensional space and, if so, what a figure in such a space would look like. For instance, what is the four-dimensional equivalent of a cube?

Many mathematicians have thought about this; in fact, a name has been invented for such a figure: it is called a *tesseract*.

What does a tesseract look like? Is it possible to make a picture of one? Since we can represent a *three*-dimensional cube in a *two*-dimensional plane (which is what we did in the picture of the cube above), perhaps a *four*-dimensional tesseract could be represented in *three*-dimensional space.

The figure below suggests a way to do this. A three-dimensional frame of a cube held in front of a light casts a two-dimensional shadow on a screen. Notice that the shadow consists of two squares, one inside the other, such that their corresponding vertices are joined by line segments. In the shadow, we can see the eight vertices of the cube, its twelve edges, and, if you look closely, even all six faces!

Now suppose that a tesseract and its "shadow" in three dimensions have the same relationship as a cube and its shadow in two dimensions. This means that the tesseract's "shadow" would consist of two cubes, one inside the other, such that their corresponding vertices would be joined by line segments.

Can you draw a diagram of this shadow? (Your drawing will actually be a *two*-dimensional diagram of the *three*-dimensional shadow of a *four*-dimensional figure!)

How many vertices, edges, "square faces," and "cubical faces" does a tesseract have?

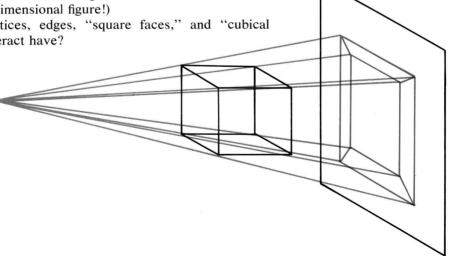

NATIONAL PARK SERVICE

Lesson 3

Prisms

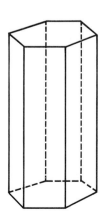

The Devil's Post Pile is a set of tall columns of rock in the Sierra Nevada mountains of California. These columns, some of which are 60 feet high, are an impressive display of geometric polyhedra in nature. Most of them are hexagonal in shape as shown in the diagram.

This polyhedron is an example of a *prism*. Notice that its top and bottom faces lie in parallel planes and that the edges joining the vertices of one face to those of the other are parallel to each other.

▶ Definition
Suppose A and B are two parallel planes, R is a polygonal region in one plane, and ℓ is a line that intersects both planes but not R. The set of all segments parallel to line ℓ that join a point of region R to a point of the other plane form a *prism*.

590

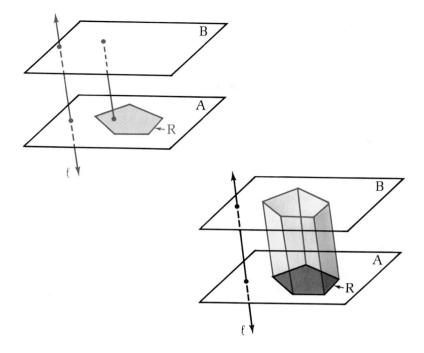

The two faces of the prism that lie in these parallel planes are called its *bases*. It can be shown that the bases of a prism are always congruent to each other. The rest of the faces of the prism, called its *lateral faces*, are parallelograms. The edges of the prism in which the lateral faces intersect are called its *lateral edges*.

Prisms are classified according to two properties: the relationship of their lateral edges to the planes containing their bases, and the shape of their bases. For example, the prism representing one of the columns of the Devil's Post Pile is a *right hexagonal prism*: its lateral edges are perpendicular to its bases and its bases are hexagons.

Long matches for lighting fireplaces are often stored in boxes that are in the shape of right hexagonal prisms. If a label is wrapped around the six lateral faces of the box, its area is called the *lateral area* of the prism. The *total area* of the prism, on the other hand, includes the areas of its two bases.

▶ Definitions
The *lateral area* of a prism is the sum of the areas of its lateral faces.

The *total area* of a prism is the sum of its lateral area and the areas of its bases.

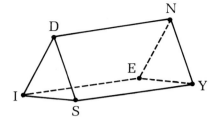

Exercises

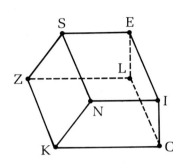

1. The word "prism" is often used to mean a triangular block of glass by which light can be broken up into a spectrum of colors. The right triangular prism shown here illustrates this meaning of the word.
   a) Name the bases of this prism.
   b) How many lateral faces does it have?
   c) Name its lateral edges.
   d) What relationship do these edges have to the bases of the prism?
   e) What kind of lateral faces does it have?

2. This figure represents an oblique trapezoidal prism.
   a) How many faces does it have?
   b) What kind of lateral faces does it have?
   c) Name the parallel edges of the prism.

3. The figure at the right shows part of a prism.
   a) The pentagon must be one of its bases. Why?
   b) Copy the figure and complete it.

4. Obtuse Ollie has decided that, since the minimum number of sides a two-dimensional polygon can have is *three*, the minimum number of sides a three-dimensional prism can have is *four*. Acute Alice disagrees, saying that four walls do *not* a prism make.
   a) What's your opinion about this?
   b) What is the minimum number of *edges* that a prism can have?
   c) What is the minimum number of *vertices* that a prism can have?

5. Give as specific a name as you can for a prism satisfying each of the following conditions.
   a) Its lateral edges are perpendicular to its bases, which are kites.
   b) Its bases each have five sides and its lateral faces are not equiangular.
   c) It has ten faces and several right angles.
   d) It has six congruent square faces.

1. Write a formula for each of the following.
   a) The total area of a cube if *e* is the length of one of its edges.
   b) The lateral area of a right prism if its height is *h* and the perimeter of one of its bases is *p*.
   c) The total area of a rectangular solid if *l*, *w*, and *h* are its dimensions.

2. If you are doing these exercises in pencil, the pencil you are using may have been a right hexagonal prism before it was first sharpened.

   Find the lateral area of a new pencil that is $7\frac{1}{4}$ inches long if one of the edges of one of its ends is $\frac{1}{8}$ inch long. Express your answer to the nearest tenth of a square inch.

3. This figure represents a piece of Limburger cheese that is in the shape of a right triangular prism. Find the amount of aluminum foil needed to wrap it up completely.

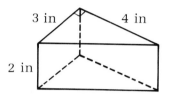

SET III

These figures are copies of pictures made by a computer.* They both represent three interlocking blocks but the one on the left is rather difficult to interpret because every edge is shown. The picture on the right is more recognizable because those edges that couldn't be seen if the blocks were actually solid have been removed. Clever computer programs have been developed to do this automatically!

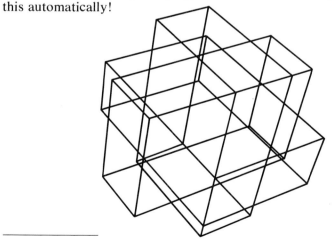
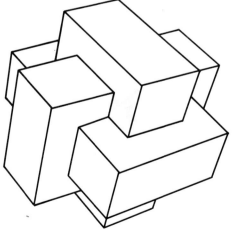

*From "Computer Displays" by Ivan E. Sutherland. Copyright © 1970 by Scientific American, Inc. All rights reserved.

The figure below also represents three intersecting prisms.

1. Can you figure out which line segments to remove so that it is easier to visualize? (A convenient way to do this is to put your paper over the figure and then draw just those edges that you think can be seen.)

2. There is a rather tricky aspect to "hidden surface" problems of this sort. Can you guess what it is?

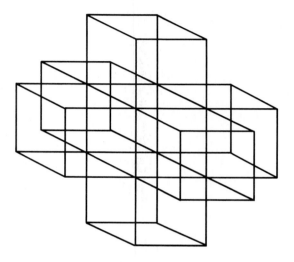

Lesson **4**

The Volume of a Prism

A fad among college students in the late fifties was cramming. Not cramming for exams, but cramming as many people into a small space as possible. The photograph above shows twenty-two fellows who have managed to crowd themselves into a telephone booth. This certainly doesn't leave very much space for each one! Just exactly how much would it be?

The measure of a region of space is called its *volume*. We will assume that the volume of a given three-dimensional region, like the area of a two-dimensional region, is a unique positive number. Since the telephone booth is in the shape of a rectangular solid, we can find its volume from its dimensions. If the base of the booth is a square 3 feet on a side and the booth is 7 feet high, how many cubes measuring 1 foot along each edge can it contain?

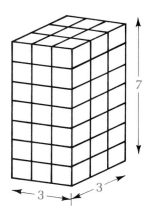

Since a layer of $3 \cdot 3 = 9$ cubes can be put on the floor of the booth and 7 such layers will reach to the ceiling, the answer to this question is $9 \cdot 7 = 63$ cubes. The volume of the phone booth is 63 cubic feet. If all twenty-two fellows were completely inside the booth, each would occupy, on the average, a space of less than 3 cubic feet!

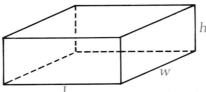

Even though it is not possible to divide *every* rectangular solid into a whole number of unit cubes, it seems reasonable that the volume of such a solid could still be found by multiplying its three dimensions. If the dimensions of the solid shown above are l, w, and h, and its volume is V, then we will assume that $V = lwh$. Since l and w are also the length and width of one of the bases of the solid, which is a rectangle, we have $B = lw$ where B is the area of one of the bases. This means that the volume of the solid can also be expressed by the formula $V = Bh$, where B is the area of a base and h is the height of the solid. Since a rectangular solid is a right rectangular prism, we now have a formula for determining the volume of one type of prism.

What if the lateral edges of a prism are *not* perpendicular to its bases, so that the prism is *oblique*? The height of such a prism is less than the length of one of its lateral edges. It is the length of one of its *altitudes*.

▶ Definition
An *altitude* of a prism is a line segment joining the planes of its bases that is perpendicular to both of them.

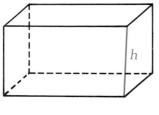

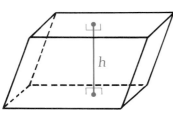

To find the volume of an oblique rectangular prism, we might imagine that it is made from a deck of playing cards. The volume of the prism is equal to the sum of the volumes of all of the cards. Now imagine pushing the cards so that they form a right rectangular prism. Its base has the same area as before, and its altitude remains unchanged. Furthermore, the cards have the same volume that they had before, so that the volume of the prism *also* remains *unchanged*. This means that the volume of an oblique rectangular prism can also be found by the formula $V = Bh$.

So far we have considered how to determine the volumes of prisms whose bases are *rectangles*. To find the volumes of prisms whose bases are other shapes, we will compare their *cross sections*.

▶ Definition
A *cross section* of a geometric solid is the intersection of a plane and the solid.

For example, one of the cross sections of the triangular prism shown on the facing page is △ABC.

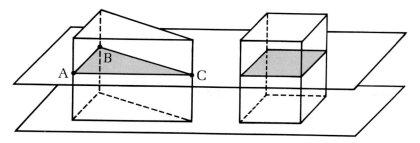

Now suppose that the bases of two prisms lie in the same plane and that every plane parallel to this plane that intersects both prisms cuts off cross sections with the same area. It seems reasonable to conclude from this that the two prisms have the same volume. Since we cannot prove this, we will state it, in more general terms, as a postulate. It was first stated in this form by an Italian mathematician, Cavalieri, who lived in the seventeenth century and was a pupil of Galileo.

▶ **Postulate 21**   (Cavalieri's Principle)
Consider two geometric solids and a plane. If every plane parallel to this plane that intersects one of the solids also intersects the other so that the resulting cross sections have the same area, then the two solids have the same volume.

It can be proved that all cross sections of a prism are congruent to its bases, and hence have the same area that they have. From this and Cavalieri's Principle, it follows that two prisms that have bases of equal area and altitudes that are equal must have equal volumes. Hence the volume of *every prism* can be found by the same formula. We will now state this conclusion as a general postulate.

▶ **Postulate 22**
The volume of any prism is Bh, where B is the area of one of its bases and h is the length of its altitude.

Formulas for the volumes of a rectangular solid and cube now follow as corollaries to this postulate.

▶ **Corollary 1**
The volume of a rectangular solid is lwh, where l, w, and h are its length, width, and height.

▶ **Corollary 2**
The volume of a cube is e^3, where e is the length of one of its edges.

Exercises

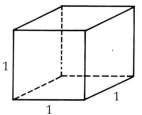

1. A cube whose edges are each 1 foot long has a volume of 1 cubic foot. Find each of the following.
   a) Its total area in square feet.
   b) The area of one of its faces in square inches.
   c) Its total area in square inches.
   d) Its volume in cubic inches.

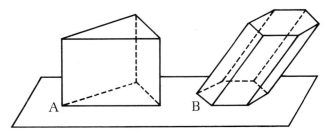

2. The bases of the right triangular prism, A, and the oblique hexagonal prism, B, shown above lie in the same plane. Each prism has an altitude 7 units long.
   a) If the area of one base of prism A is 10 square units, find its volume.

   Suppose every plane parallel to the bases of the two prisms that intersects them cuts off cross sections of equal areas.
   b) What is the volume of prism B?
   c) What is the basis for your answer to part b?

3. Three cubes have edges of lengths 1, 2, and 3 units, respectively.
   a) Find the total area of each cube.

   What happens to the total area of a cube if the length of one edge is
   b) doubled?
   c) tripled?

   d) Find the volume of each cube.

   What happens to the volume of a cube if the length of one edge is
   e) doubled?
   f) tripled?

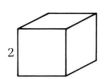
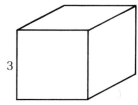

4. Every edge of the right prism shown here is the same
   length: 4 units.
   a) What is the shape of its bases?
   b) Find the exact area of one of them.
   c) What is the height of the prism?
   d) Find its exact volume.

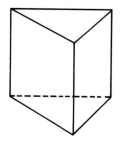

SET II

1. A right square prism is a rectangular solid with square bases.
   Suppose four right square prisms have equal volumes, 216
   cubic units each, and that the length of each edge of their
   bases is 1, 2, 3, and 6 units, respectively.
   a) Find the area of one base of each prism.
   b) Find the height of each prism.
   c) Find the total area of each prism.
   d) Of all right square prisms having a certain volume, which
      one do you think has the least area?

2. Irwin's swimming pool (pictured on page 346 and again here)
   is in the shape of a right trapezoidal prism. Find its volume.

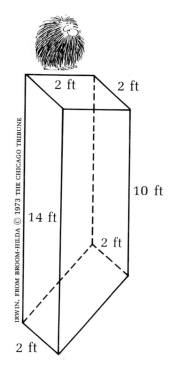

3. Several steps of a concrete staircase are shown below. Each
   step is a right triangular prism and has the dimensions
   indicated.

   If the staircase contains 24 steps in all, find the total
   volume of concrete that it contains. Express your answer in
   cubic feet.

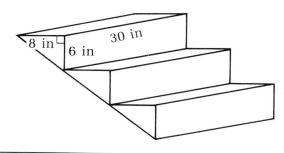

Suppose two wedges have the shape of right triangular prisms with dimensions in inches as shown in the figures below. Can you figure out which of the two wedges is larger?

BY PERMISSION OF JOHN HART AND FIELD ENTERPRISES, INC.

ELIOT ELISOFON—TIME-LIFE PICTURE AGENCY

Lesson 5

Pyramids

The largest of all man-made geometric solids was built more than four thousand years ago. It is the Great Pyramid in Egypt, the only one of the "seven wonders of the world" still in existence. This pyramid, one of about eighty such structures built by the ancient Egyptians, is comparable in height to a forty-story building and covers an area of more than 13 acres. It was put together from more than two million stone blocks, weighing between 2 and 150 tons each!

The figures below represent pyramids of several different types. Although the Egyptians consistently chose the square for the shape of the bases of their pyramids, other polygons can also be used.

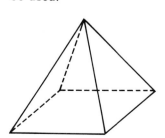 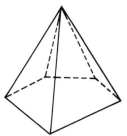 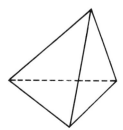

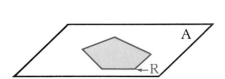
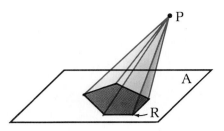

▶ Definition

Suppose A is a plane, R is a polygonal region in plane A, and P is a point not in plane A. The set of all segments that join P to a point of region R form a *pyramid*.

The face of the pyramid that lies in this plane is called its *base*. The rest of its faces are called *lateral faces* and the edges in which they intersect each other are called its *lateral edges*. Although the base of a pyramid can be any polygonal region, its lateral faces are always triangular.

The height of a pyramid is measured by the length of its *altitude*.

▶ Definition

The *altitude* of a pyramid is the perpendicular line segment joining its vertex to the plane of its base.

The Great Pyramid is an example of a *regular* pyramid.

▶ Definition

A *regular pyramid* is a pyramid whose base is a regular polygon and whose lateral edges are equal.

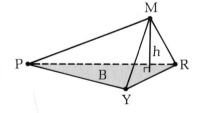

It can be proved that the altitude of such a pyramid joins its vertex to the center of its base.

The amount of space occupied by the Great Pyramid is enormous: it contains more than 91,000,000 cubic feet of rock. How can such a volume be determined? To develop a formula for the volume of a pyramid, we can imagine building a prism having the same base and altitude as the pyramid in the following way.

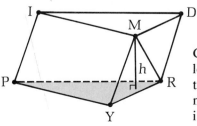

Consider pyramid PYRM with base of area B and altitude of length h. Construct $\overline{PI}$ and $\overline{RD}$ so that they are both parallel to $\overline{YM}$ and equal to it in length. Join points M, I, and D to determine the upper base of the prism. Since the volume of the prism is Bh, the volume of the pyramid is less than Bh.

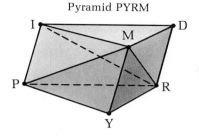

Pyramid PYRM

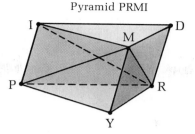

Pyramid PRMI

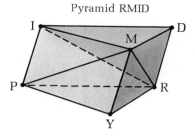

Pyramid RMID

It is possible to cut the prism we have formed into three pyramids, one of which is the original pyramid. Each one is shaded in one of the figures above. Furthermore, it can be shown that these pyramids have *equal* volumes. This means that the volume of each is $\frac{1}{3}Bh$. This result is true of all pyramids. We will state it as a theorem, even though we have not proved it.

▶ **Theorem 92**

The volume of any pyramid is $\frac{1}{3}Bh$, where B is the area of its base and h is the length of its altitude.

SET I

Exercises

1. This figure represents a regular pentagonal pyramid and its altitude.
   a) Name its vertex.
   b) What kind of base does it have?
   c) What kind of lateral faces does it have?
   d) What relationships does $\overline{HO}$ have to LIDAY?

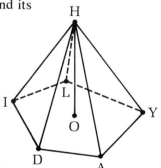

2. This figure shows part of a pyramid.
   a) Copy the figure and complete it.
   b) How many faces does it have?
   c) How many edges does it have?

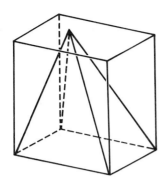

3. The figure shown here represents a pyramid contained inside a prism. (The vertex of the pyramid lies on the upper base of the prism.)
   a) How does the volume of the pyramid compare with the volume of the prism?
   b) How does the volume of the space between the walls of the pyramid and prism compare with the volume of the pyramid?

SET II

1. Find the exact volume of each of the following pyramids.
   a) Its base is a square with a perimeter of 20 units and its altitude is 9 units long.
   b) Its base is an isosceles triangle with sides of lengths 5, 5, and 8, and its altitude is $\frac{1}{2}$ unit long.
   c) Its base is a regular hexagon with sides of length 2 and its altitude is 14 units long.

2. The figure shown here represents a regular square pyramid. Each edge of its base has a length of 8 units and its altitude, $\overline{EN}$, is 7 units long. The midpoint of $\overline{FR}$ is O.
   a) Without making any calculations, arrange EO, ER, and EN in order of increasing size.
   b) Find the exact value of EO.
   c) Find the exact value of NR.
   d) Find the exact value of ER.
   e) Do your answers to parts b and d agree with your arrangement in part a?

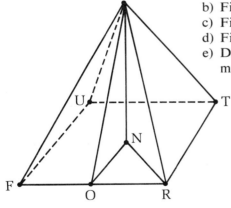

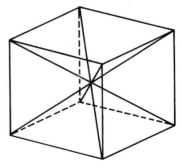

3. The cube shown above consists of six identical pyramids, each having one of the faces of the cube for its base. If an edge of the cube is 6 units long, find the volume of one of the pyramids in two different ways.

4. This figure is a pattern for the faces of a rectangular pyramid.
   a) Find the lateral area of the pyramid it represents.
   b) Find the volume of the pyramid it represents.

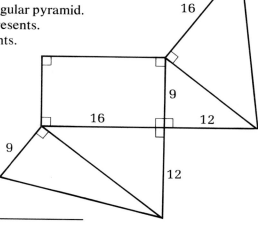

You know that a prism whose faces are all squares is called a *cube*. Eight identical cubes can be put together to form a larger cube as shown in the figure at the left below.

A pyramid whose faces are all equilateral triangles is called a *regular tetrahedron*. How many identical regular tetrahedra can be put together to form a larger tetrahedron as shown in the figure at the right below? To find out, cut out and tape several models of a regular tetrahedron and try it. Unless you're a mathematical genius, you won't be able to determine the correct answer without using a model!

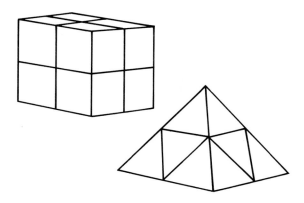

Lesson **6**

Cylinders and Cones

This picture, part of a painting titled *Euclidean Walks* by René Magritte, contains several deliberate visual tricks. One of them concerns the roof of the tower and the street extending out to the horizon. They are almost identical in appearance, yet the tower roof is shaped like a *cone*, a three-dimensional geometric solid, whereas the surface of the street is a two-dimensional figure bounded by parallel lines.

A cone is very much like a pyramid. Its base, however, is bounded by a circle rather than a polygon. And instead of having a set of flat triangular faces, it has a single curved surface called its lateral surface.

We will use the term *circular region* to mean the union of a circle and its interior. With this agreement, it is easy to define the term *cone*. We simply replace the word "polygonal" in the definition of a pyramid with the word "circular."

▶ Definition

Suppose A is a plane, R is a circular region in plane A, and P is a point not in plane A. The set of all segments that join P to a point of region R form a *cone*.

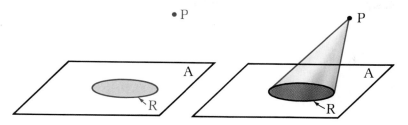

The words "base," "vertex," and "altitude" are used in the same sense with respect to cones that they are with pyramids. The line segment joining the vertex of a cone to the center of its base is called its *axis*. A cone is either *right* or *oblique* depending upon whether its axis is perpendicular or oblique to its base.

A right cone   An oblique cone

Just as cones are the circular counterparts of *pyramids*, so are cylinders the circular counterparts of *prisms*. Again by changing the word "polygonal" to "circular," the definition of a prism becomes that of a cylinder.

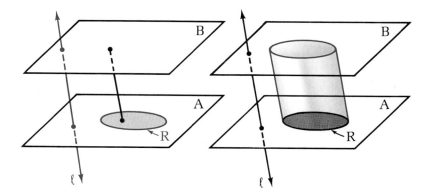

▶ Definition

Suppose A and B are two parallel planes, R is a circular region in one plane, and ℓ is a line that intersects both planes but not R. The set of all segments parallel to line ℓ that join a point of region R to a point of the other plane form a *cylinder*.

Every cylinder has three surfaces: two flat ones, which are its bases, and a curved one, which is its lateral surface. The *axis* of a cylinder is the line segment joining the centers of its bases. Cylinders, like cones, are classified as *right* or *oblique*, depending upon the direction of their axes with respect to their bases.

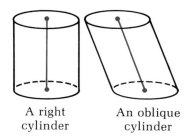

A right cylinder   An oblique cylinder

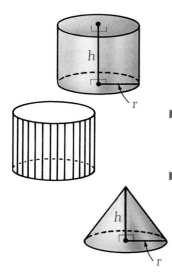

Since a cylinder can be closely approximated by a prism and a cone approximated by a pyramid, the formulas for the volumes of prisms and pyramids can be used to find the volumes of cylinders and cones as well. They are restated as the following theorems without proof.

▶ **Theorem 93**
The volume of a cylinder is $\pi r^2 h$, where r is the radius of its bases and h is the length of its altitude.

▶ **Theorem 94**
The volume of a cone is $\frac{1}{3}\pi r^2 h$, where r is the radius of its base and h is the length of its altitude.

Exercises

1. Find the exact volume of each of the following solids.
   a) A cylinder if the diameter of one of its bases is 8 units and its altitude is 111 units long.
   b) A cone if the radius of its base is 2 units and its altitude is 1119 units long.
   c) A cylinder if the sum of the areas of its bases is 302 square units and its altitude is 12 units long.

2. You know that the volume of a right cylinder is $\pi r^2 h$, where r is the radius of its bases and h is the length of its altitude. The figure below suggests a way to find the *area* of such a cylinder.
   a) Write a formula for its lateral area (the area of its lateral surface).
   b) Write a formula for its total area.

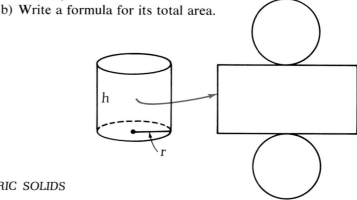

3. The radius and altitude of this cone are a and b units, respectively. Write a formula in terms of a and b for each of the following.

a) The volume of the cone.

b) The volume of a cone that has the same radius but whose altitude is twice as long.

c) The volume of a cone that has the same altitude but whose radius is twice as long.

d) The volume of a cone whose radius and altitude are both twice as long.

What happens to the volume of a cone if

e) its altitude is doubled and its radius remains unchanged?

f) its radius is doubled and its altitude remains unchanged?

g) both its radius and altitude are doubled?

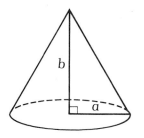

4. If a rectangle is revolved about one of its sides as an axis, it generates a right cylinder as shown in the adjoining figure.

a) If $MI = 3$ and $IL = 5$, find the exact volume of the cylinder.

b) Suppose the rectangle were revolved about $\overline{KL}$ instead. Would the right cylinder generated have the same volume?

c) What kind of solid would be generated by revolving a right triangle about one of its legs? Draw a figure to illustrate your answer.

d) In $\triangle TEA$, $TE = 9$ and $EA = 4$. Find the exact volume of the solid formed by revolving $\triangle TEA$ about $\overline{TE}$.

e) Find the exact volume of the solid formed by revolving the triangle about $\overline{EA}$.

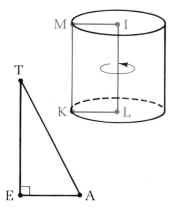

SET II

1. This "hourglass" consists of two identical cones contained in a right cylinder. The cylinder is 6 inches tall and the radius of its base is 2 inches long.

a) Find the exact volume of the shaded region of the figure.

b) Find the exact volume of the space between the cones and the cylinder in the same figure.

2. This diagram represents a conical anthill 9 inches high with a base of 31.4 inches in circumference. If the total volume of the tunnels and rooms in this anthill is 35.5 cubic inches, how much dirt does it contain?

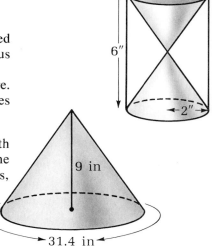

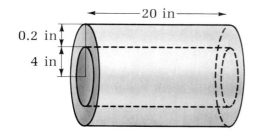

PHOTOGRAPH BY SYNDICATION INTERNATIONAL IN LAUGHING CAMERA 3,
HANNS REICH VERLAG

3. This cat (or cats?) is walking through a cylindrical pipe. If the dimensions of the pipe are as shown in the figure above, find the volume of material that it contains. Express your answer to the nearest cubic inch.

SET III

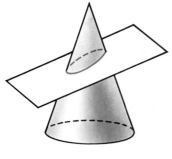

A Greek mathematician named Apollonius who lived at about the same time as Euclid observed that, if a right cone is cut by a plane as shown in this figure, the resulting cross section is not a circle but a curve called an *ellipse*.

The curve shown below is an ellipse inside of which two points, called its *foci*, have been labeled A and B. Make an accurate copy of the figure by tracing it.

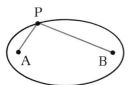

The two foci have an interesting relationship to the points on the ellipse. To find out what it is, see if you can solve the following problem. Suppose a flea at point A walks directly to some point on the curve and then walks directly to point B. An example of such a path is shown at the left. To what point should the flea walk to make his trip the shortest possible?

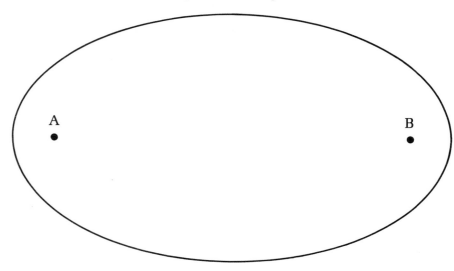

Lesson 7

Spheres

At the age of 81, Mr. Luke Roberts decided to start collecting string. This photograph shows him standing with the result of his unusual hobby, a ball of string three feet in diameter! How heavy would a ball of string this size be? To answer this, it would be helpful to know how to find the volume of a sphere.

▶ Definition

A *sphere* is the set of all points in space that are at a given distance from a given point.

The given distance is called the *radius* of the sphere and the given point is called its *center*. By the volume of a sphere, we mean the volume of the solid consisting of the sphere and its interior. The volume of a sphere is determined by its radius just as is the area of a circle.

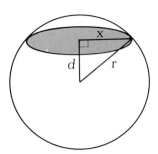

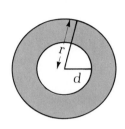

To find a formula for the volume of a sphere, we will use Cavalieri's Principle. We begin by imagining a sphere of radius r sliced by a plane at a distance d from its center. The cross section of the sphere is a circle: we will let its radius be x.

By the Pythagorean Theorem, $x^2 + d^2 = r^2$, so $x^2 = r^2 - d^2$. Since the cross section is a circle, its area is πx^2. Substituting for x^2, we have

$$\alpha_{\text{cross section}} = \pi(r^2 - d^2) = \pi r^2 - \pi d^2.$$

This result can be interpreted as the difference between the areas of two circles having radii of r and d. In the adjoining figure, this is the area of the shaded region between the two circles.

Now to apply Cavalieri's Principle to finding the volume of the sphere, we will construct a geometric solid that has this kind of cross section. It consists of a right cylinder from which two identical cones have been removed, as shown in the figure below. The cylinder has the same radius, r, and the same height, $2r$, as the sphere and is situated so that the two solids rest on the same plane. The two hollowed-out cones meet at the center of the cylinder and each shares one of its bases.

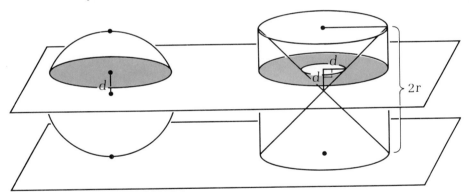

Consider a cross section of this solid at a distance d from its center. It is bounded by two circles with radii of r and d and so its area is $\pi r^2 - \pi d^2$. Since this is the same as the area of the corresponding cross section of the sphere, it follows from Cavalieri's Principle that the volumes of the two solids are the same.

We can easily find the volume of the cylindrical solid. It is equal to

$$\begin{aligned} V_{\text{cylinder}} - V_{\text{two cones}} &= \\ \pi r^2(2r) - 2[\tfrac{1}{3}\pi r^2(r)] &= \\ 2\pi r^3 - \tfrac{2}{3}\pi r^3 &= \tfrac{4}{3}\pi r^3. \end{aligned}$$

It follows that this is also the volume of the sphere.

► **Theorem 95**
  The volume of a sphere is $\frac{4}{3}\pi r^3$, where r is its radius.

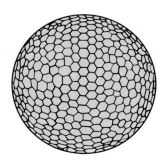

Next we will use an intuitive argument to derive a formula for the surface area of a sphere. To do this we will divide the sphere into a large number of small "pyramids." Imagine that the surface of the sphere is separated into a large number of tiny "polygons." They are not actually polygons since there are no straight line segments on the surface of a sphere. However, the smaller they are, the closer to being polygons they become.

Now imagine joining the corners of all of these "polygons" to the center of the sphere so that they become the bases of a set of "pyramids" all with a common vertex, the center of the sphere. All of the pyramids, then, have altitudes equal to the radius of the sphere.

 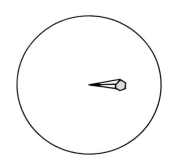

The volume of one of these pyramids is $\frac{1}{3}Br$, where B is the area of its base and r is the length of its altitude. The volume of the sphere is the sum of the volumes of all of the pyramids. If the areas of their bases are B_1, B_2, B_3, and so on, then their volumes are $\frac{1}{3}B_1r$, $\frac{1}{3}B_2r$, $\frac{1}{3}B_3r$, and so on, and

$$V_{\text{sphere}} = \frac{1}{3}B_1r + \frac{1}{3}B_2r + \frac{1}{3}B_3r + \ldots$$
$$= \frac{1}{3}r(B_1 + B_2 + B_3 + \ldots)$$

Now $\alpha_{\text{sphere}} = B_1 + B_2 + B_3 + \ldots$ so

$$V_{\text{sphere}} = \frac{1}{3}r\alpha_{\text{sphere}}.$$

Solving this equation for α_{sphere}, we get

$$\alpha_{\text{sphere}} = \frac{3V_{\text{sphere}}}{r}.$$

But $V_{\text{sphere}} = \frac{4}{3}\pi r^3$, so

$$\alpha_{\text{sphere}} = \frac{3\left(\frac{4}{3}\pi r^3\right)}{r} = 4\pi r^2.$$

As we have already noted, our argument in developing this formula has been an intuitive one. We have been talking about polygons and pyramids on and in a sphere when no such figures can really exist. So we can't consider our argument a proof. It is possible by means of the calculus, however, to derive the *same result without making any approximations* such as we have made.

▶ **Theorem 96**
The surface area of a sphere is $4\pi r^2$, where r is its radius.

Exercises

SET I

1. Two spheres have radii of 1 and 2 units, respectively.
   a) Find the exact surface area of each.
   b) Find the exact volume of each.

   On the basis of your answers to parts *a* and *b*, what do you think happens to
   c) the surface area of a sphere if its radius is doubled?
   d) the volume of a sphere if its radius is doubled?

2. The three-dimensional equivalent of a *semicircle* is a *hemisphere*. Since the circumference of a circle with radius r is $2\pi r$, the length of a semicircle with radius r is $\frac{1}{2}(2\pi r) = \pi r$.

   Suppose a hemisphere has radius r. Write a formula for each of the following.
   a) Its volume.
   b) The area of its curved surface.
   c) Its total surface area (including both its flat and curved surfaces).

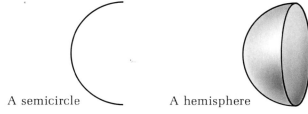

A semicircle   A hemisphere

3. The radius of the earth is approximately 4,000 miles.
   a) Find the area of the earth. Round your answer to the nearest million square miles.
   b) Find the earth's volume. Round your answer to the nearest billion cubic miles.

---

1. In our derivation of a formula for the volume of a sphere, we assumed that if a plane intersects a sphere in more than one point, the intersection is a circle. Answer the following questions to explain why this is so in terms of the figure below.

   Suppose sphere W is intersected by plane A in more than one point. Let T and R be any two points in the intersection.
   a) Why is WT = WR?

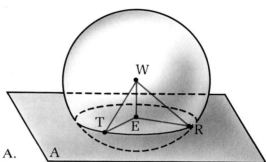

   Let $\overline{WE}$ be perpendicular to plane A.
   b) Why is $\overline{WE} \perp \overline{ET}$ and $\overline{WE} \perp \overline{ER}$?
   c) Why is $\triangle WET \cong \triangle WER$?
   d) Why is ET = ER?
   e) How does the fact that T and R represent any two points in the intersection of sphere W and plane A imply that the intersection is a circle?

2. Obtuse Ollie found the exact surface areas and volumes of spheres having the following radii: 10 inches, 20 inches, and 30 inches. He then drew the following conclusion: the volume of a sphere is always a larger number than its surface area.
   a) Do all three examples support this conclusion?
   b) Acute Alice says his conclusion is wrong. What is your opinion?

3. The surface area of a certain sphere is 36π square units.
   a) Find its radius.
   b) Find its volume.
   c) Show, using algebra, why for a given unit of measure there is a sphere of only one size having this area-volume relationship.

4. Mr. Robert's ball of string had a diameter of 3 feet. On the assumptions that one cubic inch of string weighs 0.03 pound and the ball was solid string, how much did it weigh?

© 1957 UNITED FEATURE SYNDICATE, INC.

Suppose a balloon is blown up at a steady rate. If it takes 5 seconds for the radius of the balloon to become 5 inches, how many more seconds will it take for the radius of the balloon to become 10 inches?

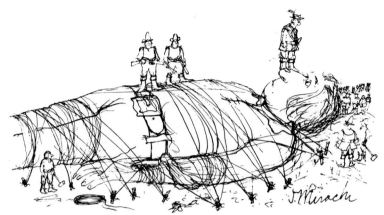

"And one more thing. Don't cross your eyes while I'm speaking with you."

Lesson **8**

Similar Solids

In *Gulliver's Travels* by Jonathan Swift, Gulliver's first voyage took him to Lilliput, a land of people "not six inches high." The emperor of Lilliput issued a decree which, among other things, specified the amount of food to which Gulliver was entitled. As Gulliver describes it:

"The Reader may please to observe that . . . the Emperor stipulates to allow me a Quantity of Meat and Drink sufficient for the Support of 1728 Lilliputians. Some time after, asking a Friend at Court how they came to fix on that determinate Number, he told me that his Majesty's Mathematicians, having taken the Height of my Body . . . and finding it to exceed theirs in the Proportion of Twelve to One, they concluded from the Similarity of their Bodies that mine must contain at least 1728 of theirs, and consequently would require as much Food as was necessary to support that Number of Lilliputians."

617

In making their calculations, the emperor's mathematicians assumed that Gulliver's body was similar to that of a Lilliputian. Instead of attempting to define what we mean by similar solids, we will simply state that they have the same shape. Hence, any two spheres are similar as are any two cubes. Two rectangular solids are similar if their corresponding dimensions are proportional; two right cones are similar if their altitudes have the same ratio as the radii of their bases; and so forth.

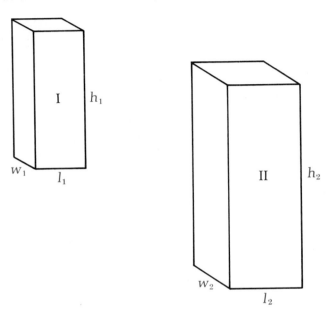

The figure shown here represents two similar rectangular solids, I and II. How do they compare in volume? If the solids are similar, their corresponding dimensions are proportional:

$$\frac{l_1}{l_2} = \frac{w_1}{w_2} = \frac{h_1}{h_2}.$$

Since $V_I = l_1 w_1 h_1$ and $V_{II} = l_2 w_2 h_2$, it follows that

$$\frac{V_I}{V_{II}} = \frac{l_1 w_1 h_1}{l_2 w_2 h_2} = \left(\frac{l_1}{l_2}\right)\left(\frac{w_1}{w_2}\right)\left(\frac{h_1}{h_2}\right).$$

By substitution,

$$\frac{V_I}{V_{II}} = \left(\frac{h_1}{h_2}\right)\left(\frac{h_1}{h_2}\right)\left(\frac{h_1}{h_2}\right) = \left(\frac{h_1}{h_2}\right)^3.$$

We have shown that the ratio of the volumes of two similar rectangular solids is equal to the cube of the ratio of their heights. This relationship is also true for other similar solids. Since Gulliver was 12 times as tall as one of the Lilliputians, the emperor's mathematicians concluded that his volume was $12^3 = 1728$ times one of theirs.

In another passage of Swift's book, the emperor's tailors made Gulliver a suit of clothes. If the suit was made of fabric having the same thickness as their own, how much material did the tailors need?

To answer this, we need to know how the surface areas of two similar solids compare. Again we will derive the relationship in terms of rectangular solids. Since $\frac{l_1}{l_2} = \frac{w_1}{w_2} = \frac{h_1}{h_2}$, it follows that $\frac{l_1 w_1}{l_2 w_2} = \frac{w_1 h_1}{w_2 h_2} = \frac{l_1 h_1}{l_2 h_2}$ by the multiplication and transitive postulates. It follows from the equal ratios theorem that

$$\frac{l_1 w_1 + w_1 h_1 + l_1 h_1}{l_2 w_2 + w_2 h_2 + l_2 h_2} = \frac{l_1 w_1}{l_2 w_2}.$$

Now $\alpha_I = 2(l_1 w_1 + w_1 h_1 + l_1 h_1)$ and $\alpha_{II} = 2(l_2 w_2 + w_2 h_2 + l_2 h_2)$, so

$$\frac{\alpha_I}{\alpha_{II}} = \frac{2(l_1 w_1 + w_1 h_1 + l_1 h_1)}{2(l_2 w_2 + w_2 h_2 + l_2 h_2)} = \frac{l_1 w_1 + w_1 h_1 + l_1 h_1}{l_2 w_2 + w_2 h_2 + l_2 h_2}.$$

Hence $\frac{\alpha_I}{\alpha_{II}} = \frac{l_1 w_1}{l_2 w_2}$. But since $\frac{l_1}{l_2} = \frac{w_1}{w_2} = \frac{h_1}{h_2}$,

$$\frac{\alpha_I}{\alpha_{II}} = \left(\frac{l_1}{l_2}\right)\left(\frac{w_1}{w_2}\right) = \left(\frac{h_1}{h_2}\right)\left(\frac{h_1}{h_2}\right) = \left(\frac{h_1}{h_2}\right)^2$$

by substitution.

The ratio of the areas of two similar rectangular solids is equal to the square of the ratio of their heights. Again this relationship is true for similar solids in general. So the emperor's tailors needed $12^2 = 144$ times as much material to make a suit for Gulliver as for themselves.

▶ **Theorem 97**
The ratio of the surface areas of two similar solids is equal to the square of the ratio of any pair of corresponding segments.

▶ **Theorem 98**
The ratio of the volumes of two similar solids is equal to the cube of the ratio of any pair of corresponding segments.

SET I

Exercises

We have proved the theorems on the areas and volumes of similar solids only for rectangular solids. In the first two exercises of this lesson, we will prove these theorems for some other solids.

1. Supply reasons for the following proof for the volumes of two similar right cones.
   a) If the two cones in the figure below are similar, then
   $$\frac{h_1}{h_2} = \frac{r_1}{r_2}. \text{ Why?}$$
   b) $V_{\mathrm{I}} = \frac{1}{3}\pi r_1^2 h_1$ and $V_{\mathrm{II}} = \frac{1}{3}\pi r_2^2 h_2$. Why?
   c) $\dfrac{V_{\mathrm{I}}}{V_{\mathrm{II}}} = \dfrac{\frac{1}{3}\pi r_1^2 h_1}{\frac{1}{3}\pi r_2^2 h_2}$. Why?
   d) $\dfrac{V_{\mathrm{I}}}{V_{\mathrm{II}}} = \left(\dfrac{r_1}{r_2}\right)^2\left(\dfrac{h_1}{h_2}\right) = \left(\dfrac{h_1}{h_2}\right)^2\left(\dfrac{h_1}{h_2}\right) = \left(\dfrac{h_1}{h_2}\right)^3$. Why?

   $\dfrac{V_{\mathrm{I}}}{V_{\mathrm{II}}} = \left(\dfrac{r_1}{r_2}\right)^3$ for the same reason.

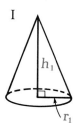
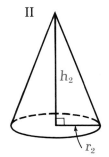

2. Show, in a similar fashion, that the ratio of the surface areas of two spheres is equal to the square of the ratio of their radii.

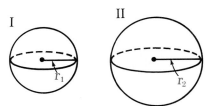

3. Two oblique triangular prisms are similar. If the lengths of the lateral edges have the ratio $\frac{2}{5}$, find each of the following.

   a) The ratio of the lengths of their altitudes.
   b) The ratio of their areas.
   c) The ratio of their volumes.

4. Two right pentagonal pyramids are similar. If their surface areas are 9 square inches and 16 square inches, find each of the following.

   a) The ratio of the lengths of the corresponding sides of their bases.
   b) The ratio of their volumes.

1. The strengths of two similar bones are proportional to their cross-sectional areas.* Suppose that the two bones shown here are similar and that the second is three times as long as the first.

    a) How much stronger than the first bone do you think the second one is?

    Suppose the bones are from two animals that have similar shapes but different sizes. The weights of two similar animals are proportional to their volumes.

    b) How many times heavier than the first animal would the second one be?

    c) Which animal has a skeleton more capable of supporting its own weight?

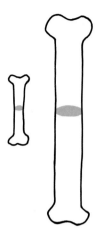

2. There is a legend that the people of Athens were troubled by a plague in 430 B.C. and that they were told by the oracle at Delphi that, to rid the city of it, they must double the size of the altar of Apollo. The altar was in the shape of a cube.

    a) Explain why the Athenians would not be obeying the command of the oracle if they built a new altar by doubling the edges of the original one.

    b) By what number would the length of the edge of Apollo's altar have to be multiplied in order to double its volume?

3. The Rin-Tin-Tin Company sells dog food in tin cans of two sizes: "large" and "colossal." The cans are 8 and 12 inches tall, respectively, and are similar in shape.

    a) If the large can sells for 24¢ and the colossal can sells for 72¢, which is a better buy?

    Suppose the walls of the two cans have the same thickness.

    b) If the large can contains 4¢ worth of metal, what is the value of the metal in the colossal can?

    c) If it costs the company 8¢ to produce the dog food in the large can, how much does it cost to produce the dog food in the colossal can?

    d) How does the selling price of each can compare with the cost of the materials in it?

    e) Which can do you think the company would prefer that people buy?

---

*A variety of applications of the properties of similar solids to the study of biology appear in a fascinating essay by the scientist J. B. S. Haldane. Titled "On Being the Right Size," the essay is in the anthology *The World of Mathematics*, edited by James R. Newman (Simon & Schuster, 1956), pp. 952–957.

SET III

Clams range in size from as little as a pinhead to more than four feet in length! If a clam 2.4 inches long weighs one ounce, how much would you expect a clam having the same shape and a length of four feet to weigh?

Lesson 9

Euler's Theorem

An American architect and sculptor, Charles Perry, has recently created several large metal sculptures based upon geometric solids. One of them, shown in this photograph, is located in the central court of the new Hyatt Regency hotel in San Francisco and is forty feet high. Its design is that of a *regular dodecahedron*, a polyhedron that has twelve faces in the shape of regular pentagons.

The regular dodecahedron is one member of a set of five geometric solids called the *regular polyhedra*.

▶ Definition
A ***regular polyhedron*** is a geometric solid, all of whose faces are congruent regular polygons, in which the same number of polygons meet at each vertex.

A tetrahedron

An octahedron

The simplest regular polyhedron is also a pyramid: it is the *tetrahedron*, in which three equilateral triangles meet at each vertex. The solid is called a "tetrahedron" because it has four faces. The prefix "tetra" comes from a Greek word meaning "four."

Two more regular polyhedra have faces that are equilateral triangles. They are the *octahedron*, in which four meet at each vertex, and the *icosahedron*, in which there are five triangles at each vertex.* The octahedron and icosahedron have eight faces and twenty faces, respectively.

There is only one regular polyhedron that has squares for its faces. It is one of the prisms, the *cube*. The fifth and last regular polyhedron is the *dodecahedron*. It has twelve regular pentagons for its faces and three of them meet at each vertex.

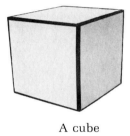

A cube

An icosahedron

A dodecahedron

Leonhard Euler, the Swiss mathematician for whom Euler diagrams are named, stated an interesting relationship involving the numbers of faces, vertices, and edges of each of these polyhedra.

	Number		
Polyhedron	Faces	Vertices	Edges
Tetrahedron	4	4	6
Cube	6	8	12
Octahedron	8	6	12
Dodecahedron	12	20	30
Icosahedron	20	12	30

From this table, it is evident that each solid has more edges than it has either faces or vertices. In fact, in all five cases the number of edges is exactly 2 less than the sum of the faces and

---

*You have seen the icosahedron before: first in boron crystals (page 511) and later in Buckminster Fuller's world map (page 553).

vertices. It is possible to show that this face-vertex-edge relationship is true for *every* polyhedron, including even those that are not regular.*

▶ **Theorem 99** (Euler's Theorem)
For every polyhedron, $F + V = E + 2$, where F, V, and E are its number of faces, vertices, and edges, respectively.

SET I

Exercises

1. The reason that there are only five regular polyhedra is based upon the fact that the sum of the measures of the angles at each vertex of a geometric solid must be less than 360°.
   a) Three of the regular polyhedra have faces that are equilateral triangles. What are their names?
   b) How many triangles meet at each vertex of each of these three polyhedra?
   c) What is the sum of the measures of the angles at each vertex of each of these polyhedra?
   d) Why is there no regular polyhedron in which six equilateral triangles meet at each vertex?
   e) What are the names of the regular polyhedra that have square and pentagonal faces?
   f) Why is there no regular polyhedron in which four squares meet at each vertex?
   g) Since each angle of a regular pentagon has a measure of 108° and 4 · 108° > 360°, there is no regular polyhedron in which four pentagons meet at each vertex. Explain why, since at least three faces must meet at each vertex of a polyhedron, there is no regular polyhedron whose faces are regular hexagons.

2. The adjoining figure represents a geometric solid that has six equilateral triangles for its faces.
   a) Is it a regular polyhedron? Explain.
   b) How many faces, vertices, and edges does it have?
   c) Does Euler's Theorem apply to it?

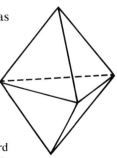

---

*A proof of this theorem is in *What is Mathematics?* by Richard Courant and Herbert Robbins (Oxford University Press, 1941), pp. 236–240.

3. The adjoining figure represents a right cylinder.
   a) How many faces, vertices, and edges does it have?
   b) Does Euler's Theorem apply to it?
   c) Is a cylinder a polyhedron?

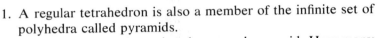

SET II

It is easy to verify that Euler's Theorem applies to all regular polyhedra because there are only five such solids and, in each case, $F + V = E + 2$.

In the following exercises, you will show that Euler's Theorem is true for other types of polyhedra.

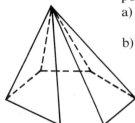

1. A regular tetrahedron is also a member of the infinite set of polyhedra called pyramids.
   a) The adjoining figure is a hexagonal pyramid. How many faces, vertices, and edges does it have?
   b) Verify that Euler's Theorem applies to it.

   Suppose the base of a pyramid has n sides; in other words, that it is an n-gon.
   c) Explain why such a pyramid has $(n + 1)$ faces altogether.
   d) How many vertices does it have?
   e) How many edges does it have?
   f) Show, by means of your answers to the previous three questions, that Euler's Theorem applies to all pyramids.

2. A cube is also a member of the infinite set of polyhedra called prisms.
   a) The adjoining figure is a pentagonal prism. How many faces, vertices, and edges does it have?
   b) Verify that Euler's Theorem applies to it.

   Suppose that each base of a prism has n sides.
   c) How many faces does such a prism have?
   d) How many vertices does it have?
   e) How many edges does it have?
   f) Show that Euler's Theorem applies to all prisms.

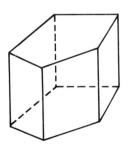

3. A pyramidal "roof" has been added to one face of the cube shown here. As a result, the solid has gained 4 new faces and has lost 1, for a net increase of 3.
   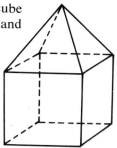
   a) How has its number of vertices changed?
   b) How has its number of edges changed?

   We know that Euler's Theorem applies to the original cube.
   c) Show that it also applies to the new solid.

   Notice that the roofed-over face had four sides. Suppose a pyramidal "roof" is added to another solid in the same way, but that the roofed-over face has n sides.
   d) How does this change the number of faces of the solid?
   e) How does this change its number of vertices?
   f) How does this change its number of edges?

   We can show that if Euler's Theorem applies to a polyhedron before one of its faces is roofed-over in this way, then it also applies afterwards.

   If F, V, and E are the number of faces, vertices, and edges of the original polyhedron and the roofed-over face had n sides, then the new polyhedron must have $(F + n - 1)$ faces, $(V + 1)$ vertices, and $(E + n)$ edges.
   g) Show that if $F + V = E + 2$, then $(F + n - 1) + (V + 1) = (E + n) + 2$.

4. One corner of the cube shown here has been cut off. As a result, the solid has gained 3 new vertices and has lost 1, for a net increase of 2.
   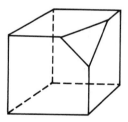
   a) How has its number of faces changed?
   b) How has its number of edges changed?

   We know that Euler's Theorem applies to the original cube.
   c) Show that it also applies to the new solid.

   Notice that the cut-off corner had three edges meeting at it. Suppose a corner is cut off another solid in the same way, but that n edges meet at it.
   d) How does this change the number of faces of the solid?
   e) How does this change its number of vertices?
   f) How does this change its number of edges?
   g) Show, in the same way that you did in Exercise 3, that, if Euler's Theorem applies to a polyhedron before one of its corners is cut off, then it also applies afterwards.

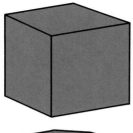

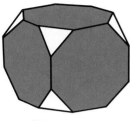

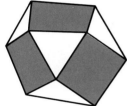

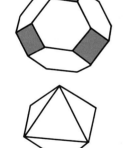

SET III

This sequence of polyhedra begins with a cube and ends with a regular octahedron. The three intermediate solids and the octahedron are the result of cutting the cube by planes passing closer and closer to its center.

First, just the corners of the cube are removed, resulting in a solid whose faces are equilateral triangles and regular octagons. It is called a *truncated cube,* the word "truncated" meaning that its corners have been cut off. The next figure, midway between the cube and octahedron, is called a *cuboctahedron.* The equilateral triangle faces are larger than before and meet at their vertices. The octagonal faces have been replaced with squares.

In the fourth solid these squares have become smaller and the triangles have been transformed into regular hexagons. It is called a *truncated octahedron,* a name whose origin is obvious when it is compared with the last solid.

Two of the solids in the sequence have the same numbers of faces, edges, and corners. Can you figure out which two they are?

"No, I can't play with you now. I'm busy truncating a cube."

COURTESY OF JOHN MCCLELLAN

Chapter 15 / Summary and Review

Basic Ideas

Postulates

21. *Cavalieri's Principle.*   Consider two geometric solids and a plane. If every plane parallel to this plane that intersects one of the solids also intersects the other so that the resulting cross sections have the same area, then the two solids have the same volume.   597
22. The volume of any prism is Bh, where B is the area of one of its bases and h is the length of its altitude.   597

Theorems

91. The length of a diagonal of a rectangular solid is $\sqrt{l^2 + w^2 + h^2}$, where l, w, and h are its dimensions. 587
Corollary. The length of a diagonal of a cube is $e\sqrt{3}$, where e is the length of one of its edges. 587
Corollary 1 to Postulate 22. The volume of a rectangular solid is lwh, where l, w, and h are its length, width, and height. 597
Corollary 2 to Postulate 22. The volume of a cube is e^3, where e is the length of one of its edges. 597

92. The volume of any pyramid is $\frac{1}{3}Bh$, where B is the area of its base and h is the length of its altitude. 603

93. The volume of a cylinder is $\pi r^2 h$, where r is the radius of its bases and h is the length of its altitude. 608

94. The volume of a cone is $\frac{1}{3}\pi r^2 h$, where r is the radius of its base and h is the length of its altitude. 608

95. The volume of a sphere is $\frac{4}{3}\pi r^3$, where r is its radius. 613

96. The surface area of a sphere is $4\pi r^2$, where r is its radius. 614

97. The ratio of the surface areas of two similar solids is equal to the square of the ratio of any pair of corresponding segments. 619

98. The ratio of the volumes of two similar solids is equal to the cube of the ratio of any pair of corresponding segments. 619

99. *Euler's Theorem.* For every polyhedron, $F + V = E + 2$, where F, V, and E are its number of faces, vertices, and edges, respectively. 625

Exercises

SET I

1. A geometry student has a mental block in the shape of a rectangular solid. If it is 3 feet long, 2 feet wide, and 4 feet high, find each of the following:
   a) the volume of the block in *cubic inches.*
   b) its total surface area in *square feet.*
   c) the exact length of one of its diagonals in feet.

2. The bases of pyramid A and cone M, at the top of the next page, both lie in plane H. Suppose that the two solids have equal altitudes and that, for each plane parallel to plane H that intersects them, the resulting cross sections have equal areas.
   a) What does this imply about the two solids?

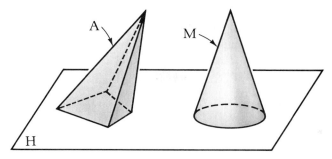

Suppose also that the altitude of cone M is 9 centimeters long and that the radius of its base is 4 centimeters long.

b) Find the volume of the cone.

c) Find the area of the base of the pyramid.

3. The figure shown here illustrates what is meant by the *projection of a point into a plane*. To project point A into plane P, we draw the line through A that is perpendicular to the plane. The point in which this line intersects plane P, point B, is the projection of A into the plane.

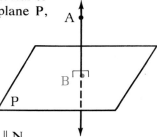

In the figure below, △BAC lies in plane O and O ∥ N.

a) Copy the figure and sketch the projection of △BAC into plane N and label it △B'A'C'.

b) What relationship does △B'A'C' seem to have to △BAC?

c) Of what kind of geometric solid are △BAC and △B'A'C' the bases?

d) In the figure, O ∥ N. Under what condition do you think the projection of a triangle lying in plane O into plane N would be a line segment?

e) Do you think the projection of a triangle into a plane could be a point?

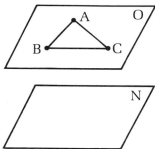

4. The base of the Great Pyramid is bounded by a square whose sides are each 230 meters long.

Suppose the small pyramid built by this Egyptian child is similar to it and that each edge of its base is 1 meter long. How do the two pyramids compare in
a) height?
b) surface area?
c) volume?

BY PERMISSION OF ED FISHER

SET II

1. The German astronomer and mathematician Johann Kepler discovered the two star-shaped polyhedra shown on the next page in about 1620. Each can be formed by adding pyramids to the faces of a *regular polyhedron.*
a) What kind of pyramids were added in each case?
b) To what regular polyhedra were they added?
c) How many points does each star have altogether?

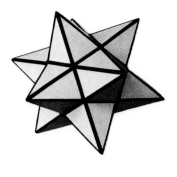

2. The Bitabak Company sells grapefruit juice in two different cans. One is in the usual shape of a right cylinder and the other is spherical in shape like a grapefruit. The diameter and height of the cylindrical can are both equal to the diameter of the spherical can.

   a) On the assumption that both cans are completely filled with grapefruit juice, exactly how would they compare in weight?

   b) If the lateral surface of each cylindrical can and the entire surface of each spherical can is painted yellow, which can requires more paint?

3. If a model of the edges of a cube were made from a piece of wire 1 yard long, what would the cube's volume be?

4. A package in the shape of a rectangular solid is tied with a ribbon so that the ribbon crosses the edges of its bases at their midpoints. If the package is 8 inches long, 6 inches wide, and 5 inches high, how long is the ribbon?

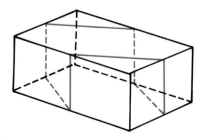

The following problem appears in an old book of puzzles by Sam Loyd, a man who invented hundreds of clever puzzles in the late nineteenth and early twentieth centuries.* Supposedly inspired by an actual attempt to fly to the moon in a balloon, the problem went like this.

A balloon for a trip to the moon is attached to a ball of wire in which the wire is one hundredth of an inch thick. If the ball of wire was originally two feet in diameter and was wound so solidly that there was no air space, what was the total length of the wire?

In his book, Mr. Loyd claimed that the problem could be solved without being concerned about the value of π. Can you do it?

---

*Originally in Sam Loyd's *Cyclopedia of Puzzles*, privately published in 1914. Included in *Mathematical Puzzles of Sam Loyd*, Volume 2, selected and edited by Martin Gardner (Dover, 1960).

Chapter **16**

NON-EUCLIDEAN
GEOMETRIES

Lesson 1

Geometry on a Sphere

Euclid defined parallel lines to be "lines that, being in the same plane and being produced indefinitely in both directions, do not meet one another in either direction." In this cartoon, Peter has attempted to demonstrate the existence of parallel lines by tracing them on the surface of the earth. Unfortunately, his two-pronged stick for drawing the lines was worn down to a nub during the long journey so Peter's friends remain skeptical upon seeing that his "parallel lines" have come together to meet in a common point at the end of the trip.

BY PERMISSION OF JOHN HART AND FIELD ENTERPRISES, INC.

Peter's rather embarrassing failure raises a number of questions about the geometry we have been studying and its relationship to the shape of the earth on which we live. For instance, our postulates about points and lines imply that lines are both straight and infinitely long. Can such lines exist on the surface of a sphere?

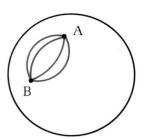

The concept of *distance between two points* is a very basic one. In a plane, it is measured along the line determined by the points. On the surface of a sphere, it must be measured along a curved path. There are many curved paths between two such points; the shortest of these paths is along the *great circle* through the two points.

► Definition

A *great circle* of a sphere is the set of points that is the intersection of the sphere and a plane containing its center.

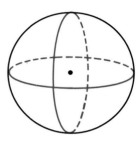

Examples of great circles on the earth are the equator and the meridians (the circles that pass through the north and south poles). Since we measure distances on the surface of the earth along these great circles, we might consider them "lines" in our development of geometry on a sphere. Of course, these lines are neither straight nor infinitely long. Their length is determined by the size of the sphere. Since we did not define the term "line" in our geometry, we might change our idea of it to conform to a great circle on a sphere.

"Lines" on a sphere

If we do, however, we need to reconsider our postulates about points and lines. Do two points determine a line on a sphere? Look at the adjoining figure. Is there exactly one line through points A and B? What about points C and D? The answer in the first case is yes, and in the second case, no, as the figures below illustrate. It is the second case in which we get into trouble, because points C and D are a pair of *polar points*.

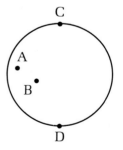

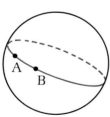

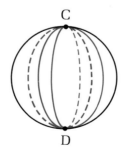

► Definition

Polar points are the points of intersection of a line through the center of the sphere with the sphere.

To get out of the difficulty of having more than one line through a pair of polar points, we might consider them to count as just one point. We can get away with this because, like the term "line," the term "point" was not defined. By agreeing that when we refer to a given *pair* of points it is understood that they are *not polar*, we have rescued our postulate that two points determine a line. For a given *pair* of points, there is exactly one line that contains them.

Our second postulate about points and lines said that a line contains at least two points. It apparently reflects the situation on a sphere without any adjustments.

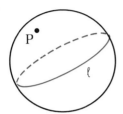

But now we come to something quite surprising. The figure shown here represents, in our new geometry, a point, P, and a line, ℓ, that does not contain it. Through P, how many lines can be drawn parallel to line ℓ?

Since our definition of parallel lines states that they lie in the same plane and do not intersect, to adapt it to our new geometry on a sphere we will evidently need to consider the sphere to be a "plane." Again, we can get away with this because we did not define the term "plane"! Now let's rephrase our question. Through P, how many lines can be drawn on the sphere that do not intersect line ℓ? The answer is *none. Every* great circle of a sphere intersects all other great circles of the sphere!

What does all this mean? It means that we have a geometry in which the Parallel Postulate we have used in the past no longer applies, for, through a point not on a line, there are now *no lines parallel to the line*!

With additional changes in our postulates concerning distance and betweenness and this new assumption that there are no parallel lines, we can develop a strange new geometry with all sorts of peculiar theorems. Although these theorems flatly contradict others that we have proved in our study of geometry, they make sense in terms of the new parallel postulate and in terms of each other! They are part of a geometry in which the Parallel Postulate has been replaced by another postulate. Such a geometry is called *non-Euclidean* and to close our study of the subject, we will become acquainted with what two of the non-Euclidean geometries are like.

Exercises

In the following exercises, we will refer to the geometry that we have been studying throughout the course as *Euclidean* geometry and to our new geometry on a sphere as *sphere* geometry. We will restrict our comparison of the two geometries to their two-dimensional versions only; this means that in each case we will assume that all "points" and "lines" lie in the same "plane."

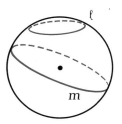

1. In sphere geometry, we picture lines as great circles of the sphere.
   a) Explain what is meant by a *great circle* of a sphere.
   b) Do one or both of the lettered curves in this figure seem to be lines?
   c) Because they have no endpoints, we say that lines in Euclidean geometry are boundless. Are lines in sphere geometry boundless?
   d) Because they are infinite in extent, we say that lines in Euclidean geometry have no lengths. Do lines in sphere geometry have lengths?

2. Points A, B, and C lie on the same great circle in the left-hand figure below.
   a) Are the three points collinear?
   b) Does B appear to be between A and C?
   c) Suppose point A moved toward the left and point C toward the right along the line, as shown in the right-hand figure. Is it always possible to tell which point is between the other two?

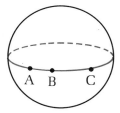 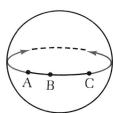

3. Early in our study of Euclidean geometry we proved that, if two lines intersect, they intersect in no more than one point.*
   a) Why does the figure below seem to show that this is not true in sphere geometry?
   b) However, the postulate upon which our proof was based is also true in sphere geometry. So, these lines intersect in just one point! Can you explain?

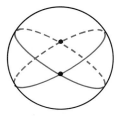

*See page 64.

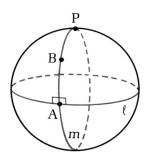

1. We will take the liberty of using such words as "angle" and "perpendicular" in sphere geometry intuitively without attempting to define them. The figure shown here illustrates two lines, ℓ and m, that intersect to form right angles. You can think of line ℓ as the "equator" and point P as the "north pole."

   How many lines do you think can be drawn through each of the following points that are perpendicular to line ℓ?
   a) Point A.
   b) Point B.
   c) Point P (called the *pole* of line ℓ).

   The number of perpendiculars to a line through a point seems to depend upon where the point is located with respect to the line.
   d) Under what condition do you think that there is exactly one line through a given point perpendicular to a given line in sphere geometry?

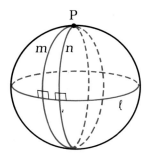

2. We proved in Euclidean geometry that, in a plane, two lines perpendicular to a third line are parallel to each other.
   a) Use the figure shown here to explain whether or not this theorem is true in sphere geometry.
   b) Referring to the same figure, do you think two lines in sphere geometry that form equal corresponding angles with a transversal are parallel?

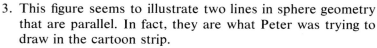

3. This figure seems to illustrate two lines in sphere geometry that are parallel. In fact, they are what Peter was trying to draw in the cartoon strip.
   a) Why do the two curves seem to be parallel?
   b) Since parallel lines do not exist on a sphere, something is wrong with our thinking about this figure. What is it?

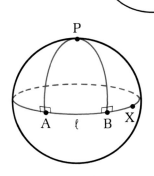

4. In the figure shown here, point P is the pole of line ℓ. Consider $\triangle$ABP with exterior $\angle$PBX.
   a) Does the Exterior Angle Theorem of Euclidean geometry apply to this triangle?
   b) How does the sum of the measures of the angles of $\triangle$ABP compare with 180°?
   c) How many right angles can a triangle have in Euclidean geometry?
   d) How many right angles do you think a triangle can have in sphere geometry?

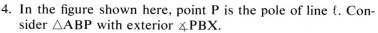

SET III

Some of the quadrilaterals whose properties we studied in Euclidean geometry do not exist in sphere geometry. For example, there are no parallelograms because there are no parallels. It is tempting to conclude from this that there are no rectangles either, since in Euclidean geometry we have proved that every rectangle is a parallelogram. Our proof, however, depended upon other Euclidean theorems that may not hold true in sphere geometry. If we consider the definition of rectangle—an equiangular quadrilateral—then it would seem that such figures *could* exist. Rectangle ABCD in the figure below is an example.

1. What kind of angles do you think this rectangle has?

2. Do you think a quadrilateral can exist in sphere geometry that has four right angles?

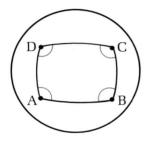

If we know that two quadrilaterals in Euclidean geometry are rectangles, then we can conclude that the angles of one are equal to the angles of the other.

3. Do you think a comparable conclusion is valid in sphere geometry?

4. Do you think there are any quadrilaterals of Euclidean geometry other than parallelograms that do not exist in sphere geometry?

HALE OBSERVATORIES

Lesson 2   The Saccheri Quadrilateral

PHOTOGRAPHY BY ROBERT M. GOTTSCHALK

We live in a mysterious universe. Before Einstein developed his theory of relativity, it was assumed that physical space is Euclidean. Although mathematicians in the nineteenth century had come to accept the idea that the geometry of Euclid is not the only one possible, it is still hard to believe that the non-Euclidean geometries might provide better models of what the universe is actually like. Yet it is these geometries upon which Einstein's theory of relativity is based.*

The first book to contain some non-Euclidean geometry was published in 1733, about two thousand years after Euclid wrote his *Elements.* It was written by an Italian priest named Girolamo Saccheri and, oddly enough, was actually intended by its author to prove that Euclidean geometry is the only logically consistent geometry possible. In fact, he named the book *Euclid Freed of Every Flaw.*

*An interesting article on this subject by P. Le Corbeiller is titled "The Curvature of Space," *Scientific American* (November 1954): 80–86; reprinted in *Mathematics in the Modern World* (W. H. Freeman and Company, 1968).

Saccheri planned to accomplish his goal by making what he thought were some false assumptions about a special quadrilateral. By reasoning indirectly from these assumptions, he felt certain that contradictions would eventually develop that could, in turn, be used to prove the Parallel Postulate. Instead of this happening, however, Saccheri ended up creating part of a new geometry without realizing it. Unfortunately for him, other mathematicians at the time paid little attention to his book and it was soon forgotten, not to be rediscovered until 1889, more than a century later. By that time, other men had independently recreated what Saccheri had developed; and it is their names that are now closely associated with non-Euclidean geometry: especially those of Nicholas Lobachevsky, a Russian, and Bernhard Riemann, a German.

Saccheri began his work with a quadrilateral that has a pair of sides perpendicular to a third side. We will call such a quadrilateral "biperpendicular."

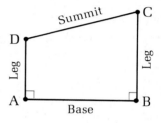

► Definition

A *biperpendicular quadrilateral* is a quadrilateral that has a pair of sides that are both perpendicular to a third side.

As you can see from the labeling of the figure, the two sides perpendicular to the same side are called the *legs*, the side to which they are perpendicular is called the *base*, and the side opposite the base is called the *summit*. The angles at A and B are called *base angles* of the quadrilateral and the angles at C and D are called *summit angles*. Furthermore, we will refer to angle C as the summit angle opposite leg $\overline{AD}$ and to angle D as the summit angle opposite leg $\overline{BC}$.

A biperpendicular quadrilateral whose legs are equal might be called "isosceles," but in honor of Saccheri it is usually called a *Saccheri quadrilateral*.

► Definition

A *Saccheri quadrilateral* is a biperpendicular quadrilateral whose legs are equal.

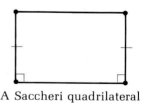

A Saccheri quadrilateral

A Saccheri quadrilateral looks very much like a rectangle; that is, the summit angles look as if they must also be right angles, so that the figure is equiangular. It is easy to prove this in Euclidean geometry. In one of the non-Euclidean geometries, however, a Saccheri quadrilateral is *not* a rectangle because, with a different postulate about parallel lines, it can be proved that its summit angles are *not* right angles!

Before we look further into this, we will state some theorems about biperpendicular quadrilaterals that are true in both Euclidean and the non-Euclidean geometries. Each of them can be proved without using the Parallel Postulate and outlines of the proofs are in the exercises. Although these theorems are not very interesting in themselves, we will use them in the next lesson to derive some of the strange results that Saccheri obtained. There is no need to memorize them.

► **Theorem 100**
The summit angles of a Saccheri quadrilateral are equal to each other.

► **Theorem 101**
The line segment joining the midpoints of the base and summit of a Saccheri quadrilateral is perpendicular to both of them.

The next two theorems are comparable to a pair of theorems about inequalities in triangles that you already know.

► **Theorem 102**
If the two legs of a biperpendicular quadrilateral are unequal, then the summit angles are unequal and the larger angle is opposite the longer side.

► **Theorem 103**
If the two summit angles of a biperpendicular quadrilateral are unequal, then the legs are unequal and the longer leg is opposite the larger angle.

Exercises

SET I

1. Which of the figures below are Saccheri quadrilaterals?

a)

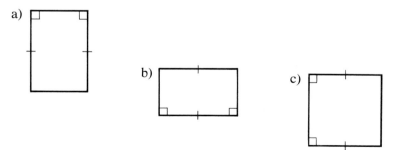

b)

c)

Answer the questions about the proofs of the following theorems.

2. *Theorem 100.* The summit angles of a Saccheri quadrilateral are equal to each other.

> Hypothesis: Saccheri quadrilateral JOEL with base $\overline{LE}$.
> Conclusion: $\angle J = \angle O$.

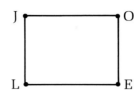

To prove this theorem, we can draw $\overline{JE}$ and $\overline{LO}$ and show that two pairs of triangles are congruent.

a) Why is JL = OE?
b) Why is $\angle JLE = \angle OEL$?
c) Why is $\triangle JLE \cong \triangle OEL$?

It follows that JE = OL.

d) Why is $\triangle JLO \cong \triangle OEJ$?
e) Why is $\angle LJO = \angle EOJ$?

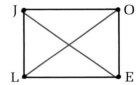

3. *Theorem 102.* If the two legs of a biperpendicular quadrilateral are unequal, then the summit angles are unequal and the larger angle is opposite the longer side.

> Hypothesis: Biperpendicular quadrilateral HOSA with base $\overline{OS}$; AS > HO.
> Conclusion: $\angle H > \angle A$.

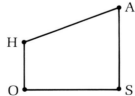

Proof.
Since HOSA is a biperpendicular quadrilateral with base $\overline{OS}$, we know that $\overline{HO} \perp \overline{OS}$ and $\overline{AS} \perp \overline{OS}$. And by hypothesis, AS > HO.

Choose point E on $\overrightarrow{SA}$ so that SE = OH and draw $\overline{HE}$.

a) Why is HOSE a Saccheri quadrilateral?
b) Why is $\angle 1 = \angle 2$?
c) Since $\angle OHA = \angle 1 + \angle 3$, $\angle OHA > \angle 1$. Why?
d) It follows that $\angle OHA > \angle 2$. Why?
e) Why is $\angle 2 > \angle A$?
f) Why is $\angle OHA > \angle A$?

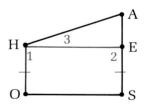

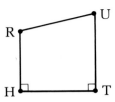

4. *Theorem 103.* If the two summit angles of a biperpendicular quadrilateral are unequal, then the legs are unequal and the longer leg is opposite the larger angle.

> Hypothesis: Biperpendicular quadrilateral RUTH with base $\overline{HT}$; $\angle R > \angle U$.
> Conclusion: UT > RH.

We will prove this theorem by the indirect method.

a) Either UT < RH, UT = RH, or UT > RH. Why?

Suppose UT < RH.

b) Then it follows that $\angle R < \angle U$. Why?
   This contradicts the hypothesis that $\angle R > \angle U$.

Suppose UT = RH.

c) What kind of quadrilateral must RUTH be?
d) It follows that $\angle R = \angle U$. Why?
   This also contradicts the hypothesis that $\angle R > \angle U$.

The only possibility remaining is that UT > RH.

SET II

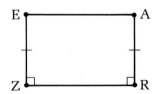

1. In Euclidean geometry, it is very easy to prove that a Saccheri quadrilateral is a parallelogram.
   a) Use the figure shown here to explain how this could be done.
   b) Why is the summit of a Saccheri quadrilateral in Euclidean geometry equal to its base?

2. Explain why, in Euclidean geometry, the summit angles of a Saccheri quadrilateral are right angles.

3. Theorem 101 can be proved without using any facts about parallel lines. Answer the following questions related to how this could be done. (Each answer will include several ideas.)
   The line segment joining the midpoints of the base and summit of a Saccheri quadrilateral is perpendicular to both of them.

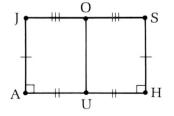

> Hypothesis: Saccheri quadrilateral JSHA with $\overline{OU}$ joining the midpoints of $\overline{JS}$ and $\overline{AH}$, respectively.
> Conclusion: $\overline{OU} \perp \overline{JS}$ and $\overline{OU} \perp \overline{AH}$.

First, we can draw $\overline{JU}$ and $\overline{SU}$.

a) How can it be shown that $JU = SU$?

Since O is the midpoint of $\overline{JS}$, we know that $JO = SO$.

b) How does it follow that $\overline{OU} \perp \overline{JS}$?

Next, we draw $\overline{OA}$ and $\overline{OH}$.

c) How can it be shown that $\triangle JOA \cong \triangle SOH$?
d) Explain, without proving any more triangles congruent, why $\overline{OU} \perp \overline{AH}$.

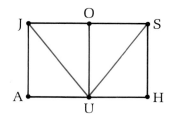

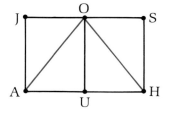

SET III

Among the many people who have tried to prove Euclid's Parallel Postulate was a Persian mathematician who was the court astronomer of the grandson of the famous Genghis Khan. His name was Nasir Eddin and he lived in the thirteenth century.

Nasir Eddin began by making the following assumption. Suppose n and ℓ are two lines such that perpendiculars to ℓ from points on n make unequal angles with n.

Then if $\angle 1$ and $\angle 3$ are acute and $\angle 2$ and $\angle 4$ are obtuse, it follows that $\overline{DI}$ must be shorter than $\overline{AE}$.

Nasir Eddin's assumption can be easily proved by means of one of the theorems in this lesson. Can you explain why?

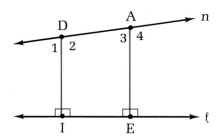

Lesson 3

The Geometries of
Lobachevsky and Riemann

Nicholas Lobachevsky.

After proving that the summit angles of a Saccheri quadrilateral are equal, Saccheri realized that, if he could prove them to be right angles, he could use this fact to prove the Parallel Postulate. This he planned to accomplish by reasoning indirectly. He would show that the assumptions that the summit angles were either acute or obtuse would lead to contradictions.

Saccheri did, in fact, eventually arrive at a contradiction from the hypothesis that the summit angles were obtuse. But to eliminate the possibility that they were acute proved to be far more difficult. In fact, instead of causing contradictions, the "acute angle" hypothesis became the beginning of one of the non-Euclidean geometries. Saccheri and his contemporaries didn't really comprehend the significance of what he had started and so it was not until later that the idea that a non-Euclidean geometry could make sense was accepted. Three men are given credit for independently realizing this: the great German mathematician Karl Friedrich Gauss, Janos Bolyai, a Hungarian, and Nicholas Lobachevsky. Since the non-Euclidean geometry based upon the "acute angle" hypothesis has often simply been called *Lobachevskian,* we will refer to it by that name.

► **The Lobachevskian Postulate**
The summit angles of a Saccheri quadrilateral are acute.

By means of this postulate, the following theorems can be proved.

► **Lobachevskian Theorem 1**
In Lobachevskian geometry, the summit of a Saccheri quadrilateral is longer than its base.

► **Lobachevskian Theorem 2**
In Lobachevskian geometry, a midsegment of a triangle is less than half as long as the third side.

Of course, these theorems contradict theorems in Euclidean geometry, *but only those based upon the Parallel Postulate.* Since the Parallel Postulate states that through a point not on a line, there is no more than one parallel to the line, these new theorems are consistent with the idea that through a point not on a line, there *is* more than one parallel to the line!

You will recall that in sphere geometry, there are *no* parallels to a line through a point not on it. However, this result contradicts other Euclidean theorems proved *before* the Parallel Postulate. In fact, it is equivalent to assuming that the summit angles of a Saccheri quadrilateral are *obtuse*, and this is why Saccheri was able to eliminate this possibility.

Nevertheless, if we are willing to change some other postulates related to distance and betweenness as well as the Parallel Postulate, a logically consistent non-Euclidean geometry can be developed in which there are no parallels at all. The German mathematician Bernhard Riemann was the first to appreciate this and we will refer to the geometry that he created as *Riemannian geometry.*

Bernhard Riemann.

The table below summarizes the basic differences between Euclidean geometry and the two non-Euclidean geometries. In each case, either statement can be proved to be a logical consequence of the other.

Statement	*Euclid*	*Lobachevsky*	*Riemann*
Through a point not on a line, there is	exactly one parallel to the line.	more than one parallel to the line.	no parallel to the line.
The summit angles of a Saccheri quadrilateral are	right.	acute.	obtuse.

Exercises

In the exercises of this and the following lesson, we will restrict our proofs to Lobachevskian geometry because in this geometry only the Parallel Postulate is changed. This means that we can use any idea considered before Chapter 7 in this book.

SET I

1. Quadrilateral BRLN is a Saccheri quadrilateral with base $\overline{NL}$; $\overline{EI}$ joins the midpoints of $\overline{BR}$ and $\overline{NL}$.
   a) What can you conclude about ∡B and ∡R?
   b) In Lobachevskian geometry, what kind of angles are they?
   c) What can you conclude about $\overline{EI}$?

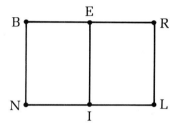

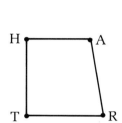

2. Quadrilateral HART is biperpendicular with base $\overline{HT}$.
   a) What can you conclude about ∡H and ∡T?
   b) If ∠A > ∠R, what can you conclude?

3. The following proof is based upon the Lobachevskian Postulate and the theorems from Lesson 2.

Lobachevskian Theorem 1. In Lobachevskian geometry, the summit of a Saccheri quadrilateral is longer than its base.

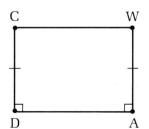

Hypothesis: Saccheri quadrilateral CWAD with base $\overline{DA}$.
Conclusion: CW > DA.

Proof.

Since CWAD is a Saccheri quadrilateral with base $\overline{DA}$, $\overline{WA} \perp \overline{DA}$. Therefore ∡A is a right angle, and hence ∠A = 90°.

a) Why is ∡W acute?
b) Why is ∠W < 90°?
c) Why is ∠W < ∠A?

Next we let O and R be the midpoints of $\overline{CW}$ and $\overline{DA}$, respectively, as shown in the figure below.

d) What permits us to do this?
e) After drawing $\overline{OR}$, how do we know that $\overline{OR} \perp \overline{CW}$ and $\overline{OR} \perp \overline{DA}$?
f) Why is OWAR a biperpendicular quadrilateral?

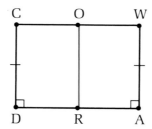

Since we have already proved that ∠W < ∠A, we know that ∠A > ∠W.

g) How does it follow that OW > RA?
h) Why is OW = ½CW and RA = ½DA?
i) Why is ½CW > ½DA?
j) Why is CW > DA?

4. We have just proved that the summit of a Saccheri quadrilateral in Lobachevskian geometry is longer than its base. The following questions are about the properties of a Saccheri quadrilateral in *Riemannian geometry*. The sphere geometry that we studied in Lesson 1 makes a good model for understanding Riemannian geometry because there are no parallel lines in either.

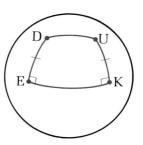

This figure shows a Saccheri quadrilateral on a sphere.

a) What kind of angles do ∡D and ∡U seem to be?
b) Does your answer agree with the table on page 649?
c) What relationship does $\overline{DU}$ seem to have to $\overline{EK}$?
d) State what you think is an appropriate theorem in Riemannian geometry to correspond to Lobachevskian Theorem 1.

1. Answer the following questions about the proof of Lobachevskian Theorem 2.

   In Lobachevskian geometry, a midsegment of a triangle is less than half as long as the third side.

   Hypothesis: △GIN with midsegment $\overline{RH}$.
   Conclusion: RH < $\frac{1}{2}$NI.

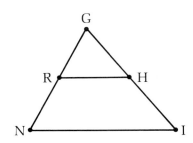 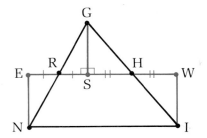

   Proof.
   Through G, draw $\overline{GS} \perp \overline{RH}$. Draw $\overleftrightarrow{RH}$ and choose points E and W on the rays opposite $\overrightarrow{RS}$ and $\overrightarrow{HS}$ so that RE = RS and HW = HS. Draw $\overline{NE}$ and $\overline{IW}$.
   a) Explain why △GSR ≅ △NER and △GSH ≅ △IWH.

   Since the triangles are congruent, EN = GS and GS = WI.
   b) Why does it follow that EN = WI?

   Angles E and W are right angles because they are equal to the right angles at S, so $\overline{NE} \perp \overline{EW}$ and $\overline{IW} \perp \overline{EW}$.
   c) Why is EWIN a Saccheri quadrilateral?
   d) Why is NI > EW?

   Since EW = ER + RS + SH + HW, ER = RS and SH = HW, it follows that EW = RS + RS + SH + SH = 2RS + 2SH = 2(RS + SH).
   e) Why is EW = 2RH?
   f) Why is NI > 2RH?
   g) Why is $\frac{1}{2}$NI > RH so that RH < $\frac{1}{2}$NI?

2. Draw a figure to represent a triangle on a sphere and add one of the triangle's midsegments.
   a) On the basis of your drawing, state what you think is an appropriate theorem in Riemannian geometry to correspond to Lobachevskian Theorem 2.
   b) Do you think a midsegment of a triangle in Riemannian geometry is parallel to the third side of the triangle? Explain.

"He keeps his place clean as a whistle."

COURTESY OF VIRGIL PARTCH

SET III

The homes in this cartoon that don't seem to be as clean as a whistle are each supported by four posts. These posts are not perpendicular to the ground but are slanted toward each other so that their tops are closer together than their bottoms. The adjoining figure illustrates the posts' relationship to each other.

Suppose that KN = ER and that KE < NR. Is it possible that the posts could be perpendicular to the floor of the home they support? Specifically, can it be that $\overline{KN} \perp \overline{KE}$ and $\overline{ER} \perp \overline{KE}$? Explain.

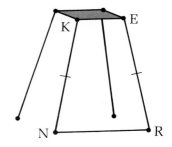

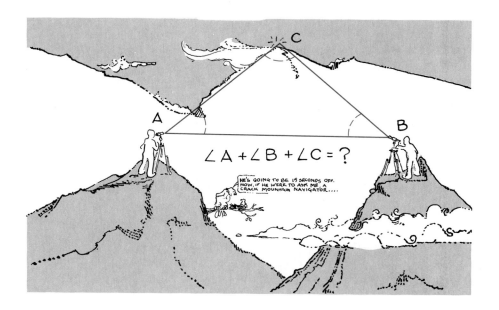

$$\angle A + \angle B + \angle C = ?$$

HE'S GOING TO BE 15 SECONDS OFF.
NOW, IF HE WERE TO ASK ME A
CRACK MOUNTAIN NAVIGATOR.....

Lesson **4**

The Triangle Angle Sum
Theorem Revisited

At about the time Lobachevsky and Bolyai were developing the non-Euclidean geometry that we have been calling Lobachevskian, the great German mathematician Karl Friedrich Gauss carried out an experiment that is illustrated by the drawing above. You know that in Euclidean geometry the sum of the measures of the angles of a triangle is 180°. According to the geometry of Lobachevsky it is *less* than 180°, whereas in Riemannian geometry it is *more* than 180°.

Which of these geometries is the one that fits physical space? In a geodesic survey, Gauss measured the angles of a giant triangle formed by light rays with its corners on three mountain tops in Germany.* The resulting sum turned out to be within just a few seconds of 180°. Even though the measurements were made as precisely and accurately as was possible at the time, the difference could have been due to experimental error. And, unfortunately, the mountain top triangle was much too small to determine if physical space is non-Euclidean, even though its

*Tord Hall, *Karl Friedrich Gauss,* translated by Albert Froderberg (The M.I.T. Press, 1970), pp. 123–124.

sides were each between 40 and 70 miles long! According to both non-Euclidean geometries, the smaller the triangle, the closer the sum of the measures of its angles is to 180°. If space is either Lobachevskian or Riemannian in nature, then a triangle with three *stars* for its corners rather than three mountain tops on the earth would be much more appropriate for deciding which it is.*

We have said that in Lobachevskian geometry the sum of the measures of the angles of a triangle is less than 180°. Furthermore, the sum of the measures of a convex quadrilateral is less than 360°. As a consequence of these facts, it can be proved that there are no scale models in this geometry! In other words, if two figures in Lobachevskian geometry have different sizes, they cannot have the same shape! And scale models cannot exist in Riemannian geometry either!

It is no wonder that with such intuition-defying results as these it took a long time for the non-Euclidean geometries to be accepted. Yet we already know enough about Lobachevskian geometry to be able to prove them as the following theorems.

► **Lobachevskian Theorem 3**

In Lobachevskian geometry, the sum of the measures of the angles of a triangle is less than 180°.

► **Corollary**

In Lobachevskian geometry, the sum of the measures of the angles of a convex quadrilateral is less than 360°.

► **Lobachevskian Theorem 4**

In Lobachevskian geometry, if two triangles are similar, they must also be congruent.

At the beginning of this course, we compared geometry to a game and said, in effect, that the name of the game depends upon the rules by which it is played.† With your brief introduction to the non-Euclidean geometries, you can probably more fully appreciate the meaning of this.

In the past century, many other geometries have been developed in addition to the three that we have studied. Man's curiosity will undoubtedly lead him to invent still others in the future. In fact, one twentieth-century mathematician has said, "When a man stops wondering and asking and playing, he is through."‡

---

*More information about how the nature of physical space might be determined is in *The Universe*, by David Bergamini, Life Nature Library series (Time-Life Books, 1962), pp. 172–179.

†See page 44.

‡In *Mathematics and the Imagination* by Edward Kasner and James R. Newman (Simon & Schuster, 1940).

Exercises

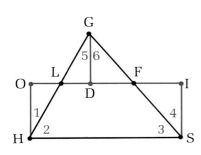

1. First we will prove Lobachevskian Theorem 3.

   In Lobachevskian geometry, the sum of the measures of the angles of a triangle is less than 180°.

   Hypothesis: △GSH.
   Conclusion: ∠G + ∠H + ∠S < 180°.

   Our proof begins with the addition of several points and lines to the triangle to form a figure like the one we used to prove Lobachevskian Theorem 2.

   Proof.
   Let L and F be the midpoints of sides $\overline{GH}$ and $\overline{GS}$ of the triangle and draw $\overleftrightarrow{LF}$. Through G, draw $\overleftrightarrow{GD} \perp \overleftrightarrow{LF}$. Choose points O and I on the rays opposite $\overrightarrow{LD}$ and $\overrightarrow{FD}$ so that LO = LD and FI = FD. Draw $\overline{HO}$ and $\overline{SI}$.
   a) Use the fact that △GDL ≅ △HOL and △GDF ≅ △SIF to explain why OISH is a Saccheri quadrilateral.
   b) Why are ∢OHS and ∢ISH acute?
   c) What can be concluded about ∠OHS + ∠ISH from the fact that ∠OHS < 90° and ∠ISH < 90°?

   Since ∠OHS = ∠1 + ∠2 and ∠ISH = ∠3 + ∠4, it follows that ∠1 + ∠2 + ∠3 + ∠4 < 180°.
   d) To which angles in the figure are ∢1 and ∢4 equal?
   e) On the basis of the last inequality and your answer to part d, explain why ∠HGS + ∠GHS + ∠GSH < 180°.

2. Use Lobachevskian Theorem 3 to decide upon answers to the following questions about triangles in *Lobachevskian geometry*.
   a) If two angles of one triangle are equal to two angles of another triangle, do you think that the third pair of angles are necessarily equal?
   b) Are the acute angles of a right triangle complementary?
   c) What can you say about the measure of each angle of an equilateral triangle?

3. Use the figure shown here to explain why the sum of the measures of the angles of a convex quadrilateral in Lobachevskian geometry is less than 360°.

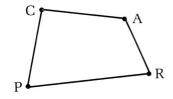

1. We will use the indirect method to prove Lobachevskian Theorem 4.

   In Lobachevskian geometry, if two triangles are similar, they must also be congruent.

   Hypothesis: △MAR ~ △LIN.
   Conclusion: △MAR ≅ △LIN.

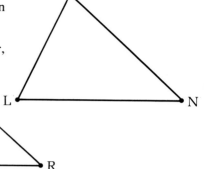

Proof.
By hypothesis, △MAR ~ △LIN.
a) Why is ∠M = ∠L, ∠A = ∠I and ∠R = ∠N?

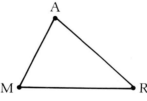

Now suppose △MAR and △LIN are not congruent. If this is the case, then AM ≠ IL and AR ≠ IN, because if either pair of these sides were equal, then the triangles would be congruent by A.S.A.

We will now copy the smaller triangle on the larger one as shown in the second pair of triangles. Choose point M′ on $\overrightarrow{IL}$ so that IM′ = AM and point R′ on $\overrightarrow{IN}$ so that IR′ = AR. Draw $\overline{M'R'}$.
b) Why is △M′IR′ ≅ △MAR?

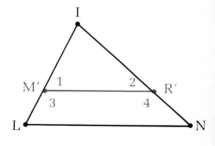

It follows that ∠1 = ∠M and ∠2 = ∠R since they are corresponding parts of these triangles.
c) Why is ∠L = ∠1 and ∠N = ∠2?
d) Why is ∠1 + ∠3 = 180° and ∠2 + ∠4 = 180°?

Adding these equations, we get ∠1 + ∠2 + ∠3 + ∠4 = 360°.
e) Why is ∠L + ∠N + ∠3 + ∠4 = 360°?
f) If we assume from the figure that LM′R′N is a convex quadrilateral, the equation in part e is impossible. Why?

Since we have arrived at a contradiction, our initial assumption that △MAR and △LIN are not congruent is false! So △MAR ≅ △LIN.

2. If the angles of one triangle are equal to the angles of another triangle, what can be concluded about the triangles in
   a) Euclidean geometry?
   b) Lobachevskian geometry?

3. In this lesson, we claimed that the sum of the measures of the angles of a triangle in *Riemannian geometry* is *more than* 180°.
   a) Draw a figure to represent a fairly large triangle on a sphere to show why this is plausible.
   b) Do you think there is any limit on the sum of the measures of the angles of a triangle in Riemannian geometry?

SET III

Obtuse Ollie is upset by all of these strange theorems from non-Euclidean geometry. A theorem of Euclidean geometry that he especially likes is the one stating that an angle inscribed in a semicircle is a right angle. His intuition tells him that the theorem is probably true in the other geometries as well.

By drawing radius $\overline{OE}$ and beginning with the four numbered angles shown here, it is possible to arrive at a definite conclusion about this. Can you do so?

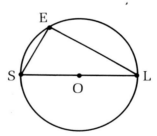 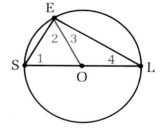

Chapter 16 / Summary and Review

Basic Ideas

Postulate

The Lobachevskian Postulate.   The summit angles of a Saccheri quadrilateral are acute.   649

Theorems

100. The summit angles of a Saccheri quadrilateral are equal to each other.   644
101. The line segment joining the midpoints of the base and summit of a Saccheri quadrilateral is perpendicular to both of them. 644
102. If the two legs of a biperpendicular quadrilateral are unequal, then the summit angles are unequal and the larger angle is opposite the longer side.   644
103. If the two summit angles of a biperpendicular quadrilateral are unequal, then the legs are unequal and the longer side is opposite the larger angle.   644
L.1. In Lobachevskian geometry, the summit of a Saccheri quadrilateral is longer than its base.   649
L.2. In Lobachevskian geometry, a midsegment of a triangle is less than half as long as the third side.   649
L.3. In Lobachevskian geometry, the sum of the measures of the angles of a triangle is less than 180°.   655
     Corollary. In Lobachevskian geometry, the sum of the measures of the angles of a convex quadrilateral is less than 360°.   655
L.4. In Lobachevskian geometry, if two triangles are similar, they must also be congruent.   655

Statement	Euclid	Lobachevsky	Riemann
Through a point not on a line, there is	exactly one parallel to the line.	more than one parallel to the line.	no parallel to the line.
The summit angles of a Saccheri quadrilateral are	right.	acute.	obtuse.
The sum of the measures of the angles of a triangle is	180°.	less than 180°.	more than 180°.

Exercises

SET I

1. Many mathematicians since the time of Euclid have tried to prove the Parallel Postulate. That is, they have tried to show by means of the other postulates of Euclidean geometry that through a point not on a line there is exactly one parallel to the line. Why has no one ever succeeded in doing this?

2. We proved in Euclidean geometry that two parallel lines are everywhere equidistant by showing that every perpendicular segment joining one line to the other has the same length.
   a) This theorem does not apply to Riemannian geometry. Why not?

   We can prove that it does not apply to Lobachevskian geometry either by using the indirect method. In the figure below, we will assume that $\ell_1 \parallel \ell_2$ and show that ℓ_1 and ℓ_2 are not everywhere equidistant.
   Let R, A, and D be three points on line ℓ_1.
   b) What permits us to draw $\overline{RM}$, $\overline{AU}$, and $\overline{DI}$, each perpendicular to line ℓ_2?

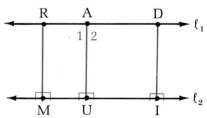

To prove that ℓ_1 and ℓ_2 are not everywhere equidistant, we will assume that they are. This means that RM = AU = DI.

c) What kind of figures are RAUM and ADIU?

d) What kind of angles are ∡1 and ∡2? (Remember that we are thinking in terms of Lobachevskian geometry.)

e) Explain how your answer to part d contradicts the fact that ∡1 and ∡2 are a linear pair.

Since we have arrived at a contradiction, our assumption is false. Hence, two parallel lines in Lobachevskian geometry are not everywhere equidistant!

3. Quadrilateral ZINC is biperpendicular and its base and legs have equal lengths.

   a) Explain why, in Lobachevskian geometry, ZINC cannot be a rhombus.

   b) Do you think a rhombus can exist in Lobachevskian geometry?

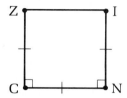

SET II

Throughout this book, you have seen a variety of pictures by the artist Maurits Escher. He has said: "By keenly confronting the enigmas that surround us, and by considering and analyzing the observations that I had made, I ended up in the domain of mathematics. . . . I often seem to have more in common with mathematicians than with my fellow artists."[*]

His print on page 662 illustrates a model devised by a French mathematician, Henri Poincaré, for visualizing the theorems of Lobachevskian geometry![†]

You will recall that we have already used a sphere as a model to make the theorems of Riemannian geometry seem plausible. Points of the plane were represented by pairs of polar points on the sphere and lines by great circles. To understand Poincaré's model, it is necessary to know what *orthogonal circles* are.

---

[*]M. C. Escher, *The Graphic Work of M. C. Escher*, rev. ed. (Meredith Press, 1967). p. 10.

[†]Another print based upon this model is on page 469 of this book.

CIRCLE LIMIT III BY M. C. ESCHER. ESCHER FOUNDATION, HAAGS GEMEENTEMUSEUM, THE HAGUE

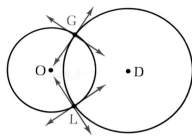

This figure represents two orthogonal circles O and D intersecting at points G and L. The tangents to the circles at these points have been drawn.

1. On the basis of this figure and your knowledge of the meaning of the prefix "ortho," define *orthogonal circles.*

In Poincaré's model of Lobachevskian geometry, points of the plane are represented by points in the interior of a circle and lines by both the diameters of the circle and the arcs of circles orthogonal to it. Examples of some lines in this model are shown here. The white arcs through the backbones of the fish in Escher's print are also "lines."

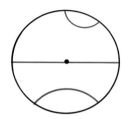

2. Through two points in a circle, there is exactly one arc of a circle orthogonal to it. What basic postulate does this illustrate?

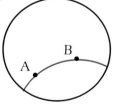

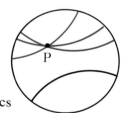

3. This figure shows an "orthogonal arc" and several such arcs through point P that do not intersect it. What idea in Lobachevskian geometry does this illustrate?

4. State a theorem or postulate in Lobachevskian geometry suggested by each of the following figures.

a)
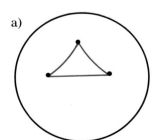

b)
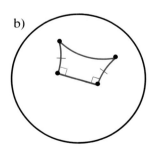

c)
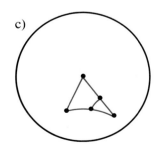

The sum of the measures of the angles of a triangle is less than, equal to, or more than 180° depending upon whether the geometry is Lobachevskian, Euclidean, or Riemannian.

Dilcue, in a flash of inspiration at the end of his study of geometry, has discovered a way to prove that the sum is 180° *without using the Parallel Postulate*! His argument goes like this.

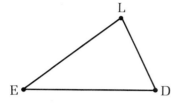

Let A be a point on side $\overline{ED}$ of $\triangle LED$ and draw $\overline{LA}$. Since we are trying to find out what the sum of the measures of the angles is, let's represent it as x. Then in $\triangle LED$,

$$\angle 1 + \angle 2 + \angle E + \angle D = x.$$

Also, in $\triangle LEA$,

$$\angle 1 + \angle 3 + \angle E = x$$

and in $\triangle LAD$,

$$\angle 2 + \angle 4 + \angle D = x.$$

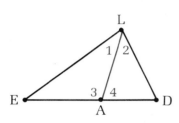

Adding the last two equations, we get

$$\angle 1 + \angle 2 + \angle 3 + \angle 4 + \angle E + \angle D = 2x.$$

Subtracting the first equation leaves

$$\angle 3 + \angle 4 = x.$$

Since $\angle 3$ and $\angle 4$ are a linear pair, they are supplementary; so

$$\angle 3 + \angle 4 = 180°.$$

Therefore, $x = 180°$ and we have proved that the sum of the measures of the angles of a triangle is equal to 180° without using the Parallel Postulate!

Dilcue's proof does not disprove the validity of the non-Euclidean geometries, however, because he has assumed something that is not true in either of them. Can you figure out what it is?

Answers

TO SELECTED EXERCISES

Chapter 1, Lesson 2

SET I

1. If you cross your eyes, then I crack up.
2. If money grew on trees, then Smokey the Bear wouldn't have to do commercials for a living.   3. If someone is a surfer, then he likes big waves.   4. If ice cream is licorice-flavored, then it has a peculiar color.   5. If a heavy object is stored in the attic of a jungle mansion, then it may crash down upon the occupants.   6. If someone lives in a grass house, then he shouldn't stow thrones.
7. If someone is a ghost, then he does not have a shadow.   Or: If someone has a shadow, then he is not a ghost.

SET II

1.

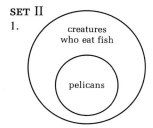

2. a and d.
3. Either of these diagrams is correct.

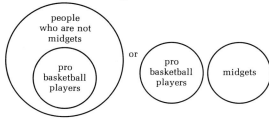

4. a and d.   5. c and d.   6. Two.

Chapter 1, Lesson 3

SET I

1. a) Inverse.   b) Converse.
c) Contrapositive.   2. a) Contrapositive.
b) Inverse.   c) Original statement.
3. a) Converse.   b) None.   c) Contrapositive.
4. a) Inverse.   b) Contrapositive.   c) Original statement.   5. If the vampires are out, the moon is full.   6. If gargling helps much, then a giraffe does not have a sore throat.
7. If we have not been receiving signals from Jupiter, it may be wise to go there.   8. If you cannot comprehend geometry, you do not know how to reason deductively.

SET II 1.

Switch Position	Bulb a	Bulb b
1	On	On
2	Off	On
3	Off	Off

2. No.   3. A statement and its converse are not logically equivalent.   4. No.
5. The fact that a statement and its inverse are not logically equivalent.   6. Yes.   7. The fact that a statement and its contrapositive are logically equivalent.   8. Reverse the positions of bulbs a and b.
9. not $a \rightarrow$ not b.

Chapter 1, Lesson 4

SET I

1. Converse error.   2. Inverse error.
3. Valid.   4. Valid.   5. Inverse error.
6. Converse error.   7. Valid.   8. Inverse error.

Chapter 1, Lesson 4 (continued)

SET II

1. I have dialed correctly.   2. No deduction possible.   3. There is no fly in your soup.   4. No deduction possible.
5. No deduction possible.   6. Fred is not a night owl.   7. No deduction possible.
8. No deduction possible.

Chapter 1, Lesson 5

SET I

1. Valid.   2. Not valid.   3. Valid.
4. Not valid.   5. Valid.   6. Valid.
7. Not valid.

SET II

1. If you buy the *Encyclopedia Cribanana*, your children will have more time to watch television.   2. No conclusion.   3. If your name is in *Who's Who*, then you're sure of where's where.   4. No conclusion.
5. No conclusion.   6. If you read *Mewsweek* magazine, then you are not fond of mice.   7. All carnations are fragrant.
8. No shoplifter is trustworthy.

Chapter 1, Lesson 6

SET I

1. No.   2. Since words are defined by means of other words, the dictionary would be of no help if it were written entirely in Spanish.   3. Laurel misunderstood the meaning of the definition given for comet.
4. If someone defined "sound" as "an auditory sensation," then the breaking flowerpot made no sound because no auditory sensation occurred.

SET II

1. No.   3. "part."

Chapter 1, Lesson 7

SET I

1. Yes. This is the contrapositive of the previous statement, so it must also be true.
2. A good definition, because if something has a low temperature it is cold.   3. A bad definition, because if a creature is a social insect it isn't necessarily an ant.   4. A bad definition, because a stringed musical instrument is not necessarily a mandolin.

5. A good definition, because a young cat is a kitten.

SET II

1. a) No. (For most people, the first sentence is true and the second is false.) b) The second sentence. c) If you get some presents, then it is your birthday.   2. a) The second diagram. b) If *a*, then *b*. c) "*a* if *b*" means the same as "if *b*, then *a*"; "*a* only if *b*" means the same as "if *a*, then *b*."   3. An automobile is a jalopy if and only if it is a junky old car. Also correct: An automobile is a junky old car if and only if it is a jalopy.
4. If a person is a goof-off, he habitually shirks responsibility. If a person habitually shirks responsibility, then he is a goof-off.
5. All equiangular quadrilaterals are rectangles.
6. All three statements must also be true.
7. It is unsatisfactory because its converse is not true. A set of points is not necessarily a line.

Chapter 1, Lesson 8

SET I

1. a) An unlimited number. b) One. c) One.
d) None.   2. Points A, B, and C; also, points A, E, and D.   3. Points lie on the same line if and only if they are collinear points. Also correct: Points are collinear if and only if they are points that lie on the same line.   4. a) *Two* points determine a line.
b) Three *collinear* points determine a line.
5. a) An unlimited number. b) One.
c) None. d) One. e) None.   6. Lines *a*, *b*, and *c*.   7. Two *intersecting* (or *concurrent*) lines determine a point.
8. Four.   9. One on line *c* and one on line *d*. Three.

SET II

1. If the thumbtack were "in line" (collinear) with the other two.   2. They must be noncollinear.   3. Coplanar.   5. The line lies in the plane.

Chapter 1, Lesson 9

SET I

1. If something takes more than two hundred hours, it takes more than eight days.   2. If you don't realize what time it is, you will be late to school.   3. If there is a total eclipse

of the sun, then the sky becomes dark. If crickets think that it is night, they start chirping. 4. If the moon were made of green cheese, there would be one giant peep for mousekind. 5. The second statement in the proof should be: If a triangle is equilateral, then it is also equiangular.

1. If someone is a baby, he is illogical. If someone is illogical, he is despised. (Originally the third statement.) If someone is despised, he cannot manage crocodiles. (Originally the second statement.) 2. Replace the second statement in the proof with its contrapositive: If the opposite sides of a quadrilateral are parallel, then it is a parallelogram. 3. If there is no Great Pumpkin, then Linus is mistaken. (This is the contrapositive of the fourth statement.) If Linus is mistaken, Lucy is pleased. If Lucy is pleased, she becomes rambunctious. If Lucy becomes rambunctious, she plays a trick on Charlie Brown. If Lucy plays a trick on Charlie Brown, he will be upset. If Charlie Brown is upset, he forgets to feed Snoopy. If Charlie Brown doesn't feed Snoopy, Snoopy won't have pie for dinner.

Chapter 1, Lesson 10

SET I

1. a) Seven is not a prime number. b) Mr. Spock likes contradictions. c) A line contains less than two points. 2. a) Its square is not odd. b) They intersect in more than one point. c) They are not parallel to each other. 3. a) It is the opposite of the conclusion of the theorem. b) A contradiction. c) That the opening statement is false. 4. a) Suppose the plant in the photograph *is* poison ivy. b) The plant has leaves in groups of three. c) The fact that the plant does not have leaves in groups of three. d) The plant in the photograph is not poison ivy.

Chapter 1, Lesson 11

SET I

1. a) Postulate 2. b) Definition of collinear points (actually, its inverse.) If three or more points do not lie on the same line, they are noncollinear. c) Postulate 3. d) Postulate 4.

2. a) Suppose the two lines intersect in more than one point. b) Postulate 1. c) That our initial assumption is false. d) Two intersecting lines intersect in no more than one point.

SET II

1. a) The previous theorem. b) Postulate 2. c) Definition of collinear points. d) Postulate 3. e) Postulate 4. 2. a) That the line and plane intersect in more than one point. b) Postulate 4. c) Our initial assumption. d) The line and plane intersect in no more than one point.

Chapter 1, Review

SET I

1. a) If a poem is a limerick, then it has five lines. b) If I perfect my perpetual motion machine, then I will make a fortune. 2. The only statement the Cat would have to agree with is c, the contrapositive. (a is the converse and b is the inverse.) 3. a) Inverse. b) Contrapositive. c) Converse. 4. a) If there's fire, then there's smoke. b) If it is a Kodak, it is an Eastman. c) If Evel Knievel didn't try to jump over the Grand Canyon, he wouldn't land in the Colorado River. 5. a) Yes. b) No. B.C. asked Peter to prove that man came from the ape. Peter proved that B.C. is not an ape.

SET II

1. a) Postulates. b) Theorems. c) Deductive reasoning. 2. We can't define every term without going around in circles. These terms are left undefined because they are especially simple. 3. This statement is good because the surface of a window is flat like a plane. It is bad because a window has edges and a plane does not. 4. Indirect proof. 5. The basis for disagreement is the meaning of *same noise*. Baby Snooks is taking it literally.

6. a)

Chapter 1, Review *(continued)*

b) If a person enjoys gambling, then he likes to take a chance.

SET III

2. If a rabbit is over six feet tall, he will not be taken seriously.

Chapter 2, Lesson 1

SET I

1. Collinear.   2. a) 6 units.   b) 5 units.
3. a) 6.   b) 5.   4. a) UG (the distance between U and G).   b) $BU = u - b$.
5. a) AT = 74.   b) NA = 10.
c) GN = 26.   d) GT = 110.

SET II

1. The distance between F and L seems to be the same as the distance between E and A, yet $-2 -(-4) = 2$, and $8 - 5 = 3$.

2. 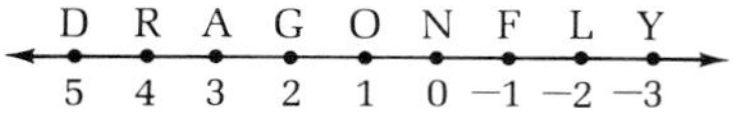
```
  D R A G O N F L Y
  5 4 3 2 1 0 -1 -2 -3
```

3. a) 12.   b) 10.   c) -11.   4. a) Either 2 or 10.

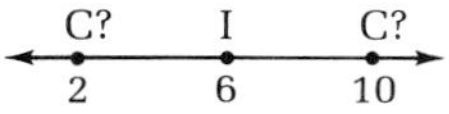

```
  C?    I    C?
  2     6    10
```

b) TI = IK.

Chapter 2, Lesson 2

SET I

1. a) They must be collinear.
b) JA + AR = JR.   c) R-A-J.
2. a) MU = UG.   b) MU + UG = MG.
3. a) A-C-N (or N-C-A).
b) AC + CN = AN.   c) No. AC = CN only if C is the midpoint of $\overline{AN}$.   4. a) $\overline{CU}$.
b) The midpoint of $\overline{PT}$.   5. a) V.
b) Since TA + VA = VT, A is between T and V.   6. a) Yes.   b) BO + OL = BL.
7. a) One.   b) Three. $\overline{TU}$, $\overline{UB}$, and $\overline{TB}$.

SET II

1. No, because it uses the word "between."
2. KE = KG − EG and EG = KG − KE.
3. a) PA + AL = PL.   b) AI + IL = AL.
c) Since (AI + IL) = AL, we can substitute (AI + IL) for AL in the first equation: PA + (AI + IL) = PL.

4. a) 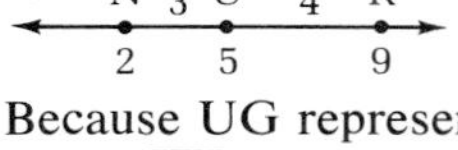
```
  N 3 U  4  R
  2   5     9
```
b) 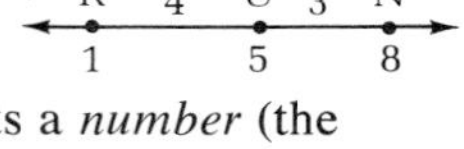
```
  R   4  U 3 N
  1      5   8
```

5. Because UG represents a *number* (the length of $\overline{UG}$). Numbers do not have midpoints.   6. a) The distance between two points is a *number*, not a line.   b) The positive difference of their coordinates.

Chapter 2, Lesson 3

SET I

1. a) Opposite rays.   b) No. An angle consists of two rays that do *not* lie on the same line.   2. a) K.   b) $\overrightarrow{KA}$, $\overrightarrow{KR}$.
c) No. Opposite rays have the same endpoint.   d) $\overline{AR}$.   3. ∡CAW, ∡WAC, ∡A.   4. a) A.   b) $\overrightarrow{AY}$ and $\overrightarrow{AP}$.
5. ∡BRY (or ∡YRB), ∡YRA (or ∡ARY), ∡BRA (or ∡ARB).

SET II

1. a) $\overrightarrow{LO}$.   b) $\overrightarrow{OL}$.   2. a) Correct.
b) Incorrect, because more than one angle has U for its vertex.   c) Correct.   d) Correct.
e) Incorrect. There is no angle in the figure with vertex G and sides $\overrightarrow{GU}$ and $\overrightarrow{GR}$.
3. a) Four.   b) Twelve.   4. a) True.
b) Not necessarily true.   c) True.   d) Not necessarily true.

Chapter 2, Lesson 4

SET I

1. Yes. Acute angles are "sharper" than obtuse angles.   2. a) Because its sides look longer.   b) She could explain that the size of an angle depends upon the "amount of opening" between its sides. Also that since the sides of an angle are rays, they are infinite and cannot be measured.   3. a) 100°, obtuse.   b) 30°, acute.   c) 90°, right.
d) 160°, obtuse.   5. a) 5,400 minutes.
b) To every angle there corresponds a unique number between 0 and 10,800.
c) 3,600.

SET II

Your measurements should be within 1° of these.
1. ∠G = 52°, ∠N = 87°, ∠U = 41°.
2. ∠B = 77°, ∠E = 94°, ∠A = 65°, ∠R = 124°.   3. ∠C = 107°, ∠A = 90°, ∠M = 138°, ∠E = 73°, ∠L = 132°.
4. a) 180°.   b) 360°.   c) 540°.   5. a) 720°.
b) 1,440°.   6. All three sides of the triangle have equal lengths.   7. The opposite sides of the figure are both equal in length and parallel.

Chapter 2, Lesson 5

1. If two angles are complementary, then the sum of their measures is 90°. If the sum of the measures of two angles is 90°, then they are complementary.   2. a) Because ∠E + ∠F + ∠G = 180°.   b) The definition of supplementary angles refers to just *two* angles, not three.   3. a) The supplement of a right angle is also a right angle.   b) The complement of an acute angle is also an acute angle.   c) An obtuse angle cannot have a complement.   4. a) 79°.   b) 89°59'.
c) 110°.   d) 179°58'50".   5. Acute angles.

SET II

1. a) ∠BOD = 80°.   b) ∠BOC = 40°.
c) 110.   d) ∠AOB = 50°.   e) 20.
2. a) The supplement is 105°; the complement is 15°.   b) The supplement is 179°; the complement is 89°.
c) 160° − 70° = 90°; 105° − 15° = 90°; 179° − 89° = 90°. The difference of the supplement and the complement in each case is the same: 90°.   d) If m = the measure of the acute angle, its supplement is $180 - m$ and its complement is $90 - m$.
$(180 - m) - (90 - m) = 180 - m - 90 + m = 90$.   3. If ∢X is the supplement of ∢Y, then ∠X + ∠Y = 180°.   If ∠X + ∠Y = 180°, then ∠X = 180° − ∠Y.   Therefore, if ∢X is the supplement of ∢Y, then ∠X = 180° − ∠Y.

Chapter 2, Lesson 6

SET I

1. a) $\overrightarrow{EL}$.   b) ∠BEL + ∠LEU = ∠BEU.
2. a)

b) $\overrightarrow{RO}$.   3. a) A linear pair.   b) They are supplementary.   c) No. We cannot write ∠OLG + ∠GLD = ∠OLD because OLD is not an angle.   4. a) ∢WHI and ∢IHT; ∢WHE and ∢EHT.   b) No.   c) $\overrightarrow{HT}$ is between $\overrightarrow{HE}$ and $\overrightarrow{HI}$.   5. ∢I, ∢INP, ∢PNK, ∢K, ∢IPK, ∢IPN, ∢NPK.
6. a) ∢BOW and ∢RON; ∢BOR and ∢WON.   b) ∢BOW and ∢WON; ∢WON

and ∢NOR; ∢NOR and ∢ROB; ∢ROB and ∢BOW.

SET II

1. Yes. (The converse of every definition is true.)   2. a) If two angles are supplementary, then they are a linear pair.
b) No.   3. ∠YRA = ∠GRA − ∠GRY.
4. a) ∠RAO + ∠OAN = ∠RAN.
b) ∠ORG + ∠GRA = ∠ORA.
c) ∠OGN − ∠OGR = ∠RGN.
d) ∠OEG + ∠GEA = 180°.
5. a) ∠BKL + ∠LKC = ∠BKC.
b) ∠LKA + ∠AKC = ∠LKC.

Chapter 2, Lesson 7

SET I

1. a) Two.   b) One.   c) None.
2. a) One.   b) An unlimited number.
3. a) An unlimited number.   b) Two.
c) One.

SET II

1. a) A line segment has exactly one midpoint.   b) The Unique Ray Corollary.
c) The Unique Point Corollary.   d) The Unique Ray Corollary.   e) The Unique Point Corollary.   f) Two points determine a line.   2. Suppose both $\overrightarrow{NA}$ and $\overrightarrow{NY}$ bisect ∢RNO. Then ∢RNO has two bisectors. But this contradicts the Angle Bisector Corollary. So both $\overrightarrow{NA}$ and $\overrightarrow{NY}$ do not bisect ∢RNO.
3. Suppose FL = FA. Then there are two different points on $\overrightarrow{FX}$ at the same distance from its endpoint. But this contradicts the Unique Point Corollary. So FL ≠ FA.

Chapter 2, Review

SET I

1. a) ∠A + ∠B = 90°.   b) ∠C < 90°.
c) ED = DF.   d) $h - g$.   e) I-J-K.
f) L-N-M.   g) O-Q-P or O-P-Q.   h) S-R-T.
2. a) Vertex: W; sides: $\overrightarrow{WO}$ and $\overrightarrow{WL}$.
b) If no other angle has the same vertex.
c) ∢W or ∢VWX.
3. a) ∠SNW + ∠WNA = ∠SNA.
b) ∠SNW = ∠WNA.   4. a) 36°.
b) 72°.   c) 108°.   5. a) Linear pair, supplementary.   b) Vertical angles.
6. a) 150°.   b) Right angles.   c) No.   d) No.
7. a) A line segment has exactly one midpoint.   b) The Unique Point Corollary.
c) Two points determine a line.

SET II

1. a) 48. b) −6. c) 35. d) 18. e) 35.
f) −52. 2. a) b) c)
d) e)

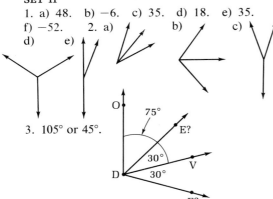

3. 105° or 45°.

4. J must be −10; Y could be either 2 or −8.

$$\underset{-10}{\text{J}} \qquad \underset{-3}{\text{A}} \qquad \underset{2}{\text{Y}}$$

or

$$\underset{-10\,-8}{\text{J Y}} \qquad \underset{-3}{\text{A}}$$

5. 4,900 minutes.

SET III

1. If the three rays are coplanar and
∠RCO + ∠OCW = ∠RCW. 2. If the
three points are collinear and
RO + OW = RW. 3. Because three
collinear points do not determine a plane.

Chapter 3, Lesson 1

SET I

1. a) If ∠E = ∠D, then ∠D = ∠E.
b) If JA = CK, then 2JA = 2CK. c) If
$DO^2 = UG^2$, then DO = UG. d) AL = AL.
e) If DA = VE, then DA − LE = VE − LE.
f) If MI = KE and KE = NT, then
MI = NT. 2. a) DA − CK = EL.
b) 3∠P + ∠T = 90°. c) $BR^2 − 1 = DY^2$.
d) $\dfrac{PH}{GA} = \dfrac{NE}{RY}$. 3. a) Subtraction.
b) Division. 4. a) Substitution.
b) Subtraction. c) Symmetric. d) Division.
5. a) Substitution. b) Subtraction.
c) Multiplication. d) Square roots.

SET II

1. a) Definition of betweenness of points.
b) Symmetric postulate. c) Square roots

postulate. d) Subtraction postulate.
2. a) Reflexive postulate. b) Definition of
midpoint. c) Transitive postulate.
d) Substitution postulate.
3. a) Multiplication postulate. b) Definition
of betweenness of rays. c) Addition
postulate. d) Definition of angle bisection.
4. a) Transitive postulate. b) Reflexive
postulate. c) Division postulate.
d) Substitution postulate.

Chapter 3, Lesson 2

SET I

1. a) Definition of betweenness of points.
b) Subtraction postulate. 2. a) Addition
postulate. b) Definition of betweenness of
points. c) Substitution or transitive postulate.
d) Definition of betweenness of points.
e) Transitive postulate. 3. a) The
midpoint of a line segment divides it into
segments that are half as long (Theorem 1).
b) Hypothesis. c) Substitution postulate.
d) Same as a. e) Substitution postulate.
4. a) Definition of betweenness of rays.
b) Substitution postulate. c) Hypothesis.
d) Same as b. e) Subtraction postulate.

SET II

1. $\overrightarrow{\text{MA}}$ bisects ∡WMR. (Hypothesis.)
2. ∠1 = ∠2. (The bisector of an angle divides
it into two equal angles.)
3. ∠1 + ∠2 = ∠WMR. (Definition of
betweenness of rays.)
4. ∠1 + ∠1 = ∠WMR. (Substitution
postulate.) 5. 2∠1 = ∠WMR. (Substitution
postulate.) 6. ∠1 = $\frac{1}{2}$∠WMR.
(Multiplication or division postulate.)
7. ∠2 = $\frac{1}{2}$∠WMR. (Substitution postulate.)

Chapter 3, Lesson 3

SET I

1. a) ∠O = ∠T. b) Supplements of the same
angle are equal. 2. a) If two angles are
equal, then they are vertical. b) No.
c) If two angles are not equal, then they are
not vertical. d) Yes. 3. a) ∠N = ∠R.
b) Complements of equal angles are equal.
4. a) ∠I = ∠N. b) Vertical angles are
equal. c) ∠I + ∠N = 90°. d) The sum of
the measures of two complementary angles
is 90°. e) ∠I = 45°, ∠N = 45°.

SET II

1. a) Vertical angles are equal. b) The bisector of an angle divides it into two equal angles. c) Transitive postulate.   2. a) A right angle has a measure of 90°.
b) Definition of betweenness of rays.
c) Substitution postulate. d) If the sum of the measures of two angles is 90°, they are complementary.   3. a) If two angles are complementary, the sum of their measures is 90°. b) Substitution postulate.
c) Substitution postulate. d) Subtraction postulate.

Chapter 3, Lesson 4

SET I

1. a) $\overleftrightarrow{IO} \perp \overleftrightarrow{WA}$. b) Because IO and WA represent numbers (distances in the figure). Numbers are not perpendicular. c) Yes. Two segments are perpendicular if they are contained in perpendicular lines.   2. *Proof.*
1. ∢N and ∢Y are right angles. (Hypothesis.)
2. ∠N = 90°, ∠Y = 90°. (A right angle has a measure of 90°.)   3. ∠N = ∠Y. (Substitution postulate.)   3. In the figure, the two right angles form a linear pair. This is not implied by Theorem 6.   4. b) Yes.
c) Symmetric. e) No.

SET II

1. The definition says only that two perpendicular lines form *one* right angle.
2. a) Because ∢1 and ∢3 are vertical angles, and vertical angles are equal. b) ∢1 and ∢2 are a linear pair, so they are also supplementary. c) Since ∠1 + ∠2 = 180°, and ∠1 = 90°, it follows that ∠2 = 90°. So ∢2 must be a right angle. d) ∢2 and ∢4 are vertical angles.   3. *Reasons.* 2. If two angles are a linear pair, then they are supplementary. 3. If two angles are supplementary, then the sum of their measures is 180°. 4. Hypothesis.
5. Substitution postulate. 6. Division postulate. 7. If an angle has a measure of 90°, then it is a right angle. 8. Substitution postulate.

Chapter 3, Lesson 5

SET I

1. Given.   2. If two angles have a common side and their other sides are opposite rays,

they are a linear pair.   3. If two angles are a linear pair, they are supplementary.
4. If two angles are supplementary, the sum of their measures is 180°.   5. Definition of betweenness of rays.   6. Substitution postulate.   7. Given.   8. If two rays are perpendicular, they form a right angle.
9. A right angle has a measure of 90°.
10. Substitution postulate. (Step 9 into step 6.)
11. Subtraction postulate.   12. If the sum of the measures of two angles is 90°, the angles are complementary.

Chapter 3, Review

SET I

1. a) ∠H = ∠E. Transitive. b) TR. Division.
c) HA = HT + TE. Substitution. d) ∠ETA. Symmetric.   2. a) Linear pair, supplementary. b) ∠JOR + ∠ROE = ∠JOE.
c. $\overrightarrow{OE}$ bisects ∢ROK. d) $\overrightarrow{OR} \perp \overrightarrow{OJ}$.
3. a) Vertical angles are equal. b) If two angles are supplementary, the sum of their measures is 180°. c) Substitution.
d) Division. e) An angle with a measure of 90° is a right angle. f) If two lines form a right angle, they are perpendicular.   4. If their measures are 90°, they are right angles and cannot be complementary. The sentence should say: "Two angles are complementary if *the sum* of their measures is 90°."
5. If two lines are perpendicular, they form four right angles. If the two angles in a linear pair are equal, then each is a right angle.

SET II

1. *Proof.* 1. $\overline{CU} \perp \overline{LB}$. (Given.) 2. ∢CUL and ∢CUB are right angles. (Perpendicular lines form right angles.) 3. ∠CUL = ∠CUB. (Any two right angles are equal.)   2. *Proof.*
1. ∢1 and ∢2 are a linear pair. (Given.)
2. ∢1 and ∢2 are supplementary. (The angles in a linear pair are supplementary.) 3. ∢J and ∢A are supplementary. (Given.)
4. ∠1 = ∠J. (Given.) 5. ∠2 = ∠A. (Supplements of equal angles are equal.)
3. *Proof.* 1. ∢KIG and ∢GIN are complementary. (Given.) 2. ∠KIG + ∠GIN = 90°. (If two angles are complementary, the sum of their measures is 90°.)
3. ∠KIN = ∠KIG + ∠GIN. (Betweenness of rays.) 4. ∠KIN = 90°. (Transitive.)

Chapter 3, Review (continued)

5. ⊾KIN is a right angle. (An angle with a measure of 90° is a right angle.)   6. $\overline{IK} \perp \overline{IN}$. (If two lines form a right angle, they are perpendicular.)

Chapter 4, Lesson 1

SET I

1. a) Isosceles.   b) $\overline{SA}$ and $\overline{SP}$.   c) $\overline{AP}$.
d) ⊾S.   e) ⊾A and ⊾P.   2. a) Scalene, obtuse.   b) $\overline{BO}$.   c) $\overline{BO}$ and $\overline{BA}$.   d) ⊾O.
e) ⊾O and ⊾A.   3. a) Right.
b) $\overline{VI}$ and $\overline{IE}$.   c) $\overline{VE}$.   d) Yes.   e) No.
4. a) True.   b) True.   c) True.   d) True.
e) False.   5. a) △CAO, △ABO, △ABR, △CBO, △RAO.   b) Three. (△CBO, △ABO, △ABR.)   c) Two. (△CAO, △RAO.)
d) None.

SET II

3. a) Yes.   b) Every equilateral triangle is isosceles since it has at least two equal sides.
c) No.

Chapter 4, Lesson 2

SET I

1. Correspondences a, b, and d.
2. a) A ↔ D, L ↔ N, M ↔ O.
b) ALM ↔ DNO.   c) LAM ↔ NDO.
d) ODN ↔ MAL.   3. a) CA = HE.
b) EW = AS.   c) SC = WH.   d) ∠C = ∠H.
e) ∠W = ∠S.   4. a) KO = KA, OL = AL, KL = KL, ∠1 = ∠2, ∠O = ∠A, ∠3 = ∠4.
b) KO = AK, OL = KL, KL = AL, ∠1 = ∠A, ∠O = ∠2, ∠3 = ∠4.
5. a) △PIS ≅ △HIO.   b) Two triangles congruent to a third triangle are congruent to each other.   6. a) Yes. Every triangle is congruent to itself; for example, △PEA ≅ △PEA.   b) Yes. If △PEA ≅ △NUT, then △NUT ≅ △PEA.

SET II

1. a) No.   b) Yes.   c) Congruent angles are angles that have equal measures.
d) ⊾I ≅ ⊾L iff ∠I = ∠L.   2. a) $\overline{HE}$ and $\overline{AZ}$, $\overline{HL}$ and $\overline{AL}$, $\overline{LE}$ and $\overline{LZ}$.   b) ⊾1 and ⊾6, ⊾2 and ⊾5, ⊾3 and ⊾4.   c) △EHL and △ZAL, △EHA and △ZAH, △EHZ and △ZAE.   3. a) That ∠1 = ∠6, ∠2 = ∠5, and EL = ZL.   b) Six.

Chapter 4, Lesson 3

SET I

1. a) $\overline{ON}$.   b) ⊾E.   c) ⊾O and ⊾E.
d) $\overline{NO}$ and $\overline{NE}$.   2. a) $\overline{WO}$.   b) ⊾W.
c) ⊾T and ⊾O.   d) $\overline{TW}$ and $\overline{TO}$.
3. a) No. The angle must be *included* by the sides.   b) No. We do not have an A.A.A. congruence postulate.   c) No. The sides are *not included* by the angles.   4. a) S.A.S.
b) S.S.S.   c) A.S.A.   d) No conclusion possible.   e) A.S.A.   f) S.S.S.   g) No conclusion possible.   h) No conclusion possible.   i) S.A.S.   j) No conclusion possible.

Chapter 4, Lesson 4

SET I

1. a) UI = TA.   b) ∠G = ∠R.   c) UI = TA and GI = TR.   2. a) Yes. Corresponding parts of congruent triangles are equal.
b) No. These are not corresponding parts of the triangles. (You can't tell this by looking at the triangles; you have to look at the congruence correspondence.)   c) Yes. These are corresponding angles of the triangles.
3. a) Given.   b) The midpoint of a line segment divides it into two equal segments.
c) Given.   d) Reflexive postulate.   e) S.S.S. postulate.   f) Corresponding parts of congruent triangles are equal.   g) Given.
h) If two angles have a common side and their other sides are opposite rays, they are a linear pair.   i) If two angles in a linear pair are equal, they are right angles.   j) If two lines form right angles, they are perpendicular.

Chapter 4, Lesson 5

SET I

1. a) Given.   b) If two angles are a linear pair, they are supplementary.   c) Supplements of equal angles are equal.   d) Given.
e) Reflexive postulate.   f) S.A.S. postulate.
g) Corresponding parts of congruent triangles are equal.   h) If an angle is divided into two equal angles, it is bisected.   2. a) Given.
b) Given.   c) Reflexive postulate.
d) A.S.A. postulate.   e) Corresponding parts of congruent triangles are equal.
f) Betweenness of points.   g) Subtraction postulate.   h) Substitution postulate.
i) Substitution postulate.

Chapter 4, Lesson 6

1. a) Hypothesis. b) If a triangle is equilateral, all its sides are equal. c) If two sides of a triangle are equal, the angles opposite them are equal. d) Same as step b.
e) Same as step c. f) Transitive postulate.
g) If a triangle has three equal angles, it is equiangular.   2. a) Hypothesis.
b) Reflexive postulate. c) Symmetric postulate. d) A.S.A. postulate.
e) Corresponding parts of congruent triangles are equal.   3. a) That △OCT is isosceles and that ∠C = ∠O. b) That △OCT is equiangular, that it is equilateral, and that OC = CT = OT. c) Yes. The contrapositive of Theorem 11 says: If two sides of a triangle are not equal, the angles opposite them are not equal.

Chapter 4, Lesson 7

SET I

1. a) AO = AB, because if two angles of a triangle are equal, the sides opposite them are equal. b) △OAG ≅ △BAR, because if two triangles are congruent to a third triangle, they are congruent to each other.
c) △GOA ≅ △RBA, because of the A.S.A. postulate. d) △GOB ≅ △RBO, because of the S.A.S. postulate. (OB = OB.)
2. a) Given. b) If two angles of a triangle are equal, the sides opposite them are equal. c) Given. d) Substitution postulate.
e) Reflexive postulate. f) Given. g) The midpoint of a line segment divides it into two equal segments. h) S.S.S. postulate.
i) Corresponding parts of congruent triangles are equal. j) If two angles in a linear pair are equal, they are right angles. k) If two lines form right angles, they are perpendicular.

Chapter 4, Lesson 8

SET I

1. a) Steps 4–6 would be eliminated, and in statements 7 and 8 ∡CEA and ∡CEB would be changed to ∡CDA and ∡CDB. b) No.
2. a) E is equidistant from O and V.
b) NO = NV. c) The perpendicular bisector of $\overline{OV}$.   3. By construction, KI = KL and IN = LN. Since KN = KN (reflexive), △KIN ≅ △KLN (S.S.S.). Therefore,

∠IKN = ∠LKN (corresponding parts of congruent triangles are equal). Since $\overrightarrow{KN}$ divides ∡IKL into two equal angles, it bisects it.   4. By construction, HE = RT, HA = RH', and EA = TH'. Therefore, △HAE ≅ △RH'T (S.S.S.). So ∠R = ∠H.

Chapter 4, Review

SET I

1. a) Scalene. b) $\overline{UM}$. c) △COL ≅ △ENI.
d) Yes, because we have proved that two triangles congruent to a third triangle are congruent to each other.   2. a) △OHC and △CRO; △OHI and △CRD; △ODR and △CIH. b) △OHC and △CRO. Either S.S.S. or S.A.S. (OC = OC). c) Either $\overline{OI}$ and $\overline{CD}$ (S.A.S.) or ∡OHI and ∡CRD (A.S.A.).
3. No, because ∡1 and ∡2 are not angles of the same triangle.   4. a) Yes; bisect the angle and then bisect each of the angles that result. b) No, because if this could be done, the angle could be trisected.

SET II

1. *Proof.* 1. DA = DI. (Given.) 2. ∠A = ∠1. (If two sides of a triangle are equal, the angles opposite them are equal.) 3. ∡1 and ∡2 are vertical angles. (Given.) 4. ∠1 = ∠2. (Vertical angles are equal.) 5. ∠A = ∠2. (Transitive postulate).   2. *Proof.* 1. $\overrightarrow{LT}$ bisects ∡ULP and $\overrightarrow{LP}$ bisects ∡TLI. (Given.)
2. ∠3 = ∠4 and ∠4 = ∠5. (If an angle is bisected, it is divided into two equal angles.)
3. ∠3 = ∠5. (Transitive postulate.)
4. ∠1 = ∠2. (Given.) 5. LT = LP. (If two angles of a triangle are equal, the sides opposite them are equal.) 6. LU = LI. (Given.) 7. △TUL ≅ △PIL. (S.A.S. postulate.)   3. *Proof.* 1. ∠1 = ∠2. (Given.) 2. AR = AR. (Reflexive postulate.)
3. ∠5 = ∠6. (Given.) 4. ∡3 and ∡5 are supplementary, ∡4 and ∡6 are supplementary. (If two angles are a linear pair, they are supplementary.) 5. ∠3 = ∠4. (Supplements of equal angles are equal.)
6. △ASR ≅ △AER. (A.S.A. postulate.)
7. SR = ER. (Corresponding parts of congruent triangles are equal.) 8. ∠7 = ∠8. (If two sides of a triangle are equal, the angles opposite them are equal.)   4. *Proof.*
1. ∠1 = ∠2. (Given.) 2. LF = FX. (If two angles of a triangle are equal, the sides

Chapter 4, Review *(continued)*

opposite them are equal.)   3. $\angle 3 = \angle 4$.
(Given.)   4. FA = FA. (Reflexive postulate.)
5. $\triangle$LFA $\cong$ $\triangle$XFA. (S.A.S. postulate.)
6. LA = XA. (Corresponding parts of
congruent triangles are equal.)   7. $\triangle$LAX is
isosceles. (If a triangle has two equal sides, it
is isosceles.)

Chapter 5, Lesson 1

SET I
1. a) Yes. b) Point E. c) Itself. d) Line ℓ
is the perpendicular bisector of $\overline{TR}$ and $\overline{IE}$.
2.

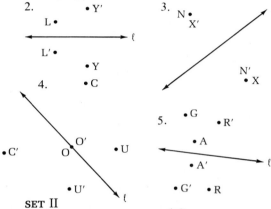

SET II
1. a) Points I and O. b) $\overleftrightarrow{IO}$ is the
perpendicular bisector of $\overline{LN}$. c) Because of
the answer to part b and the definition of the
reflection of a point through a line.

Chapter 5, Lesson 2

SET I
1. a) Reflection of a set of points through a
line preserves distance. b) Reflection of a set
of points through a line preserves collinearity.
c) Reflection of a set of points through a line
preserves angle measure. d) Reflection of a
set of points through a line preserves
betweenness. e) A triangle and its reflection
through a line are congruent.
2. a) Counterclockwise. b) $\triangle$NIT or $\triangle$ITN.
c) Clockwise, because reflecting a triangle
through a line reverses its orientation.

SET II
1. a) Line ℓ is the perpendicular bisector of
$\overline{BO}$ and $\overline{EN}$. b) S.A.S. c) Corresponding
parts of congruent triangles are equal.
d) $\angle 3$ and $\angle 4$ are the respective complements

of $\angle 1$ and $\angle 2$. e) Complements of equal
angles are equal. f) S.A.S.
g) Corresponding parts of congruent triangles
are equal.   2. a) Hypothesis. b) Definition
of betweenness of points. c) Reflection of a
set of points through a line preserves
collinearity. d) Reflection of a set of points
through a line preserves distance.
e) Substitution. f) Definition of
betweenness of points.   3. a) Hypothesis.
b) Two points determine a line.
c) Reflection of a set of points through a line
preserves distance. d) S.S.S.
e) Corresponding parts of congruent triangles
are equal.

Chapter 5, Lesson 3

SET I
1. Figures a and c.   2. Figure b.
4. a) One. b) Six. c) None.

SET II
1. a) By hypothesis, line ℓ bisects $\angle$BSC. If
an angle is bisected, it is divided into two
equal angles. b) S.A.S. (SI = SE, $\angle 1 = \angle 2$,
and ST = ST.) c) Corresponding parts of
congruent triangles are equal. d) They are
equidistant from its endpoints. e) In a
plane, two points equidistant from the
endpoints of a line segment determine the
perpendicular bisector of the line segment.
f) One point is the reflection through a line of
another point if the line is the perpendicular
bisector of the segment that joins the two
points.

Chapter 5, Lesson 4

SET I
1. a) A translation. b) A triangle and its
reflection through a line are congruent.
c) Two triangles congruent to a third triangle
are congruent to each other.   2. a) $\triangle$GHI.
b) $\triangle$JKL. c) $\triangle$MNO. d) $\triangle$DEF.
e) $\triangle$JKL. f) $\triangle$DEF.   3. a) A reflection.
b) A translation.

SET II
1. c) PR = AG and AG = UE (reflection of a
set of points through a line preserves
distance), so PR = UE (transitive). d) They
seem to be parallel.   2. d) They seem to be
parallel and to have equal lengths.   3. f) $2x$.
5. c) No.

Chapter 5, Lesson 5

SET I

1. a) A rotation.   b) The center of the rotation.   c) BN = EN and EN = RN (reflection of a set of points through a line preserves distance), so BN = EN = RN (transitive).   2. a) Six.   b) Two.
c) Three.

SET II

1. a) 40°.   d) ∠MNL = 80°. It is twice the answer to part a.   2. d) The circle that contains E also contains N and U; the circle that contains D also contains B and R; the circle that contains I also contains G.
e) Point H.   f) The center of the rotation.
3. b) 90°.   c) 180°.   4. c) 100°.
5. a) The magnitude of the rotation.
b) Reflection of a set of points through a line preserves angle measure.

Chapter 5, Lesson 6

SET I

1. a) I is the midpoint of each of these segments.   b) Yes.   c) Betweenness and distance.   2. a) A rotation with A as its center.   b) Yes.   c) No.
3. a)

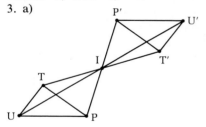

b) Yes.   c) No.   4. c) Yes.
5. a)

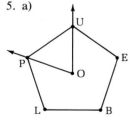

b) ∠POU = 72°.   c) 144°.   d) No.

SET II

1. a) Six.   b) Yes.   c) Three.   d) No.

Chapter 5, Review

SET I

1. a) The points on the line.   b) No points.
c) The center of the rotation.
2.   3.

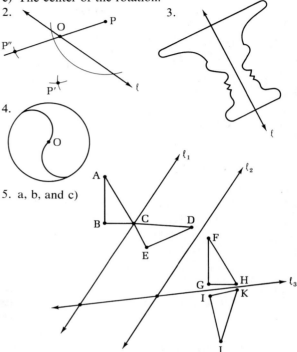

4.

5. a, b, and c)

d) A translation because it is the composite of two reflections through parallel lines.   e) A rotation because it is the composite of two reflections through intersecting lines.

6. Possible answers:

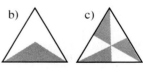

b)   c)

SET II

1. Translations; line symmetry.
2. Rotations and translations; point and rotational symmetry.   3. Translations and rotations; rotational symmetry.
4. Translations and reflections; no symmetry.

SET III

1. The following playing cards have point symmetry. Diamonds: all 13 cards. Clubs, hearts, and spades: 2, 4, 10, jack, queen, king.   2. No playing cards have line symmetry.

Chapter 6, Lesson 1

SET I

1. a) Hypothesis. b) Addition.
c) Hypothesis. d) Addition. e) Transitive.
2. a) Hypothesis. b) Subtraction.
c) Hypothesis. d) Substitution. e) Addition.
3. a) ∠S > ∠T. b) EA < AL.
4. a) Multiplication. b) Transitive. c) The "whole greater than its part" theorem.
d) Substitution.   5. a) Three possibilities postulate. b) Subtraction. c) Addition theorem. d) Transitive.

Chapter 6, Lesson 2

SET I

1. a) ∡TUN. b) ∡N and ∡A.
c) ∠TUN > ∠N and ∠TUN > ∠A.
2. a) Linear pair; supplementary. b) Vertical angles; equal. c) No, because it does not form a linear pair with an angle of the triangle.
d) Two. e) Six.   3. a) True. b) False.
c) True. d) True. e) True.   4. a) ∡5 and ∡O. b) ∡7. c) ∡1, ∡S, ∡6, and ∡O.

Chapter 6, Lesson 3

SET I

1. a) NU > NR. b) NU < NR and NU = NR. c) If two sides of a triangle are unequal, the angles opposite them are unequal and the larger angle is opposite the longer side.
d) The hypothesis that ∠R > ∠U. e) If two sides of a triangle are equal, the angles opposite them are equal. f) The hypothesis that ∠R > ∠U. g) NU > NR.
2. a) ∠Y = ∠L. b) FY < LY.
c) ∠F > ∠Y.   3. a) Yes, by A.S.A.
b) AL. c) WK. d) Yes, because △WAL ≅ △LKW. Corresponding parts of congruent triangles are equal.   4. a) RD.
b) DE. c) Yes, DE > RD. d) No.

Chapter 6, Lesson 4

SET I

1. a) Two points determine a line. b) The Unique Point Corollary. c) Two points determine a line. d) If two sides of a triangle are equal, the angles opposite them are equal.
e) Betweenness of rays. f) The "whole greater than its part" theorem. g) Substitution (steps d and f). h) If two angles of a triangle

are unequal, the sides opposite them are unequal and the longer side is opposite the larger angle. i) Betweenness of points.
j) Substitution (steps h and i). k) Substitution (steps b and j).   2. a) EA + AS < ES contradicts the Triangle Inequality Theorem.
b) Yes. c) No.   3. a) Yes. b) No.
c) No.

Chapter 6, Review

SET I

1. a) The "whole greater than its part" theorem. b) Transitive. c) An exterior angle of a triangle is greater than either remote interior angle. d) If two angles of a triangle are unequal, the sides opposite them are unequal and the longer side is opposite the larger angle. e) The sum of the lengths of two sides of a triangle is greater than the length of the third side. f) Addition theorem. g) Three possibilities postulate.
2. a) Given. b) An exterior angle of a triangle is greater than either remote interior angle.
c) Given. d) If two sides of a triangle are equal, the angles opposite them are equal.
e) Substitution (steps b and d). f) An exterior angle of a triangle forms a linear pair with an angle of the triangle. g) The two angles in a linear pair are supplementary.
h) The sum of the measures of two supplementary angles is 180°. i) Subtraction.
j) Substitution (steps e and i). k) Addition.
l) Division. m) Substitution (steps d and l).
n) An angle with a measure less than 90° is acute.   3. a) An exterior angle of a triangle is greater than either remote interior angle.
b) Substitution. c) Same as part a.
d) Transitive. e) The fact that ∠1 = ∠2.

SET II

1. *Proof.* 1. BT > BO. (Given.)
2. ∠1 > ∠3. (If two sides of a triangle are unequal, the angles opposite them are unequal, and the larger angle is opposite the longer side.) 3. ∠3 > ∠L. (An exterior angle of a triangle is greater than either remote interior angle.) 4. ∠1 > ∠L. (Transitive.)
2. *Proof.* 1. Draw NI. (Two points determine a line.) 2. NA + AI > NI. (The sum of the lengths of two sides of a triangle is greater than the length of the third side.)
3. NA + AI + IL > NI + IL. (Addition.)

4. NI + IL > NL. (Same as step 2.)
5. NA + AI + IL > NL. (Transitive.)
3. Reflect point C through any one of the four sides of the table and hit ball A so that it moves toward the reflection. After striking the cushion of the table, the ball will move in the direction of ball C.

SET III

If the roads are straight, we can conclude by the Triangle Inequality Theorem that WB < 4 + 3, BF < 3 + 5, and FW < 5 + 4. Hence WB + BF + FW < 7 + 8 + 9 = 24. Frodo's trip must have been less than 24 kilometers in length.

Chapter 7, Lesson 1

SET I

1. a) Line e.   b) $\angle 2$ and $\angle 6$, $\angle 4$ and $\angle 8$, and $\angle 3$ and $\angle 7$.   c) $\angle 3$ and $\angle 5$.   d) $\angle 3$ and $\angle 8$.   2. a) $t \parallel h$. If two lines form equal corresponding angles with a transversal, the lines are parallel.   b) $b \parallel e$. If two lines form equal alternate interior angle with a transversal, the lines are parallel.
c) No.   3. a) Hypothesis.   b) Vertical angles are equal.   c) Transitive.   d) If two lines form equal corresponding angles with a transversal, the lines are parallel.
4. a) Hypothesis.   b) If two angles form a linear pair, they are supplementary.
c) Supplements of the same angle are equal.
d) If two lines form equal alternate interior angles with a transversal, the lines are parallel.
5. a) Hypothesis.   b) Perpendicular lines form right angles.   c) Any two right angles are equal.   d) If two lines form equal corresponding angles with a transversal, the lines are parallel.

Chapter 7, Lesson 2

SET I

1. According to Theorem 24, there is exactly one line through O that is perpendicular to $\overleftrightarrow{HM}$. The figure shows two.
3. a) Hypothesis.   b) Perpendicular lines form right angles.   c) Any two right angles are equal.   d) An exterior angle of a triangle is greater than either remote interior angle.
e) Substitution.   f) If two angles of a triangle are unequal, then the sides opposite them are unequal and the longer side is opposite the larger angle.

SET II

2. a) Three.   b) No.   c) In a plane, two points equidistant from the endpoints of a line segment determine the perpendicular bisector of the line segment.   3. b) They are parallel to each other.   c) In a plane, two lines perpendicular to a third line are parallel to each other.   5. b) The three lines should appear to be concurrent.

Chapter 7, Lesson 3

SET I

1. The blanket is longer from side to side!
2. Through a point not on a line, there is at least one line parallel to the line. (Theorem 26).   Through a point not on a line, there is no more than one line parallel to the line. (The Parallel Postulate.)   3. a) Two points determine a line.   b) In a plane, through a point on a line there is exactly one line perpendicular to the line.   c) Through a point not on a line, there is exactly one line perpendicular to the line.   d) An angle has exactly one bisector.   e) The Unique Point Corollary.   f) Through a point not on a line, there is exactly one line parallel to the line.
g) A line segment has exactly one midpoint.
h) The Unique Ray Corollary.   4. a) Lines e and d intersect.   b) They are both parallel to line a.   c) The Parallel Postulate.

SET II

1. a) An angle is being constructed at P that is equal to the corresponding angle at Q.
b) If two lines form equal corresponding angles with a transversal, then the lines are parallel.

Chapter 7, Lesson 4

SET I

2. a) Hypothesis.   b) If two parallel lines are cut by a transversal, the corresponding angles are equal.   c) If two angles are a linear pair, they are supplementary.   d) If two angles are supplementary, the sum of their measures is 180°.   e) Substitution.   f) If the sum of the measures of two angles is 180°, they are supplementary.   3. a) Hypothesis.   b) If two parallel lines are cut by a transversal, the corresponding angles are equal.
c) Hypothesis.   d) Perpendicular lines form

Chapter 7, Lesson 4 *(continued)*

right angles.   e) A right angle has a measure of 90°.   f) Substitution.   g) An angle with a measure of 90° is a right angle.   h) If two lines form a right angle, they are perpendicular.

Chapter 7, Lesson 5

SET I

1. a) Hypothesis.   b) The sum of the measures of the angles of a triangle is 180°.
c) Substitution.   d) Hypothesis.
e) Substitution.   f) Subtraction.
2. a) Hypothesis.   b) The sum of the measures of the angles of a triangle is 180°.
c) An exterior angle of a triangle forms a linear pair with one of the angles of the triangle.   d) If two angles are a linear pair, then they are supplementary.   e) If two angles are supplementary, the sum of their measures is 180°.   f) Substitution.
g) Subtraction.   3. a) Since the sum of the measures of the angles of a triangle is 180° and the right angle of a right triangle has a measure of 90°, the sum of the measures of the other two angles must be 90°. Therefore, they are complementary.   b) Since an equilateral triangle is also equiangular, each of its angles has the same measure. One third of 180° is 60°.   4. a) 63°.   b) 150°.   c) 120°.
d) 45°.   e) 29°.   f) 80°.   g) 60°.   h) 88°.

Chapter 7, Lesson 6

SET I

1. a) NT = MG.   b) ∠N = ∠M.
2. *Proof.*   1. △GAR and △LIC with ∠G = ∠L and ∠R = ∠C. (Hypothesis.)   2. ∠A = ∠I. (If two angles of one triangle are equal to two angles of another, then the third pair of angles are equal.)   3. AR = IC. (Hypothesis.)
4. △GAR ≅ △LIC. (A.S.A.)   3. a) S.S.A. or A.S.S.   b) We know they are congruent by H.L.   c) Two sides and the angle opposite one of them in one triangle are equal to the corresponding parts of the other triangle, yet the triangles are obviously not congruent.

Chapter 7, Review

SET I

1. a) Hypothesis.   b) Through a point not on a line, there is exactly one line parallel to the line.   c) If two parallel lines are cut by a transversal, the alternate interior angles are equal.   d) If two parallel lines are cut by a transversal, the corresponding angles are equal.   e) Betweenness of rays.
f) Substitution.

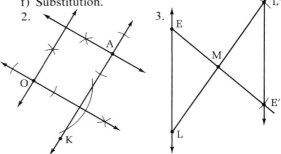

a) They seem to be parallel.   b) Since E′ and L′ are the respective reflections through M of E and L, EM = E′M and LM = L′M. Also, ∠EML = ∠E′ML′ since vertical angles are equal. Therefore, △EML ≅ △E′ML′ (S.A.S.) so ∠LEM = ∠L′E′M. It follows that $\overrightarrow{EL} \parallel \overrightarrow{E'L'}$ since they form equal alternate interior angles with a transversal.
4. a) No.   b) No.   c) Yes.   5. a) 115°.
b) 25°.   c) 40°.

SET II

1. a) $\overline{SP} \perp \overline{PU}$.   b) In a plane, if a line is perpendicular to one of two parallel lines, it is also perpendicular to the other.   c) ∡1 and ∡2 are complementary.   d) Since $\overline{ER} \perp \overline{PU}$, ∡PCR is a right angle and △PCR is a right triangle. The acute angles of a right triangle are complementary.   e) One point is the reflection through a line of another point if the line is the perpendicular bisector of the segment that joins the two points.
f) ∠2 = ∠3.   g) Reflection of a set of points through a line preserves angle measure.

The reasons have been omitted in the following proofs.
2. *Proof.*   1. AP > AL.   2. ∠ALP > ∠P.
3. $\overline{AL} \parallel \overline{ME}$.   4. ∠E = ∠ALP.
5. ∠E > ∠P.   3. *Proof.*   1. $\overrightarrow{MA}$ bisects ∡PML.   2. ∠PMA = ∠AML.   3. PM = PA.
4. ∠PMA = ∠PAM.   5. ∠PAM = ∠AML.
6. $\overline{PA} \parallel \overline{ML}$.   4. *Proof.*   1. $\overline{AN} \perp \overline{NE}$, $\overline{AS} \parallel \overline{NE}$.   2. $\overline{AN} \perp \overline{AS}$.   3. ∡SAN and ∡ANE are right angles.   4. ∠SAN = ∠ANE.
5. ∠S = ∠E.   6. AN = AN.
7. △SAN ≅ △ENA.

Chapter 8, Lesson 1

SET I

1. a) Three of the points are collinear. b) Two segments intersect in more than their endpoints. c) The four points are not coplanar.   2. a) 58°. b) 116°. c) 143°. 3. a) I and IV. b) III. c) IV. d) II. e) III.

Chapter 8, Lesson 2

SET I

1. a) Hypothesis. b) In a plane, two lines perpendicular to a third line are parallel to each other. c) If both pairs of opposite sides of a quadrilateral are parallel, it is a parallelogram. d) The opposite sides of a parallelogram are equal.   2. a) Hypothesis. b) The opposite sides of a parallelogram are parallel. c) If two parallel lines are cut by a transversal, the interior angles on the same side of the transversal are supplementary. 3. a) Hypothesis. b) The consecutive angles of a parallelogram are supplementary. c) Supplements of the same angle are equal. 4. a) Hypothesis. b) The opposite sides of a parallelogram are parallel. c) If two parallel lines are cut by a transversal, the alternate interior angles are equal. d) The opposite sides of a parallelogram are equal. e) A.S.A. f) Corresponding parts of congruent triangles are equal. g) If a segment is divided into two equal parts, it is bisected.   5. a) GE. b) ∠AGE. c) ∢ADE and ∢AGE. d) AS.

Chapter 8, Lesson 3

SET I

1. a) S.S.S. b) If two lines form equal alternate interior angles with a transversal, the lines are parallel. c) If both pairs of opposite sides of a quadrilateral are parallel, then the quadrilateral is a parallelogram. 2. a) If two parallel lines are cut by a transversal, the alternate interior angles are equal. b) S.A.S. c) If both pairs of opposite sides of a quadrilateral are equal, then the quadrilateral is a parallelogram.   3. a) The sum of the measures of the angles of a quadrilateral is 360°. b) Substitution. c) Division. d) If two lines form supplementary interior angles on the same

side of a transversal, then the lines are parallel. e) Substitution.   f) If both pairs of opposite sides of a quadrilateral are parallel, then the quadrilateral is a parallelogram. 4. a) S.A.S. b) If two lines form equal alternate interior angles with a transversal, the lines are parallel. c) If two sides of a quadrilateral are both parallel and equal, then the quadrilateral is a parallelogram. 5. Show that both pairs of opposite sides are equal; both pairs of opposite angles are equal; one pair of opposite sides is both parallel and equal; the diagonals bisect each other.

Chapter 8, Lesson 4

SET I

1. a) A rhombus is an equilateral quadrilateral (or, all four sides of a rhombus are equal). b) If both pairs of opposite sides of a quadrilateral are equal, the quadrilateral is a parallelogram.   2. a) Points R and S are equidistant from F and O. b) $\overline{RS} \perp \overline{FO}$. c) In a plane, two points equidistant from the endpoints of a line segment determine the perpendicular bisector of the line segment. 3. a) Isosceles triangles. Two of their sides are equal because they are also sides of the rhombus. b) Right triangles. Since the diagonals of a rhombus are perpendicular to each other, they form right angles. c) Two. d) Yes.   4. a) False. b) False. c) True. d) True. e) False. f) False.

Chapter 8, Lesson 5

SET I

1. *Proof.* 1. RUTH is a rectangle. (Hypothesis.) 2. ∠R = ∠T and ∠H = ∠U. (If a quadrilateral is a rectangle, it is equiangular.) 3. RUTH is a parallelogram. (If both pairs of opposite angles of a quadrilateral are equal, then the quadrilateral is a parallelogram.) 2. a) All rectangles are parallelograms. b) The opposite sides of a parallelogram are equal. c) All four angles of a rectangle are right angles. d) S.A.S.   3. a) Equal, perpendicular. b) Equal, right angles, supplementary. c) Parallelograms, quadrilaterals. d) Rectangles, rhombuses, parallelograms, quadrilaterals. e) Equal, bisectors of each other. f) Equal, perpendicular, bisectors of each other.

Chapter 8, Lesson 5 (continued)

4. a)

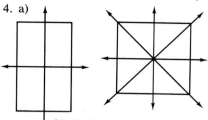

b) Both figures have point symmetry.

3. If the diagonals of a parallelogram are equal, then the parallelogram is a rectangle.

Chapter 8, Lesson 6

SET I

1. a) $\overline{FD} \parallel \overline{OR}$. b) FO = DR.
c) $\angle O = \angle R$. d) $\angle F$ and $\angle O$ are supplementary.   2. a) If the base angles of a trapezoid are equal, then its legs are equal.
b) Hypothesis: BUIK is a trapezoid; $\angle K = \angle I$. Conclusion: BK = UI.
c) Through a point not on a line, there is exactly one line parallel to the line. d) If two parallel lines are cut by a transversal, the corresponding angles are equal. e) Transitive.
f) If two angles of a triangle are equal, the sides opposite them are equal. g) If both pairs of opposite sides of a quadrilateral are parallel, it is a parallelogram. h) The opposite sides of a parallelogram are equal.
i) Substitution.

Chapter 8, Lesson 7

SET I

1. a) $\overline{MI} \parallel \overline{HL}$. b) $\angle IMU = \angle H$.
c) HE = EL. d) ME = $\frac{1}{2}$UL. e) MULE is a trapezoid. f) MILE is a parallelogram.
2. a) Given. b) Two points determine a line.
c) A line segment that joins the midpoints of two sides of a triangle is a midsegment of the triangle. d) A midsegment of a triangle is parallel to the third side. e) Given. f) If two lines are parallel, every perpendicular segment joining one line to the other has the same length.   3. a) A line segment has exactly one midpoint. b) A midsegment of a triangle is parallel to the third side.
c) Through a point not on a line, there is exactly one line parallel to the line.

SET II

1. a) A parallelogram. b) A midsegment of a triangle is parallel to the third side. c) In a plane, two lines parallel to a third line are parallel to each other. d) A midsegment of a triangle is half as long as the third side.
e) Substitution. f) If two sides of a quadrilateral are both parallel and equal, then the quadrilateral is a parallelogram.
2. a) Again quadrilateral HOIE seems to be a parallelogram. b) Yes.

Chapter 8, Review

SET I

1. a) Parallelograms, rectangles, rhombuses, squares, trapezoids. b) All of them.
c) Parallelograms (and hence, rectangles, rhombuses, and squares). d) Rhombuses and squares. e) Rectangles and squares.
2. a) The consecutive angles of a parallelogram are supplementary. b) If an angle is bisected, it is divided into angles half as large. c) Multiplication (or division).
d) Substitution. e) $\angle 3 = 90°$. f) If two lines form a right angle, they are perpendicular.
3. a) A line segment has exactly one midpoint. b) GM = PS (the legs of an isosceles trapezoid are equal), $\angle M = \angle S$ (the base angles of an isosceles trapezoid are equal), and MU = SU (U is the midpoint of $\overline{MS}$). So $\triangle GMU \cong \triangle PSU$ (S.A.S.). c) In a plane, two points equidistant from the endpoints of a line segment determine the perpendicular bisector of the line segment.
d) In a plane, if a line is perpendicular to one of two parallel lines, it is also perpendicular to the other. e) One point is the reflection through a line of another point if the line is the perpendicular bisector of the segment that joins the two points. f) A set of points has line symmetry if there is a line such that the reflection through the line of each point of the set is also a point of the set.

SET II

1. If MICA is a kite, MI = MA and CI = CA (a kite is a quadrilateral that has two pairs of equal consecutive sides with no side common to both pairs). Hence M and C are equidistant from I and A. Therefore, $\overline{MC} \perp \overline{IA}$ (in a plane, two points equidistant from the

endpoints of a line segment determine the perpendicular bisector of the line segment).

The reasons have been omitted in the following proofs.
2. *Proof.* 1. $\overline{UR}$ and $\overline{RZ}$ are midsegments of $\triangle AQT$. 2. $\overline{UR} \parallel \overline{QT}$ and $\overline{RZ} \parallel \overline{AQ}$. 3. $\angle A = \angle ZRT$ and $\angle URA = \angle T$. 4. R is the midpoint of $\overline{AT}$. 5. $AR = RT$. 6. $\triangle AUR \cong \triangle RZT$. 3. *Proof.* 1. TALC is a parallelogram. 2. $\angle T = \angle L$. 3. $\angle 1 = \angle 2$. 4. $AC = AC$. 5. $\triangle TAC \cong \triangle LAC$. 6. $TC = CL$. 7. $TA = CL$ and $TC = AL$. 8. $TA = CL = TC = AL$. 9. TALC is a rhombus. 4. Suppose $\overline{RE}$ and $\overline{AL}$ do bisect each other. Then quadrilateral RAEL is a parallelogram (if the diagonals of a quadrilateral bisect each other, the quadrilateral is a parallelogram). Hence $\overleftrightarrow{RA} \parallel \overleftrightarrow{LE}$ (the opposite sides of a parallelogram are parallel). But $\overleftrightarrow{RA}$ and $\overleftrightarrow{LE}$ intersect at M. So the assumption that $\overline{RE}$ and $\overline{AL}$ can bisect each other is false.

Chapter 9, Lesson 1
SET I
2. a) $\alpha\triangle PRE = \alpha\triangle NIC$ because congruent triangles have equal areas. b) $\alpha PNCE = \alpha\triangle PRE + \alpha RICE + \alpha\triangle NIC$ because of the Area Addition Postulate. 3. a) No. b) No. 4. a) Given. b) The opposite sides of a parallelogram are equal. c) The opposite angles of a parallelogram are equal. d) S.A.S. e) Congruent triangles have equal areas. f) Area Addition Postulate. g) Substitution. h) Division.

SET II
1. a) Two. b) Yes. c) X-P-Y.

Chapter 9, Lesson 2
SET I
1. a)

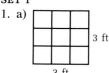

3 ft

3 ft

b) 1 square foot is the area of a square whose sides are each 1 foot long. Since 1 foot = 12 inches, 1 square foot = $(12 \text{ inches})^2 = 144$ square inches. 2. a) 17,000 square feet. b) 2,448,000 square inches. c) $1,888\frac{8}{9}$

square yards. 3. a) Because $7^2 = 49$, and 50 is slightly larger than 49. b) 49.7025 square inches. 4. a) 1.41 centimeters. b) 4.48 centimeters. c) 14.1 centimeters. d) 44.8 centimeters. 5. a) The perimeter of the larger is twice that of the smaller. b) The area of the larger is 4 times that of the smaller.

SET II
1. a) 25 square feet. b) 16 square feet. c) 21 square feet. d) 9 square feet. 2. 24 square units. 3. 64 square units. 4. 52 square units. 5. 10,028 square feet.

Chapter 9, Lesson 3
SET I
1. a) Through a point not on a line, there is exactly one line parallel to the line. b) A parallelogram. c) bh. d) $\triangle YTC \cong \triangle HCT$. e) Area Addition Postulate. f) Division. 2. a) The area of a triangle is half the product of the lengths of any base and corresponding altitude. b) Substitution. 3. a) $\alpha\triangle DIN = \frac{1}{2}ah$ and $\alpha\triangle GHY = \frac{1}{2}bh$. b) Division. c) $\dfrac{\alpha\triangle DIN}{\alpha\triangle GHY} = \dfrac{a}{b}$. 4. a) If two lines are parallel, every perpendicular segment joining one line to the other has the same length. b) Triangles with equal bases and equal altitudes have equal areas. 5. a) $\alpha\triangle KHE = 2\alpha\triangle THC$. b) If two triangles have equal altitudes, then the ratio of their areas is equal to the ratio of the lengths of their bases.

SET II
1. a) A rectangle. b) The area of a rectangle is the product of the lengths of its base and altitude. c) $\alpha ABCD = \alpha ABEF$ because they are both equal to $\alpha 1 + \alpha 2$. d) The part numbered 2 is a triangle instead of a trapezoid. e) $\alpha ABCD = \alpha BGFE$ because they are both equal to $\alpha 1 + \alpha 2 + \alpha 3$.

Chapter 9, Lesson 4
SET I
1. a) 30 square units. b) 77 square units. c) 22 square units. d) 9 square units. 2. a) 30 square units. b) 30 square units. 3. a) The length of the altitude to its base.

SET II

1. 90 square units.   2. 61 square units.
3. 35 square units.   4. a) LA = x and
AT = $2y$.   b) $\alpha\triangle PLO = \frac{1}{2}xy$; $\alpha ATOL = \frac{3}{2}xy$.
c) $\alpha\triangle PLO = \frac{1}{3}\alpha ATOL$.
5. b) $\alpha\triangle BAO = \frac{1}{2}xy$; $\alpha BACN = 2xy$.
c) $\alpha\triangle BAO = \frac{1}{4}\alpha BACN$.   6. b) LE′L′E
seems to be a parallelogram.   d) OE′ = b and
KL′ = a.   e) $\alpha LE′L′E = h(a+b)$.   f) $\alpha LOKE$
$= \frac{1}{2}\alpha LE′L′E$.   g) $\alpha LOKE = \frac{1}{2}h(a+b)$.

Chapter 9, Lesson 5

SET I

1. a) 13.   b) 10.   c) 17.   3. a) 30.
b) $2\sqrt{10}$.   c) $4\sqrt{3}$.   d) $10\sqrt{2}$.
4. a) In a right triangle, the square of the
hypotenuse is equal to the sum of the squares
of the legs.   b) Hypothesis.   c) Substitution.
d) Square roots postulate.   e) S.S.S.
f) ∡C is a right angle since $\angle C = \angle T$.
Therefore, $\triangle ROC$ is a right triangle.
5. a) It is illogical because the words
"hypotenuse" and "legs" in the hypothesis
imply that the triangle is a right triangle.
b) The longest side of the triangle (the one
whose square is equal to the sum of the
squares of the other two sides) is the
hypotenuse.   6. Sets b and c.

SET II

1. a) $\alpha\triangle ARL = \alpha\triangle LFE = \frac{1}{2}ab$;
$\alpha\triangle RLE = \frac{1}{2}c^2$.   b) A trapezoid.
c) $\alpha AREF = \frac{1}{2}(a+b)^2$.
d) $\alpha AREF = \alpha\triangle ARL + \alpha\triangle LFE + \alpha\triangle RLE$.
2. a) $b-a$.   b) $(b-a)^2 = b^2 - 2ab + a^2$.
c) $4(\frac{1}{2}ab) = 2\ ab$.   3. a) Line symmetry.
b) One seems to be the reflection of the other
through the line. They also seem to be
congruent and to have the same area.
c) One of the cut-out quadrilaterals would be
turned over.   d) Point symmetry.   e) The
squares on the legs of the original right triangle
and two congruent right triangles.   f) The
square on the hypotenuse of the original right
triangle and two more right triangles congruent
to the previous pair.

Chapter 9, Lesson 6

SET I

1. a) 6 square units.   b) 6 square units.
2. a) 60 square units.   b) $30\sqrt{5}$ square units.

c) $14\sqrt{26}$ square units.   4. a) $\sqrt{3}$ square
units.   b) $25\sqrt{3}$ square units.   c) $5\sqrt{3}$
square units.

SET II

1. $12\sqrt{3}$ square units.   2. $24 + 2\sqrt{66}$ square
units.   3. 12.

Chapter 9, Review

SET I

1. a) 625 square units.   b) 15 square units.
c) $16\sqrt{3}$ square units.   2. a) Yes, because
$29^2 = 20^2 + 21^2$.   b) ∡E and ∡U are
complementary, and $\angle E > \angle U$.
3. a) 40.   b) 180 square units.
4. a) Although 1 yard = 3 feet, 1 square
yard = 9 square feet. Ollie should have
divided 288 by 9. Furthermore, he should
have *multiplied* the result by 10 to find the
cost.   b) $320.

SET II

1. 46 square units.   2. 11 square units.
3. $100 - 25\sqrt{3}$ square units.   4. 12 square
units.   5. 15 square units.   6. 86 square
units.   7. Since N is the midpoint of $\overline{EA}$,
EN = NA. Also, since CRAE is a
parallelogram, $\overleftrightarrow{CR} \parallel \overleftrightarrow{EA}$. If altitudes were
drawn from points C and R to $\overleftrightarrow{EA}$, they would
have equal lengths (if two lines are parallel,
every perpendicular segment joining one line
to the other has the same length). Therefore,
$\alpha\triangle CEN = \alpha\triangle RNA$ because triangles with
equal bases and equal altitudes have equal
areas.   8. Since HE = HO and RE = RO,
points H and R are equidistant from points
E and O. Hence, $\overline{HR} \perp \overline{EO}$ (in a plane, two
points equidistant from the endpoints of a line
segment determine the perpendicular bisector
of the line segment). Therefore, $\overline{HN}$ is an
altitude of $\triangle EHO$ and $\overline{RN}$ is an altitude of
$\triangle ERO$. $\alpha\triangle EHO = \frac{1}{2}EO \cdot HN$ and
$\alpha\triangle ERO = \frac{1}{2}EO \cdot RN$. $\alpha HERO = \alpha\triangle EHO$
$+ \alpha\triangle ERO = \frac{1}{2}EO \cdot HN + \frac{1}{2}EO \cdot RN$
$= \frac{1}{2}EO(HN + RN) = \frac{1}{2}HR \cdot EO$.
9. a) $\alpha MBAN = \alpha DBEF$. Since congruent
triangles have equal areas, $\alpha\triangle MBC$
$= \alpha\triangle ABE$; $\alpha MBAN = 2\alpha\triangle MBC$
$= 2\alpha\triangle ABE = \alpha DBEF$.   b) Adding the two
equations, we get $\alpha HACL + \alpha MBAN$
$= \alpha DCGF + \alpha DBEF$. By the Area Addition
Postulate, $\alpha CBEG = \alpha DCGF + \alpha DBEF$.
Substituting, $\alpha HACL + \alpha MBAN = \alpha CBEG$.

Chapter 10, Lesson 1

1. a) $\frac{7}{11}$ and $\frac{21}{33}$. b) 21. c) 11 and 21.
d) $11 \cdot 21 = 231; 7 \cdot 33 = 231.$ 2. a) 2.7.
b) 9. c) 6. d) -1. 3. a) 3.14. b) 3.14.
c) Suppose $\frac{22}{7} = \frac{355}{113}$. Then, by the means-
extremes theorem, $22 \cdot 113 = 7 \cdot 355$. But
$22 \cdot 113 = 2486$ and $7 \cdot 355 = 2485$.
Therefore, $\frac{22}{7} \neq \frac{355}{113}$. 4. a) No. b) 48
inches; $\frac{48}{16}$ or $\frac{3}{1}$. c) $1\frac{1}{3}$ feet; $\frac{4}{1\frac{1}{3}}$ or $\frac{3}{1}$. d) No.

SET II
1. Multiplication. 2. *Proof.* 1. $\frac{a}{b} = \frac{c}{d}$.
(Hypothesis.) 2. $bc = ad$. (The means-
extremes theorem.) 3. $\frac{bc}{ac} = \frac{ad}{ac}$. (Division.)
4. $\frac{b}{a} = \frac{d}{c}$. (Substitution.)

Chapter 10, Lesson 2

SET I
1. a) Hypothesis. b) Definition of mean
proportional. c) The means-extremes
theorem. d) The square roots postulate.
2. a) 10. b) 6. c) $7\sqrt{3}$. 3. a) Because
the extremes are the same number.
b) Symmetric. c) The mean proportional
between them. 4. a) $x = 12, y = \frac{1}{3}$.
b) $x = 10, y = 2$.

SET II
1. a) Means-extremes. b) Upsidedownable.
c) Interchangeable. d) Denominator addition.
2. a) $\frac{3}{7}$. b) $\frac{n}{9}$. c) $\frac{6}{5}$. d) $\frac{5}{6}$. e) $\sqrt{13}$.
3. a) Hypothesis. b) Transitive.
c) Multiplication. d) Addition.
e) Substitution. f) Division.
4. Substitution. 5. a) $\frac{x}{y}$ $\left(\text{or } \frac{1}{2}\right)$.
b) $\frac{a+b+c+d}{14} = \frac{a}{2}$.

Chapter 10, Lesson 3

SET I
1. a) $\frac{WN}{NC}$. b) $\frac{RO}{CO}$. c) $\frac{UB}{RU}$. d) $\frac{RE}{RL}$.

2. a) No; he should have written $\frac{x}{10 - x} = \frac{4}{8}$.
b) No. c) $\frac{x}{10} = \frac{4}{12}$. 3. a) 2.4. b) 10.
c) 8. d) 7.5.

SET II
1. a) Hypothesis. b) If a line parallel to one
side of a triangle intersects the other sides in
different points, it divides the sides in the same
ratio. c) Denominator addition theorem.
d) Betweenness of points. e) Substitution.
f) Upsidedownable theorem.
2. *Proof.* 1. $\frac{PO}{OU} = \frac{PD}{DN}$. (Step b.)
2. $\frac{OU}{PO} = \frac{DN}{PD}$. (Upsidedownable theorem.)
3. $\frac{OU + PO}{PO} = \frac{DN + PD}{PD}$. (Denominator addition theorem.) 4. $\frac{PU}{PO} = \frac{PN}{PD}$. (Substitution.)
5. $\frac{PO}{PU} = \frac{PD}{PN}$. (Upsidedownable theorem.)

Chapter 10, Lesson 4

SET I
1. a) Interchangeable. b) Upsidedownable.
c) Denominator addition. d) Equal ratios.
2. a) IG. b) N. c) NI. d) Y.
3. $\frac{CO}{RA} = \frac{ON}{AD} = \frac{CN}{RD}; \frac{12}{8} = \frac{6}{x} = \frac{8}{y}; x = 4, y = 5\frac{1}{3}$.

SET II
1. a) Given. b) A midsegment of a triangle
is half as long as the third side. c) A
midpoint divides a segment into segments
half as long. d) Division. e) Substitution.
f) A midsegment of a triangle is parallel to the
third side. g) If two parallel lines are cut by
a transversal, the corresponding angles are
equal. h) Reflexive. i) If the corresponding
sides of two triangles are proportional and the
corresponding angles are equal, the triangles
are similar.

Chapter 10, Lesson 5

SET I
1. a) Hypothesis. b) Corresponding angles
of similar triangles are equal. c) Substitution.
d) A.A. 2. a) $\angle K = \angle NRO$ (they are
right angles) and $\angle O = \angle O$. b) RO.
c) 18 feet. 3. a) $\angle FLI = \angle TNI$ (they are

Chapter 10, Lesson 5 (continued)

right angles) and $\angle FIL = \angle NIT$ (they are vertical angles.)   b) 15 feet.

Chapter 10, Lesson 6

SET I

1. a) $6\frac{2}{3}$.   b) $7\frac{1}{2}$.   2. a) Statement 6; $\angle BAC = \angle EDF$.   b) $\angle BAG$ and $\angle BAC$ are supplementary, $\angle EDH$ and $\angle EDF$ are supplementary. (If two angles form a linear pair, they are supplementary.) $\angle BAG = \angle EDH$. (Supplements of equal angles are equal.)

Chapter 10, Lesson 7

SET I

1. Proportions a, c, and d.   2. a) 5. b) $13\frac{1}{2}$.   c) $2\frac{6}{7}$.   d) $6\frac{2}{3}$.   3. a) Two points determine a line.   b) Through a point not on a line, there is exactly one line parallel to the line.   c) If a line parallel to one side of a triangle intersects the other two sides in different points, it divides the sides in the same ratio.   d) Hypothesis.   e) Substitution. f) Upsidedownable theorem. g) Multiplication.   h) If two sides of a triangle are equal, the angles opposite them are equal.   i) If two parallel lines are cut by a transversal, the corresponding angles are equal.   j) If two parallel lines are cut by a transversal, the alternate interior angles are equal.   k) Substitution.   l) If an angle is divided into two equal angles, it is bisected.

Chapter 10, Lesson 8

SET I

1. a) Hypothesis.   b) Corresponding sides of similar triangles are proportional.   c) Equal ratios theorem.   d) The perimeter of a triangle is the sum of the lengths of its sides. e) Substitution.   2. a) Hypothesis. b) Through a point not on a line, there is exactly one line perpendicular to the line. c) A perpendicular line segment from one vertex of a triangle to the line of the opposite side is an altitude of the triangle. d) Corresponding altitudes of similar triangles have the same ratio as the corresponding sides.   e) The area of a triangle is half the product of the lengths of any base and corresponding altitude.   f) Division.

g) Substitution.   3. a) $\frac{2}{3}$. b) $\rho\triangle TJA = 48$ and $\rho\triangle DER = 72$; $\frac{\rho\triangle TJA}{\rho\triangle DER} = \frac{48}{72} = \frac{2}{3}$.   c) $\alpha\triangle TJA = 84$ and $\alpha\triangle DER = 189$; $\frac{\alpha\triangle TJA}{\alpha\triangle DER} = \frac{84}{189} = \frac{4}{9} = \left(\frac{2}{3}\right)^2$.

SET II

1. a) $\frac{25}{49}$.   b) $\frac{1}{5}$.   c) 24.   d) 72.

Chapter 10, Review

SET I

1. Conditions a, b, and c.   2. a) 12. b) 49.   c) No.   3. a) $\frac{4}{5}$.   b) 30.   c) 50.

SET II

The reasons have been omitted in the following proofs.
1. *Proof.* $\overleftrightarrow{EH} \parallel \overleftrightarrow{AC}$.   2. $\angle PEH = \angle PAC$ and $\angle PHE = \angle PCA$.   3. $\triangle PEH \sim \triangle PAC$.
4. $\frac{EH}{AC} = \frac{PH}{PC}$.   2. *Proof.*   1. Draw $\overline{AI}$.
2. $\overline{AP} \parallel \overline{OR} \parallel \overline{CI}$.
3. $\frac{AO}{OC} = \frac{AT}{TI}$ and $\frac{AT}{TI} = \frac{PR}{RI}$.
4. $\frac{AO}{OC} = \frac{PR}{RI}$.   5. $\frac{AO}{PR} = \frac{OC}{RI}$.
3. *Proof.*   1. QUIN is a rectangle.   2. QUIN is a parallelogram.   3. $\overline{QN} \parallel \overline{UI}$.   4. $\angle NCE = \angle EIU$ and $\angle CNE = \angle EUI$.
5. $\triangle CEN \sim \triangle IEU$.   6. $\frac{CE}{IE} = \frac{EN}{EU}$.
7. $CE \cdot EU = NE \cdot EI$.

SET III

The area relationship is also true for other sets of similar figures.

Chapter 11, Lesson 1

SET I

1. a) $\overline{OO'} \perp \overline{DV}$.   b) The altitude to its hypotenuse.   c) Their projections on the hypotenuse.
2. a) $\triangle IAD \sim \triangle DAL \sim \triangle IDL$.
b) $\frac{IA}{DA} = \frac{AD}{AL}$.   c) $\frac{IL}{ID} = \frac{ID}{IA}$.   d) $\frac{IL}{DL} = \frac{DL}{AL}$.
e. Proportion b illustrates Corollary 1; proportions c and d illustrate Corollary 2.
3. a) ZT and TS.   b) $\overline{ZT}$.   c) ZS and ZT. d) ES.   4. a) 4.   b) 6.   c) $2\sqrt{10}$.   d) 1.6.

1. a) Because it is determined by lines that form right angles. c) It is parallel to the line. d) It is perpendicular to the line.
2. a) Through a point not on a line, there is exactly one line parallel to the line. b) In a plane, if a line is perpendicular to one of two parallel lines, it is perpendicular to the other. c) The perpendicular segment from a point to a line is the shortest segment joining them. d) In a plane, two lines perpendicular to a third line are parallel to each other. e) If two lines are parallel, every perpendicular segment joining one line to the other has the same length. f) Substitution.

Chapter 11, Lesson 2

SET I
1. $7^2 + 24^2 = 625, 25^2 = 625.$ 2. Yes.
3. There are no such numbers in the table.
4. *Group 1*: 3-4-5, 6-8-10, 9-12-15, 12-16-20, 15-20-25, 18-24-30, 21-28-35, 24-32-40, 27-36-45, 30-40-50. *Group 2*: 5-12-13, 10-24-26, 15-36-39. *Group 3*: 7-24-25, 14-48-50. *Group 4*: 8-15-17, 16-30-34. *Group 5*: 9-40-41. *Group 6*: 12-35-37. *Group 7*: 20-21-29. 5. a) Hypothesis.
b) The square of the hypotenuse of a right triangle is equal to the sum of the squares of the legs. c) Multiplication. d) Substitution. e) If the square of one side of a triangle is equal to the sum of the squares of the other two sides, the triangle is a right triangle.

SET II
1. $a = \sqrt{2}, b = \sqrt{3}, c = 2, d = \sqrt{5}, e = \sqrt{6}.$
3. a) $\angle N = \angle E$ and $\angle 1 = \angle 2.$
b) Corresponding sides of similar triangles are proportional. c) $100 - x.$ d) $ND = 60,$ $ED = 40.$ e) $AD = 75, DS = 50.$
f) 125 feet.

Chapter 11, Lesson 3

SET I
1. a) $\dfrac{1}{\sqrt{2}}.$ b) $\dfrac{1}{2}.$ 2. a) $\dfrac{2}{1}.$ b) $\dfrac{\sqrt{3}}{1}.$
c) $\dfrac{2}{\sqrt{3}}.$ 3. a) $5\sqrt{2}.$ b) 7. c) 4.
d) $3\sqrt{2}.$ e) $e = 8, d = 4\sqrt{3}.$ f) $n = 5,$ $d = 5\sqrt{3}.$ g) $n = \sqrt{3}, e = 2\sqrt{3}.$
4. a) $8\sqrt{2}.$ b) $4\sqrt{2}.$ c) $\dfrac{9}{2}\sqrt{3}.$ d) $6\sqrt{3}.$

1. a) 12. b) $12\sqrt{2}.$ c) $12\sqrt{3}.$ 2. a) 10.
b) $10\sqrt{2}.$ c) $10\sqrt{3}.$ 3. a) Perimeter, 40 units; area, 50 square units. b) Perimeter, $20\sqrt{2}$ units; area, $25\sqrt{2}$ square units.
4. a) $36\sqrt{3}$ square units. b) $18\sqrt{3} + 54$ square units.

Chapter 11, Lesson 4

SET I
1. a) $\tan D = \dfrac{3}{4}.$ b) $\tan A = \dfrac{12}{5}.$
c) $\tan N = \dfrac{15}{8}.$ d) $\tan E = \sqrt{3}.$
3. a) 27.5. b) 5.6. c) 20. 4. a) 40°.
b) 55°. c) 30°.

SET II
1. a) .9996. b) 1. c) $\tan H = \dfrac{s}{y}$ and
$\tan U = \dfrac{y}{s}.$ $(\tan H)(\tan U) = \dfrac{s}{y} \cdot \dfrac{y}{s} = \dfrac{sy}{sy} = 1.$
2. a) $\angle HCW = \angle HCO + \angle OCW$ (betweenness of rays), so $\angle HCW > \angle OCW$ (the "whole greater than its part" theorem).
b) $\tan \angle HCW = \dfrac{HW}{CW}$ and $\tan \angle OCW = \dfrac{OW}{CW}.$
Since $HW = HO + OW$, $HW > OW$. Dividing by CW, $\dfrac{HW}{CW} > \dfrac{OW}{CW}.$ By substitution, $\tan \angle HCW > \tan \angle OCW.$

Chapter 11, Lesson 5

SET I
1. a) $\dfrac{a}{s}.$ b) $\dfrac{t}{s}.$ c) $\dfrac{a}{t}.$ d) $\dfrac{i}{r}.$ e) $\dfrac{e}{r}.$ f) $\dfrac{i}{e}.$
2. a) 47°. b) 22°. c) 39°. d) 20°. e) 58°.
f) 85°. 3. a) Sine. b) Tangent.
c) Cosine. d) Tangent. e) Sine.
f) Cosine. 4. a) 5.3. b) 2.1.
c) 42°. d) 60°.

SET II
1. 81 feet. 2. 73 feet. 3. a) 37 yards.
b) 155 yards.

Chapter 11, Review

SET I
1. a) $3\sqrt{2}.$ b) 10. c) $6\sqrt{3}.$ d) $7\sqrt{2}.$
e) $4\sqrt{3}.$ f) $2\sqrt{30}.$ 2. a) 50 feet. b) 20 feet. c) $\angle W = 37°, \angle X = 53°.$ 3. a) $5\sqrt{3}.$
b) $5\sqrt{2}.$ c) 5. d) 0. e) The pattern is $5\sqrt{3}, 5\sqrt{2}, 5\sqrt{1}, 5\sqrt{0}.$

Chapter 11, Review (continued)

SET II

1. 139 feet.   2. 46°.   3. 9°.   4. In a right triangle, the leg opposite one acute angle is the adjacent leg with respect to the other: $\sin G = \dfrac{UM}{GU}$ and $\cos U = \dfrac{UM}{GU}$. Therefore, $\sin G = \cos U$.   5. a) $2\sqrt{5}$.   b) $14\sqrt{5}$.   c) 4.   d) $4\sqrt{5}$.

SET III

The insect would be 27 centimeters from the candle.

Chapter 12, Lesson 1

SET I

2. a) J.   b) Two.   c) Radii.   d) A chord. e) Isosceles.   f) It has two equal sides because all radii of a circle are equal. g) Yes.   h) $\angle G = 60°$ because $\angle G = \angle I$. Hence, $\angle J = 60°$ and $\triangle JIG$ is equiangular. An equiangular triangle is also equilateral. 3. a) Hypothesis.   b) Perpendicular lines form right angles.   c) Two points determine a line.   d) A triangle that contains a right angle is a right triangle.   e) All radii of a circle are equal.   f) Reflexive.   g) H.L. h) Corresponding parts of congruent triangles are equal.   i) If a line segment is divided into two equal parts, it is bisected.

SET II

5. a) Two points determine a line.   b) A line through the center of a circle that bisects a chord that is not a diameter is perpendicular to it.   c) Through a point not on a line, there is exactly one line perpendicular to the line.

Chapter 12, Lesson 2

SET I

1. a) One.   b) Two.   c) None. 2. a) All radii of a circle are equal.   b) If two sides of a triangle are equal, the angles opposite them are equal.   c) Through a point not on a line, there is exactly one line perpendicular to the line. *Or,* In a plane, two lines perpendicular to a third line are parallel to each other. *Or,* An exterior angle of a triangle is greater than either remote interior angle.   4. a) Yes.   b) Yes.   c) Yes. d) Yes.   e) No.

Chapter 12, Lesson 3

SET I

1. a) $\overset{\frown}{DE}$ and $\overset{\frown}{EN}$.   b) $\overset{\frown}{MAK}$ and $\overset{\frown}{MKA}$. c) A semicircle.   d) $\overset{\frown}{MK}$.   2. a) 90°. b) 130°.   c) 230°.   d) 320°. 3. a) Hypothesis.   b) A minor arc is equal in measure to its central angle.   c) Substitution. d) Reverse statements a and c. 4. a) Two points determine a line.   b) All radii of a circle are equal.   c) Hypothesis. d) S.S.S.   e) Corresponding parts of congruent triangles are equal.   f) In a circle, equal central angles have equal minor arcs.

Chapter 12, Lesson 4

SET I

1. a) $\overset{\frown}{AD}$.   b) $\overset{\frown}{AJD}$.   c) $\angle D = \frac{1}{2}m\overset{\frown}{AJ}$. d) $\angle J = \angle A$.   3. a) $\angle N = 35°$. b) $\angle R = 35°$.   c) $m\overset{\frown}{NR} = 60°$. d) $m\overset{\frown}{GN} = 120°$.   e) $\angle GAN = 60°$. f) $\angle GAR = 90°$.   4. a) Two points determine a line.   b) Addition. c) Betweenness of rays.   d) Arc addition postulate.   e) Substitution.   5. The proof would be like that for the second figure but would involve subtraction rather than addition.

Chapter 12, Lesson 5

SET I

1. a) Hypothesis.   b) Two points determine a line.   c) An exterior angle of a triangle is equal in measure to the sum of the measures of the two remote interior angles.   d) An inscribed angle is equal in measure to half its intercepted arc.   e) Substitution. 3. a) 10°.   b) 50°.   c) 30°.   d) 75°.

SET II

1. a) Yes.   b) A secant angle whose vertex is on the circle.

Chapter 12, Lesson 6

SET I

1. a) $5\sqrt{3}$.   b) 8.   c) 7.   2. RA = RM and RM = RS, because the tangent segments to a circle from an external point are equal. So, RA = RS by the transitive postulate. 3. a) An angle inscribed in a semicircle is a right angle.   b) Two lines that form a right angle are perpendicular.   c) If a line is perpendicular to a radius at its outer endpoint, then it is tangent to the circle.

Chapter 12, Lesson 7

1. a) Hypothesis. b) Two points determine a line. c) Reflexive. d) Inscribed angles that intercept the same arc are equal.
e) A.A. f) Corresponding sides of similar triangles are proportional. g) Means-extremes theorem.   2. If two secants contain chords $\overline{AB}$ and $\overline{CD}$ in a circle and intersect in a point that is not on the circle, then
$AP \cdot PB = CP \cdot PD$.   3. a) 4. b) $2\sqrt{6}$.
c) 10. d) 35.

Chapter 12, Lesson 8

SET I

1. ($OS' = 18$, $OA' = 12$, $OB' = 9$,
$OL' = 7.2$, $OE' = 6$.)   2. a) Toward the center of the inversion circle. b) It is the point itself. c) The center of the circle.
d) $OP = 0$. There is no number for OP' such that $0 \cdot OP' = r^2$.   3. a) $\overline{LT}$.
b) Through a point on a line, there is exactly one perpendicular to the line. c) $\overline{OT}$.
d) Two points determine a line. e) $\overline{TL'}$.
f) Same reason as b.

SET II
1. a)

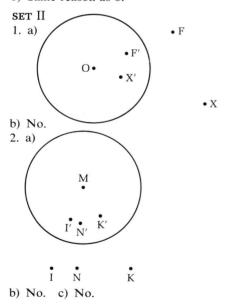

b) No.
2. a)

b) No. c) No.

Chapter 12, Lesson 9

SET I

1. a) The upper semicircle, $\widehat{N'R'O}$, except for point O. b) Counterclockwise around

$\widehat{N'R'O}$. c) On $\widehat{N'R'O}$, very close to point O.
2. a) It is a circle tangent to the inversion circle (except for point O.) b) It is a circle that intersects the inversion circle in two points (except for point O.) c) No. If the line passes through the center of the inversion circle, its inverse is itself (except for point O.)

SET II
1. a)

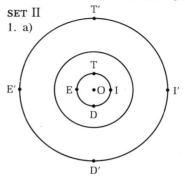

Chapter 12, Review

SET I

1. a) All radii of a circle are equal. b) Arc addition postulate. c) If a line through the center of a circle is perpendicular to a chord, it bisects it. d) A central angle is equal in measure to its intercepted arc. e) A tangent to a circle is perpendicular to the radius drawn to the point of contact. f) If two minor arcs of a circle are equal, their chords are equal.
g) An inscribed angle is equal in measure to half its intercepted arc. h) The Intersecting Chords Theorem.   2. a) 40°. b) 16.5.
c) 100°. d) 9. e) 3. f) 1.25. g) 160°.

SET II

1. $\angle V$ and $\angle G$ are inscribed angles intercepting $\widehat{LGA}$ and $\widehat{LVA}$;
$\angle V = \frac{1}{2}m\widehat{LGA}$ and $\angle G = \frac{1}{2}m\widehat{LVA}$;
$\angle V + \angle G = \frac{1}{2}m\widehat{LGA} + \frac{1}{2}m\widehat{LVA}$
$= \frac{1}{2}(m\widehat{LGA} + m\widehat{LVA}) = \frac{1}{2}360° = 180°$;
therefore, $\angle V$ and $\angle G$ are supplementary.
The same argument can also be applied to $\angle L$ and $\angle A$.

The reasons have been omitted in the following proofs. 2. *Proof.* 1. $\overline{NI} \parallel \overline{ER}$ in circle G. 2. Draw $\overline{NR}$. 3. $\angle N = \angle R$.
4. $\angle N = \frac{1}{2}m\widehat{IR}$, $\angle R = \frac{1}{2}m\widehat{NE}$. 5. $\frac{1}{2}m\widehat{IR}$
$= \frac{1}{2}m\widehat{NE}$. 6. $m\widehat{IR} = m\widehat{NE}$.   3. *Proof.*
1. Draw $\overline{RN}$. 2. $\overline{RI}$ is a diameter of circle H.
3. $\widehat{RNI}$ is a semicircle. 4. $\angle RNI$ is a right

Chapter 12, Review (continued)

angle. 5. $\overline{RN} \perp \overline{IE}$. 6. $\overline{RN}$ bisects $\overline{IE}$.
7. IN = NE. 4. If $\overleftrightarrow{DE}$ and $\overleftrightarrow{NU}$ are not
parallel, $\overleftrightarrow{DE}$ and $\overleftrightarrow{NU}$ intersect. Extend them
to intersect in point P. Then PD = PN and
PE = PU. Since PD = PE + ED and
PN = PU + UN, PE + ED = PU + UN. So
ED = UN by subtraction.

5.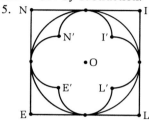

SET III

The radius of the saucer is 25 centimeters.

Chapter 13, Lesson 1

SET I

1. a) They are concyclic. b) Its circumcircle.
c) Its circumcenter. d) They are inscribed
angles of the circle. e) It is inscribed in the
circle. 3. a) It lies on the triangle. (It is
the midpoint of the hypotenuse.) b) It lies
outside the triangle. 4. b) They lie on the
perpendicular bisector of the line segment
that joins the two points.

SET II

1. a) If two sides of a triangle are equal, then
the angles opposite them are equal.
b) Substitution. c) If two angles in a linear
pair are equal, then they are right angles.
d) Two lines that form a right angle are
perpendicular. e) Through a point not on a
line, there is exactly one perpendicular to
the line. 2. a) Chord $\overline{AC}$ would be a
diameter of the circle. b) No. Because it
is a diameter, *every* line through P bisects $\overline{AC}$.
c) Yes. Every line that bisects $\overline{AC}$ must pass
through P because it is the midpoint of $\overline{AC}$.

Chapter 13, Lesson 2

SET I

1. a) The opposite angles of a cyclic
quadrilateral are supplementary.
b) Supplements of the same angle are equal.
c) An exterior angle of a triangle is greater
than either remote interior angle.

d) Addition. e) Betweenness of rays.
f) Substitution. g) Step b in which
$\angle DCB = \angle DEB$. 2. a) $50°$. b) $60°$.
c) $90°$. d) $80°$. 3. Since every angle of a
rectangle is a right angle, the opposite angles
of the rectangle are supplementary. Therefore,
every rectangle is cyclic.

Chapter 13, Lesson 3

SET I

1. a) N. b) A. c) N. d) A.
2. a) To determine the length of the radius.
c) No; it is always inside the triangle.
3. a) Yes. b) No. c) Yes. d) No.
e) Yes.

SET II

2. a) $\triangle JON \cong \triangle JEN$ (S.S.S.) Therefore,
$\angle OJN = \angle EJN$ and $\angle ONJ = \angle ENJ$ and
hence $\overleftrightarrow{JN}$ bisects $\angle OJE$ and $\angle ONE$.
b) $\triangle JES \cong \triangle JOS$ and $\triangle SEN \cong \triangle SON$
(S.A.S.) Hence $\angle JES = \angle JOS$ and
$\angle SEN = \angle SON$. Since $\angle JOS = \angle SON$,
$\angle JES = \angle SEN$. Therefore, $\overleftrightarrow{SE}$ bisects
$\angle JEN$.

Chapter 13, Lesson 4

SET I

1. a) $\overline{CT}, \overline{MB}, \overline{EA}$. b) $\dfrac{CB}{BE} \cdot \dfrac{ET}{TM} \cdot \dfrac{MA}{AC} = 1$.
c) $\dfrac{MT}{TE} \cdot \dfrac{EB}{BC} \cdot \dfrac{CA}{AM} = 1$. 2. a) Yes. b) Yes.
c) $\dfrac{6}{7}$. d) 3.

SET II

1. a) An angle bisector of a triangle divides
the opposite side into segments that have the
same ratio as the other two sides.
b) $\dfrac{YV}{VH} = \dfrac{NY}{NH}$. c) $\dfrac{HE}{EN} = \dfrac{YH}{YN}$.
d) $\dfrac{NR}{RY} \cdot \dfrac{YV}{VH} \cdot \dfrac{HE}{EN} = \dfrac{NH}{HY} \cdot \dfrac{NY}{NH} \cdot \dfrac{YH}{YN} = 1$.
e) It follows from Ceva's Theorem that these
three segments are concurrent.

Chapter 13, Lesson 5

SET I

1. b) No. d) The orthocenter of a triangle is
inside, on, or outside the triangle, depending
upon whether the triangle is acute, right, or
obtuse. The orthocenter of a right triangle is
the vertex of its right angle. 2. b) The four
points are collinear. c) The circumcenter,

incenter, centroid, and orthocenter would all be the same point.   3. a) If the opposite sides of a quadrilateral are parallel, it is a parallelogram.   b) The opposite sides of a parallelogram are equal.   c) Substitution. d) An altitude of a triangle is a line segment from a vertex perpendicular to the line of the opposite side.   e) In a plane, if a line is perpendicular to one of two parallel lines, it is also perpendicular to the other.   f) The perpendicular bisectors of the sides of a triangle are concurrent.

SET II
1. a) ∠UGE = ∠AGI (reflexive), and ∠GAI = ∠GUE (all right angles are equal). b) Corresponding sides of similar triangles are proportional.

Chapter 13, Lesson 6
SET I
1.

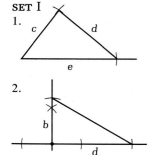

2.

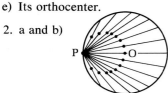

3. a) $\overline{BA}$ was bisected.   b) A semicircle was drawn with the midpoint of $\overline{BA}$ as its center and $\frac{1}{2}BA$ as its radius.   c) An angle inscribed in a semicircle is a right angle.

Chapter 13, Review
SET I
1. a) Its circumcenter.   b) Each side of △RHI is a midsegment of △ATC; a midsegment of a triangle is parallel to the third side.   c) In a plane, if a line is perpendicular to one of two parallel lines, it is perpendicular to the other.   d) Its altitudes. e) Its orthocenter.

2. a and b)

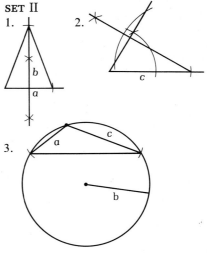

c) They lie on a circle with $\overline{PO}$ as its diameter.

3.

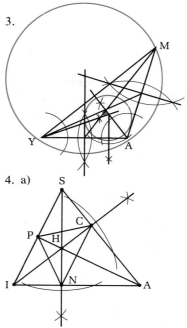

4. a)

b) $\overleftrightarrow{SN}$, $\overleftrightarrow{IC}$, and $\overleftrightarrow{AP}$ bisect the angles of △PCN.   c) $\overleftrightarrow{PC}$, $\overleftrightarrow{PN}$, and $\overleftrightarrow{CN}$.   5. Since $\overline{OC} \perp \overline{CN}$ and $\overline{OR} \perp \overline{RN}$, ∢C and ∢R are right angles. Therefore, ∢C and ∢R are supplementary. CORN is cyclic since a pair of its opposite angles are supplementary.

SET II
1.   2.

3.

4. 3.2.   5. 125°.

SET III
Grandpa Dilcue's new farm has one-seventh of the area of his original farm.

Chapter 14, Lesson 1

SET I

1. Figure b is not a pentagon because two of its segments intersect in more than their endpoints. Figure d is not a pentagon because two of its segments with a common endpoint are collinear.   2. c) No.   3. a) A concave dodecagon. b) Yes. c) The second. d) The first. e) No.
4. a) A polygon is *cyclic* iff there exists a circle that contains all of its vertices.
b) No. c) No.

SET II

1. a) Three and four, respectively. b) 540° and 720°.

c)

No. of sides of polygon	3	4	5	6
No. of diagonals from one vertex	0	1	2	3
No. of triangles formed	1	2	3	4
Sum of the measures of the polygon's angles	180°	360°	540°	720°

d) $n - 2$. e) Sum $= (n - 2)180°$.
2. a) No. b) Yes.   3. a) Two polygons are *congruent* iff there is a correspondence between their vertices such that the corresponding sides and corresponding angles of the polygons are equal. b) Yes. c) Yes.
4. a) Two polygons are *similar* iff there is a correspondence between their vertices such that the corresponding sides of the polygons are proportional and the corresponding angles are equal. b) Corresponding sides of similar polygons are proportional. c) Equal ratios theorem. d) Substitution.

e) $\dfrac{\alpha \text{DWARF}}{\alpha \text{D}'\text{W}'\text{A}'\text{R}'\text{F}'} = \left(\dfrac{\text{DW}}{\text{D}'\text{W}'}\right)^2$.

Chapter 14, Lesson 2

SET I

1. a) No. c) No. e) Yes.   2. a) Its center. b) An apothem. c) If a line through the center of a circle is perpendicular to a chord, it also bisects it. d) A radius.
e) A central angle.   3. a) The center of the square. b) Yes, MINT can be rotated through less than 360° so that it coincides with itself.   4. a) A regular *n*-gon has *n* lines of symmetry. b) In a regular polygon with an odd number of sides, each symmetry line is both the bisector of an angle and the perpendicular bisector of a side. In a regular

polygon with an even number of sides, each symmetry line is either of these but not both.
c) Only regular polygons with even numbers of sides have point symmetry.

SET II

1. b) Equilateral, and hence, equiangular.
c) 60°. d) 360° − 5(60°) = 60°.
e) In △OPG, ∠OPG = 60° and ∠GOP = ∠GPO. It follows that △OPG is equiangular, and hence, equilateral.
f) Because $\overline{\text{OG}}$, like the other sides of the hexagon, is equal in length to its radius.
g) Because each of its angles has a measure of 120°.   4. a) It decreases.
b) Equilateral triangle, 120°; square, 90°; regular pentagon, 72°; regular hexagon, 60°.
c) Central angle $= \dfrac{360°}{n}$.

Chapter 14, Lesson 3

SET I

1. a) 42. b) $p = 2(6 \sin 30°)7 = 84(.500)$
$= 42$. c) 33. d) $\dfrac{3}{1}$ or 3.   2. a) A 30°-60° right triangle. b) 6. c) $3\sqrt{3}$. d) $6\sqrt{3}$.
e) $18\sqrt{3}$. f) $18(1.732) = 31.176$.
g) $p = 2(3 \sin 60°)6 = 36(.866) = 31.176$.
3. a) $\dfrac{360}{9} = 40°$. b) 20°. c) $\sin 20° = \dfrac{\text{AY}}{10}$;
AY $= 10 \sin 20° = 10(.342) = 3.42$. d) 6.84.
e) $p = 9(6.84) = 61.56$. f) $p = 2(9 \sin 20°)10 = 180(.342) = 61.56$.

SET II

1. a) $\overline{\text{OD}}$ is the hypotenuse of isosceles right △OND; the hypotenuse of an isosceles right triangle is $\sqrt{2}$ times the length of one leg.
b) Sin ∠ODN $= \dfrac{\text{ON}}{\text{OD}}$, so $\sin 45° = \dfrac{1}{\sqrt{2}}$.
c) $p_{\text{ADIS}} = 2(4 \sin 45°)\sqrt{2} = 8$.
d) DI $= 2$DN $= 2$; $p_{\text{ADIS}} = 4$ DI $= 4(2) = 8$.
2. a) 51.96. b) 60. c) 62.16. d) No.
3. 3.15.

Chapter 14, Lesson 4

SET I

1. a) 36 square units. b) 3. c) $3\sqrt{2}$.

e) $\alpha_{\text{regular } n\text{-gon}} = 4\left(\dfrac{1}{\sqrt{2}}\right)\left(\dfrac{1}{\sqrt{2}}\right)(3\sqrt{2})^2 = 36$.

2. a) $25\sqrt{3}$ square units. b) $150\sqrt{3}$ square units. c) 259.8. d) $600(.4330) = 259.8$.

3. a) $\alpha_1 = v \cos \dfrac{180}{n} r_1{}^2$ and $\alpha_2 = v \cos \dfrac{180}{n} r_2{}^2$;

$$\dfrac{\alpha_1}{\alpha_2} = \dfrac{v \cos \dfrac{180}{n} r_1{}^2}{v \cos \dfrac{180}{n} r_2{}^2} = \dfrac{r_1{}^2}{r_2{}^2} = \left(\dfrac{r_1}{r_2}\right)^2.$$ b) The ratio

of the perimeters of two regular polygons that have the same number of sides is equal to the ratio of their radii. c) The area of the first is four times the area of the second. d) The perimeter of the first is twice the perimeter of the second.

SET II
1. a) 68 (approximately). b) 356 square units (approximately). 2. b) $\alpha_{\text{regular } n\text{-gon}} = 3.1392 r^2$.

Chapter 14, Lesson 5
SET I
1. a) 5.0001. b) .0001. c) 5.0000001 (the seventh term). d) Yes. e) 5.
2. a) 1.9999, 1.99999, 1.999999. b) 32, 64, 128. c) 7, 7, 7. d) $1, \dfrac{1}{3}, \dfrac{1}{9}$. e) 0, 1, 0.
3. a) Sequences b and e. b) Sequence a, 2; sequence c, 7; sequence d, 0.
4. a) $1, \dfrac{1}{2}, \dfrac{1}{3}, \dfrac{1}{4}, \dfrac{1}{5}$. b) It becomes smaller and smaller. c) 0. 5. a) $\dfrac{1}{2}, \dfrac{2}{3}, \dfrac{3}{4}, \dfrac{4}{5}, \dfrac{5}{6}$.
b) It gets closer and closer to 1. c) 1.

SET II
1. a) It decreases. b) 0. c) 0. d) 0.
2. a) It increases. b) 1. c) 1.
3. a) 3.1390, 3.1409, 3.1414, 3.1415.
b) 3.14. c) 3.14. d) π.

Chapter 14, Lesson 6
SET I
1. a) $\pi = \dfrac{c}{2r}$. b) The diameter of a circle is

twice the length of its radius. c) $\pi = \dfrac{c}{d}$.

d) π is the ratio of the circumference of a circle to its diameter. 2. a) 30 cubits.
b) 10 cubits. c) 3. 3. a) 3.14286.
b) $3\frac{1}{7}$. 4. a) 9.8596. b) 9.9225.
c) Smaller. 5. Six decimal places.

SET II
1. a) Circle F, 2π units; circle I, 4π units; circle R, 6π units. b) The ratio of the

circumferences of two circles is equal to the ratio of their radii. c) Circle F, π square units; circle I, 4π square units; circle R, 9π square units. d) The ratio of the areas of two circles is equal to the square of the ratio of their radii. 2. a) 27π square units.
b) $64 - 16\pi$ square units. c) $25\pi - 48$ square units. 3. 7.5 inches.
4. a) 30 centimeters. b) Approximately 31.4 centimeters. c) Approximately 65 square centimeters. d) Approximately 78.5 square centimeters. 5. Approximately 1.6 million miles each day.

Chapter 14, Lesson 7
SET I
1. a) 135°. b) 8π units. c) 3π units.
d) 16π square units. e) 6π square units.
2. a) π centimeters. b) π centimeters.
c) $\dfrac{\pi}{2}$ centimeters. d) No. e) No.

SET II
1. a) $\dfrac{25}{9}\pi$, or approximately 9 square inches.
b) $\dfrac{200}{9}\pi$, or approximately 70 square inches.
2. $\pi - 2$ square units. 3. 3 inches.
4. a) $\dfrac{100}{3}\pi - 25\sqrt{3}$ square units.
b) $10\pi - 20$ square units. c) $8\pi - 16$ square units.

Chapter 14, Review
SET I
1. 50π, or approximately 157 yards.
2. a) 2.1. b) 2.01. c) 2.000001. d) 2.
3. a) 4. b) None. c) If n is even, it has $\frac{1}{2}n$ pairs of parallel sides; if n is odd, it has no pairs of parallel sides. 4. a) It decreases.
b) It increases. c) The area of the circle.
5. a) The decagon. b) The area of the nonagon is about 289 square units and the area of the decagon is about 294 square units.

SET II
1. The reasons have been omitted in the following proof. 1. DRAGON is a regular hexagon. 2. DN = RA and NO = AG.
3. $\angle N = \angle A$. 4. $\triangle DNO \cong \triangle RAG$.
5. DO = RG. 6. DR = OG. 7. DRGO is a parallelogram. 8. $\overline{DR} \parallel \overline{OG}$. 2. a) 3

Chapter 14, Review *(continued)*

units. b) 3.14 units. c) 0.14 units. d) 3.12 units. e) 3.14 units. f) 0.02 units.
3. a) 5π square units. b) $75\sqrt{3} - 25\pi$ square units.

SET III

1. The members of the middle class.
2. The priests. 3. The equilateral triangle: radius, 2.3 units; area, 6.9 square units. The square: radius, 2.1 units; area, 9 square units. Conclusion: As n, the number of sides of a regular n-gon having a constant perimeter increases, its radius decreases and its area increases.

Chapter 15, Lesson 1

SET I

1. a) A ∥ B, B ∥ C, A ∥ C, E ∥ F. b) A ⊥ D, A ⊥ E, A ⊥ G, B ⊥ D, B ⊥ E, B ⊥ F, B ⊥ G, C ⊥ D, C ⊥ E, C ⊥ F, C ⊥ G, D ⊥ E, D ⊥ F, E ⊥ G, F ⊥ G. c) x ∥ A, x ∥ G, y ∥ A, y ∥ B, y ∥ F, z ∥ F, z ∥ G.
d) x ⊥ E, x ⊥ F, y ⊥ G, z ⊥ A, z ⊥ B, z ⊥ C. 2. a) Yes. b) Yes. c) No.
d) Yes. e) No. 3. a) $\overleftrightarrow{FD}$ is perpendicular to every line in A that passes through point O. b) ∡FOL, ∡FOI, ∡FOR, ∡DOL, ∡DOI, ∡DOR.

Chapter 15, Lesson 2

SET I

1. a) 8 vertices and 12 edges. b) R.
c) $\overline{AM}$, $\overline{AR}$, $\overline{CI}$, and $\overline{CE}$. d) $\overline{IC}$, $\overline{WR}$, and $\overline{BE}$. e) $\overline{WB}$, $\overline{BE}$, $\overline{MI}$, and $\overline{IC}$.
2. a)

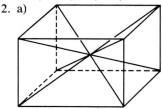

b) They are concurrent, have equal lengths, and bisect each other. c) 12 (one with each edge). 3. a) $\angle ILE = 45°$ and $\angle ELB = 45°$ because $\triangle ILE$ and $\triangle ELB$ are isosceles right triangles. b) No. $\overrightarrow{LE}$ is not between $\overrightarrow{LI}$ and $\overrightarrow{LB}$ because the three rays are not coplanar.
c) 60°. ($\triangle ILB$ is equilateral and hence, equiangular.) d) $8\sqrt{2}$; approximately 11.28 units. e) $8\sqrt{3}$; approximately 13.84 units.

Chapter 15, Lesson 3

SET I

1. a) $\triangle DIS$ and $\triangle NEY$. b) 3. c) $\overline{DN}$, $\overline{IE}$, and $\overline{SY}$. d) They are perpendicular to them.
e) Rectangles. 2. a) 6. b) Parallelograms.
c) $\overline{ZK}$ ∥ $\overline{SN}$ ∥ $\overline{EI}$ ∥ $\overline{LC}$; $\overline{SE}$ ∥ $\overline{ZL}$ ∥ $\overline{NI}$ ∥ $\overline{KC}$.
3. a) Because the lateral faces of a prism are always parallelograms.
b)

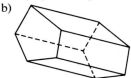

4. a) Alice is right; the minimum number of faces is five. b) 9. c) 6. 5. a) A right kite prism! b) An oblique pentagonal prism. c) An octagonal prism. d) A cube.

SET II

1. a) $6e^2$. b) ph. c) $2(lw + wh + lh)$.
2. 5.4 square inches. 3. 36 square inches of foil.

Chapter 15, Lesson 4

SET I

1. a) 6 square feet. b) 144 square inches.
c) 864 square inches. d) 1728 cubic inches.
2. a) 70 cubic units. b) 70 cubic units.
c) Cavalieri's Principle. 3. a) 6, 24, and 54 square units, respectively. b) It is quadrupled. c) It is multiplied by 9.
d) 1, 8, and 27 cubic units, respectively.
e) It is multiplied by 8. f) It is multiplied by 27. 4. a) Equilateral triangles.
b) $4\sqrt{3}$ square units. c) 4 units. d) $16\sqrt{3}$ cubic units.

SET II

1. a) 1, 4, 9, and 36 square units, respectively.
b) 216, 54, 24, and 6 units, respectively.
c) 866, 440, 306, and 216 square units, respectively. d) The cube having that volume. 2. 48 cubic feet. 3. 10 cubic feet.

Chapter 15, Lesson 5

SET I

1. a) H. b) A regular pentagon.
c) Isosceles triangles. d) $\overline{HO}$ is perpendicular to the plane of LIDAY at its center.

2. a)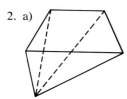

b) 5. c) 8.   3. a) It is one third the volume of the prism. b) It is twice the volume of the pyramid.

SET II
1. a) 75 cubic units. b) 2 cubic units.
c) $28\sqrt{3}$ cubic units.
2. a) EN < EO < ER. b) $\sqrt{65}$ units.
c) $4\sqrt{2}$ units. d) 9 units. e) Yes;
$7 < \sqrt{65} < 9$.   3. The volume of one of the pyramids is 36 cubic units.
4. a) 360 square units. b) 576 cubic units.

Chapter 15, Lesson 6

SET I
1. 1776π cubic units. b) 1492π cubic units.
c) 1812 cubic units.   2. a) $2\pi rh$.
b) $2\pi rh + 2\pi r^2$.   3. a) $\frac{1}{3}\pi a^2 b$.
b) $\frac{2}{3}\pi a^2 b$. c) $\frac{4}{3}\pi a^2 b$. d) $\frac{8}{3}\pi a^2 b$. e) It is doubled. f) It is quadrupled. g) It is multiplied by 8.   4. a) 45π cubic units.
b) No. (Its volume would be 75π cubic units.) c) A right cone. d) 48π cubic units.
e) 108π cubic units.

SET II
1. a) 8π cubic inches. b) 16π cubic inches.
2. 200 cubic inches.   3. 103 cubic inches.

Chapter 15, Lesson 7

SET I
1. a) 4π square units and 16π square units.
b) $\frac{4}{3}\pi$ cubic units and $\frac{32}{3}\pi$ cubic units.
c) It is quadrupled.   d) It is multiplied by 8.
2. a) $\frac{2}{3}\pi r^3$. b) $2\pi r^2$. c) $3\pi r^2$.   3. a) 201 million square miles. b) 268 billion cubic miles.

SET II
1. a) All radii of a sphere are equal. b) If a line is perpendicular to a plane, it is perpendicular to every line in the plane that passes through the point of intersection.
c) H.L. d) Corresponding parts of congruent triangles are equal. e) The fact that ET = ER implies that every point in the intersection is equidistant from point E. The

set of all points in a plane that are equidistant from a given point in the plane is a circle.
2. a) Yes.   3. a) 3 units. b) 36π cubic units.   4. Approximately 730 pounds.

Chapter 15, Lesson 8

SET I
1. a) If two right cones are similar, their altitudes have the same ratio as the radii of their bases. b) The volume of a cone is $\frac{1}{3}\pi r^2 h$, where r is the radius of its base and h is the length of its altitude. c) Division.
d) Substitution.   3. a) $\frac{2}{5}$. b) $\frac{4}{25}$. c) $\frac{8}{125}$.
4. a) $\frac{3}{4}$. b) $\frac{27}{64}$.

SET II
1. a) 9 times as strong. b) 27 times as heavy.
c) The smaller animal.   2. b) $\sqrt[3]{2}$.
3. a) The colossal can. b) 9¢. c) 27¢.
d) It is twice the cost of the materials in each case. e) The company wouldn't care.

Chapter 15, Lesson 9

SET I
1. a) The regular tetrahedron, regular octahedron, and regular icosahedron.
b) 3, 4, and 5, respectively. c) 180°, 240°, and 300°, respectively. d) Because $6 \cdot 60° = 360°$. e) The cube and regular dodecahedron. f) Because $4 \cdot 90° = 360°$.
g) Each angle of a regular hexagon has a measure of 120°; $3 \cdot 120° = 360°$.   2. a) No, because the same number of triangles do not meet at each vertex. b) 6 faces, 5 vertices, and 9 edges. c) Yes; $6 + 5 = 9 + 2$.
3. a) 3 faces, 0 vertices, and 2 edges.
b) No; $3 + 0 \neq 2 + 2$. c) No, because a polyhedron is a solid bounded by parts of intersecting *planes*.

SET II
1. a) 7 faces, 7 vertices, and 12 edges.
b) $7 + 7 = 12 + 2$. c) Because it has n lateral faces and 1 base. d) $n + 1$. e) $2n$.
f) $(n + 1) + (n + 1) = 2n + 2$.
2. a) 7 faces, 10 vertices, and 15 edges.
b) $7 + 10 = 15 + 2$. c) $n + 2$. d) $2n$.
e) $3n$. f) $(n + 2) + 2n = 3n + 2$.
3. a) It has increased by 1. b) It has increased by 4. c) $9 + 9 = 16 + 2$. d) It increases by $n - 1$. e) It increases by 1.

Chapter 15, Lesson 9 *(continued)*

f) It increases by n.   4. a) It has increased by 1.   b) It has increased by 3.
c) $7 + 10 = 15 + 2$.   d) It increases by 1.
e) It increases by $n - 1$.   f) It increases by n.

Chapter 15, Review

SET I
1. a) 41,472 cubic inches.   b) 52 square feet.
c) $\sqrt{29}$ feet.   2. a) That they have equal volumes.   b) 48π cubic centimeters.
c) 16π square centimeters.

3. a)

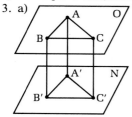

b) The two triangles seem to be congruent.
c) A right triangular prism.   d) If the planes are perpendicular.   e) No.   4. a) The Great Pyramid is 230 times as high.   b) The surface area of the Great Pyramid is 52,900 times that of the small one.   c) The volume of the Great Pyramid is 12,167,000 times that of the small one.

SET II
1. a) Pentagonal pyramids and triangular pyramids, respectively.   b) A dodecahedron and an icosahedron, respectively.   c) 12 and 20 points, respectively.   2. a) The cylindrical can would be 1.5 times as heavy as the spherical can.   b) The amount of paint would be the same for both cans.
3. 27 cubic inches.   4. 40 inches.

SET III
The wire would be 92,160,000 inches long.

Chapter 16, Lesson 1

SET I
1. a) A great circle of a sphere is the intersection of the sphere and a plane that contains its center.   b) Curve m seems to be a line; curve ℓ does not.   c) Yes.   d) Yes.
2. a) Yes.   b) Yes.   c) No.
3. a) These lines (great circles) seem to intersect in two points.   b) The two points are polar, so are counted as just one.

SET II
1. a) One.   b) One.   c) An unlimited number.   d) If the point is not the pole of the line.   2. a) The theorem is not true because there are no parallel lines in sphere geometry; $m \perp \ell$ and $n \perp \ell$, yet m and n intersect at P.
b) No.   3. a) Because they do not intersect.   b) One or both of the curves is not a line.   4. a) No, because $\angle PBX = \angle PAB$.
b) It is greater than 180°.   c) Only one.
d) Three.

Chapter 16, Lesson 2

SET I
1. a) Figures a and c.   2. a) The legs of a Saccheri quadrilateral are equal.   b) Since $\overline{JL} \perp \overline{LE}$ and $\overline{OE} \perp \overline{LE}$, $\angle JLE$ and $\angle OEL$ are both right angles.   c) S.A.S.   d) S.S.S.
e) Corresponding parts of congruent triangles are equal.   3. a) A biperpendicular quadrilateral whose legs are equal is a Saccheri quadrilateral.   b) The summit angles of a Saccheri quadrilateral are equal.   c) The "whole greater than its part" theorem.
d) Substitution.   e) An exterior angle of a triangle is greater than either remote interior angle. f) Transitive.   4. a) The three possibilities postulate.   b) If the two legs of a biperpendicular quadrilateral are unequal, then the summit angles are unequal and the larger angle is opposite the longer side.   c) A Saccheri quadrilateral.   d) The summit angles of a Saccheri quadrilateral are equal.

Chapter 16, Lesson 3

SET I
1. a) $\angle B = \angle R$.   b) Acute.   c) $\overline{EI} \perp \overline{BR}$ and $\overline{EI} \perp \overline{NL}$.   2. a) $\angle H$ and $\angle T$ are right angles.   b) $TR > HA$.   3. a) The summit angles of a Saccheri quadrilateral in Lobachevskian geometry are acute.   b) An acute angle has a measure of less than 90°.
c) Substitution.   d) A line segment has exactly one midpoint.   e) The line segment joining the midpoints of the base and summit of a Saccheri quadrilateral is perpendicular to both of them.   f) A quadrilateral that has a pair of sides perpendicular to a third side is biperpendicular.   g) If the two summit angles of a biperpendicular quadrilateral are unequal, then the legs are unequal and the longer leg is opposite the larger angle.   h) The midpoint of

a line segment divides it into segments half as long. i) Substitution. j) Multiplication.
4. a) Obtuse. c) $\overline{DU}$ seems to be shorter than $\overline{EK}$. d) In Riemannian geometry, the summit of a Saccheri quadrilateral is shorter than its base.

SET II

1. a) $\triangle GSR \cong \triangle NER$ because $ER = RS$, $GR = RN$, and $\angle ERN = \angle GRS$ (S.A.S.) $\triangle GSH \cong \triangle IWH$ for the same reason. b) $EN = GS$ and $GS = WI$, (corresponding parts of congruent triangles are equal), so $EN = WI$ (transitive). c) It is a biperpendicular quadrilateral with equal legs.
d) In Lobachevskian geometry, the summit of a Saccheri quadrilateral is longer than its base.
e) Since $EW = 2(RS + SH)$ and $RS + SH = RH$ (betweenness of points), $EW = 2RH$ (substitution). f) Substitution.
g) Division (or multiplication).

Chapter 16, Lesson 4

SET I

1. a) Since $\overline{GD} \perp \overline{LF}$, $\angle GDL$ and $\angle GDF$ are right angles. $\angle O = \angle GDL$ and $\angle I = \angle GDF$ (corresponding parts of congruent triangles are equal), so $\angle O$ and $\angle I$ are right angles. Therefore, $\overline{HO} \perp \overline{OI}$ and $\overline{SI} \perp \overline{OI}$, which means that OISH is a biperpendicular quadrilateral. $OH = GD$ and $GD = IS$ (corresponding parts of congruent triangles are equal), so $OH = IS$ (transitive). b) In Lobachevskian geometry, the summit angles of a Saccheri quadrilateral are acute. c) $\angle OHS + \angle ISH < 180°$.
d) $\angle 1 = \angle 5$ and $\angle 4 = \angle 6$. e) Since $\angle 1 + \angle 2 + \angle 3 + \angle 4 < 180°$, $\angle 5 + \angle 2 + \angle 3 + \angle 6 < 180°$ (substitution). $\angle HGS = \angle 5 + \angle 6$ (betweenness of rays), $\angle GHS = \angle 2$, and $\angle GSH = \angle 3$. Therefore, $\angle HGS + \angle GHS + \angle GSH < 180°$ (substitution). 2. a) No. b) No. (The sum of their measures is less than 90°.)
c) Each angle has a measure of less than 60°.
3. (Draw a diagonal of the quadrilateral to form two triangles.)

SET II

1. a) Corresponding angles of similar triangles are equal. b) S.A.S. c) Transitive.

($\angle L = \angle M$ and $\angle M = \angle 1$, $\angle N = \angle R$ and $\angle R = \angle 2$.) d) The angles form linear pairs and are therefore supplementary.
e) Substitution. f) In Lobachevskian geometry, the sum of the measures of the angles of a convex quadrilateral is less than 360°. 2. a) The triangles are similar.
b) The triangles are congruent.

Chapter 16, Review

SET I

1. Because the Parallel Postulate cannot be proved by means of the other postulates of Euclidean geometry. 2. a) There are no parallel lines in Riemannian geometry.
b) Through a point not on a line, there is exactly one line perpendicular to the line.
c) Saccheri quadrilaterals. d) Acute. (In Lobachevskian geometry, the summit angles of a Saccheri quadrilateral are acute.) e) If $\angle 1$ and $\angle 2$ are acute, $\angle 1 < 90°$ and $\angle 2 < 90°$, so $\angle 1 + \angle 2 < 180°$. But, since $\angle 1$ and $\angle 2$ are a linear pair, they are supplementary so that $\angle 1 + \angle 2 = 180°$.
3. a) Suppose ZINC is a rhombus. Then $ZI = CN$ because a rhombus is equilateral. But in Lobachevskian geometry, $ZI > CN$ because the summit of a Saccheri quadrilateral is longer than its base. Therefore, ZINC cannot be a rhombus. b) Yes. (It does not have any right angles, however.)

SET II

1. Two circles are orthogonal iff they intersect in two points and their tangents at the points of intersection are perpendicular to each other.
2. Two points determine a line. 3. The fact that through a point not on a line, there is more than one line parallel to the line.
4. a) The sum of the measures of the angles of a triangle is less than 180°. b) The summit angles of a Saccheri quadrilateral are acute.
c) A midsegment of a triangle is less than half as long as the third side.

SET III

Dilcue assumed that the sum of the measures of the angles of every triangle is the same number.

Index